# Neutrosophic Multi-Criteria Decision Making

# Neutrosophic Multi-Criteria Decision Making

Special Issue Editors

**Florentin Smarandache**

**Jun Ye**

**Yanhui Guo**

MDPI • Basel • Beijing • Wuhan • Barcelona • Belgrade

*Special Issue Editors*

Florentin Smarandache
University of New Mexico
USA

Jun Ye
Shaoxing University
China

Yanhui Guo
University of Illinois at Springfield
USA

*Editorial Office*
MDPI
St. Alban-Anlage 66
Basel, Switzerland

This is a reprint of articles from the Special Issue published online in the open access journal *Axioms* (ISSN 2075-1680) from 2017 to 2018 (available at: http://www.mdpi.com/journal/axioms/special_issues/neutrosophic_mcdm)

For citation purposes, cite each article independently as indicated on the article page online and as indicated below:

LastName, A.A.; LastName, B.B.; LastName, C.C. Article Title. *Journal Name* **Year**, *Article Number*, Page Range.

**ISBN 978-3-03897-288-4 (Pbk)**
**ISBN 978-3-03897-289-1 (PDF)**

# Contents

# About the Special Issue Editors

**Florentin Smarandache**, polymath, professor of mathematics, scientist, writer, and artist. He got his M. Sc. in Mathematics and Computer Science from the University of Craiova, Romania, and his Ph. D in Mathematics from the State University of Kishinev and pursued Post-Doctoral studies in Applied Mathematics at Okayama University of Sciences, Japan. He is the founder of neutrosophic set, logic, probability, and statistics and, since 1995, has published hundreds of papers on neutrosophic physics, superluminal and instantaneous physics, unmatter, absolute theory of relativity, redshift and blueshift due to the medium gradient and refraction index besides the Doppler effect, paradoxism, outerart, neutrosophy as a new branch of philosophy, Law of Included Multiple-Middle, degree of dependence and independence between the neutrosophic components, refined neutrosophic over-under-off-set, neutrosophic overset, neutrosophic triplet and duplet structures, DSmT, and so on in numerous peer-reviewed international journals and books and he has presented papers and plenary lectures in many international conferences around the world.

**Jun Ye** is currently a professor in the Department of Electrical and Information Engineering, Shaoxing University, China. He has more than 30 years of experience in teaching and research. His research interests include soft computing, fuzzy decision-making, intelligent control, robotics, pattern recognitions, rock mechanics, and fault diagnosis. He has published more than 200 papers in journals.

**Yanhui Guo**, Assistant Professor, received his B.S. degree in Automatic Control from Zhengzhou University, China, his M.S. degree in Pattern Recognition and Intelligence System from Harbin Institute of Technology, China, and his Ph.D. degree from Utah State University, Department of Computer Science, USA. He was a research fellow in the Department of Radiology at the University of Michigan and an assistant professor in St. Thomas University. Dr. Guo is currently an assistant professor in the Department of Computer Science at the University of Illinois at Springfield. Dr. Guo has published more than 100 research papers, has completed more than 10 grant-funded research projects, and has worked as an associate editor for different international journals and as a reviewer for top journals and conferences. His research area includes computer vision, machine learning, data analytics, computer-aided detection/diagnosis, and computer-assisted surgery.

# Preface to "Neutrosophic Multi-Criteria Decision Making"

The notion of a neutrosophic quadruple BCK/BCI-number is considered in the first article ("Neutrosophic Quadruple BCK/BCI-Algebras", by Young Bae Jun, Seok-Zun Song, Florentin Smarandache, and Hashem Bordbar), and a neutrosophic quadruple BCK/BCI-algebra, which consists of neutrosophic quadruple BCK/BCI-numbers, is constructed. Several properties are investigated, and a (positive implicative) ideal in a neutrosophic quadruple BCK-algebra and a closed ideal in a neutrosophic quadruple BCI-algebra are studied. Given subsets A and B of a BCK/BCI-algebra, the set NQ(A,B), which consists of neutrosophic quadruple BCK/BCI-numbers with a condition, is established. Conditions for the set NQ(A,B) to be a (positive implicative) ideal of a neutrosophic quadruple BCK-algebra are provided, and conditions for the set NQ(A,B) to be a (closed) ideal of a neutrosophic quadruple BCI-algebra are given.

Techniques for the order of preference by similarity to ideal solution (TOPSIS) and elimination and choice translating reality (ELECTRE) are widely used methods to solve multi-criteria decision-making problems. In the second research article ("Decision-Making with Bipolar Neutrosophic TOPSIS and Bipolar Neutrosophic ELECTRE-I"), Muhammad Akram, Shumaiza, and Florentin Smarandache present the bipolar neutrosophic TOPSIS method and the bipolar neutrosophic ELECTRE-I method to solve such problems. The authors use the revised closeness degree to rank the alternatives in the bipolar neutrosophic TOPSIS method. The researchers describe the bipolar neutrosophic TOPSIS method and the bipolar neutrosophic ELECTRE-I method by flow charts, also solving numerical examples by the proposed methods and providing a comparison of these methods.

In the third article ("Interval Neutrosophic Sets with Applications in BCK/BCI-Algebra", by Young Bae Jun, Seon Jeong Kim and Florentin Smarandache), the notion of (T(i,j),I(k,l),F(m,n))-interval neutrosophic subalgebra in BCK/BCI-algebra is introduced for i,j,k,l,m,n infoNumber 1,2,3,4, and properties and relations are investigated. The notion of interval neutrosophic length of an interval neutrosophic set is also introduced, and the related properties are investigated.

The bipolar neutrosophic set is an important extension of the bipolar fuzzy set. The bipolar neutrosophic set is a hybridization of the bipolar fuzzy set and the neutrosophic set. Every element of a bipolar neutrosophic set consists of three independent positive membership functions and three independent negative membership functions. In the fourth paper ("Cross-Entropy Measures of Bipolar and Interval Bipolar Neutrosophic Sets and Their Application for Multi-Attribute Decision-Making"), Surapati Pramanik, Partha Pratim Dey, Florentin Smarandache, and Jun Ye develop cross-entropy measures of bipolar neutrosophic sets and prove their basic properties. They also define cross-entropy measures of interval bipolar neutrosophic sets and prove their basic properties. Thereafter, they develop two novel multi-attribute decision-making strategies based on the proposed cross-entropy measures. In the decision-making framework, the authors calculate the weighted cross entropy measures between each alternative and the ideal alternative to rank the alternatives and choose the best one, solving two illustrative examples of multi-attribute decision-making problems and comparing the obtained result with the results of other existing strategies to show the applicability and effectiveness of the developed strategies. At the end, the main conclusion and future scope of research are summarized.

Soft sets (SSs), neutrosophic sets (NSs), and rough sets (RSs) are different mathematical models for handling uncertainties, but they are mutually related. In the fifth research paper ("Multi-Attribute Decision-Making Method Based on Neutrosophic Soft Rough Information"), Muhammad Akram, Sundas Shahzadi, and Florentin Smarandache introduce the notions of soft rough neutrosophic sets (SRNSs) and neutrosophic soft rough sets (NSRSs) as hybrid models for soft computing. The researchers describe a mathematical approach to handle decision-making problems in view of NSRSs and also present an efficient algorithm of the proposed hybrid model to solve decision-making problems.

Neutrosophic sets (NSs) handle uncertain information, while fuzzy sets (FSs) and intuitionistic fuzzy sets (IFs) fail to handle indeterminate information. Soft set theory, neutrosophic set theory, and rough set theory are different mathematical models for handling uncertainties and they are mutually related. The neutrosophic soft rough set (NSRS) model is a hybrid model combining neutrosophic soft sets with rough sets. Muhammad Akram, Hafsa M. Malik, Sundas Shahzadi, and Florentin Smarandache apply neutrosophic soft rough sets to graphs in the sixth research paper ("Neutrosophic Soft Rough Graphs with Application"), introducing the idea of neutrosophic soft rough graphs (NSRGs) and describing different methods for their construction. The authors consider the application of NSRG in decision-making problems, developing, in particular, efficient algorithms to solve decision-making problems.

In practical situations, one often has to handle programming problems involving indeterminate information. Building on the concepts of indeterminacy I and neutrosophic number (NN) (z = p+ qI for p,q infoNumber R), the seventh paper ("Neutrosophic Number Nonlinear Programming Problems and Their General Solution Methods under Neutrosophic Number Environments", by Jun Ye, Wenhua Cui, and Zhikang Lu) introduces some basic operations of NNs and concepts of NN nonlinear functions and inequalities. These functions and/or inequalities contain indeterminacy I and naturally lead to a formulation of NN nonlinear programming (NN-NP). These techniques include NN nonlinear optimization models for unconstrained and constrained problems and their general solution methods. Additionally, numerical examples are provided to show the effectiveness of the proposed NN-NP methods. It is obvious that the NN-NP problems usually yield NN optimal solutions, but not always. The possible optimal ranges of the decision variables and NN objective function are indicated when the indeterminacy I is considered for possible interval ranges in real situations. A neutrosophic number (a + bI) is a significant mathematical tool to deal with indeterminate and incomplete information which generally exists in real-world problems, where a and bI denote the determinate component and indeterminate component, respectively. Kalyan Mondal, Surapati Pramanik, Bibhas C. Giri, and Florentin Smarandache define score functions and accuracy functions for ranking neutrosophic numbers in the eighth paper, entitled "NN-Harmonic Mean Aggregation Operators-Based MCGDM Strategy in a Neutrosophic Number Environment". The authors then define a cosine function to determine the unknown weight of the criteria. The researchers define the neutrosophic number harmonic mean operators and prove their basic properties. Then, they develop two novel multi-criteria group decision-making (MCGDM) strategies using the proposed aggregation operators, solving a numerical example to demonstrate the feasibility, applicability, and effectiveness of the two proposed strategies.

Sensitivity analysis with the variation of "I" on neutrosophic numbers is performed to demonstrate how the preference for the ranking order of alternatives is sensitive to the change of "I". The efficiency of the developed strategies is ascertained by comparing the results obtained from the proposed strategies with the results obtained from the existing strategies in the literature.

A rough neutrosophic set model is a hybrid model, which deals with vagueness by using the lower and upper approximation spaces. In the ninth research paper ("Rough Neutrosophic Digraphs with Application"), Sidra Sayed, Nabeela Ishfaq, Muhammad Akram, and Florentin Smarandache apply the concept of rough neutrosophic sets to graphs, introducing rough neutrosophic digraphs and describing methods of their construction. Moreover, the researchers present the concept of self-complementary rough neutrosophic digraphs and, finally, consider an application of rough neutrosophic digraphs in decision-making.

The notion of a neutrosophic positive implicative N-ideal in BCK-algebras is introduced, and several properties are investigated in the tenth paper ("Neutrosophic Positive Implicative N-Ideals in BCK-Algebras", by Young Bae Jun, Florentin Smarandache, Seok-Zun Song, and Madad Khan). Relations between a neutrosophic N-ideal and a neutrosophic positive implicative N-ideal are discussed. Characterizations of a neutrosophic positive implicative N-ideal are considered. Conditions for a neutrosophic N-ideal to be a neutrosophic positive implicative N-ideal are provided. An extension property of a neutrosophic positive implicative N-ideal based on the negative indeterminacy membership function is discussed.

Hough transform (HT) is a useful tool for both pattern recognition and image processing communities. In regard to pattern recognition, it can extract unique features for the description of various shapes, such as lines, circles, ellipses, etc. In regard to image processing, a dozen of applications can be handled with HT, such as lane detection for autonomous cars, blood cell detection in microscope images, and so on. As HT is a straightforward shape detector in a given image, its shape detection ability is low in noisy images. To alleviate its weakness in noisy images analysis and improve its shape-detection performance, in the eleventh paper ("Neutrosophic Hough Transform"), ¨Umit Budak, Yanhui Guo, Abdulkadir S¸eng ¨ur, and Florentin Smarandache propose neutrosophic Hough transform (NHT). As it was proved earlier, neutrosophy theory-based image processing applications were successful in noisy environments. To this end, the Hough space is initially transferred into the NS domain by calculating the NS membership triples (T, I, and F). An indeterminacy filtering is constructed where the neighborhood information is used in order to remove the indeterminacy in the spatial neighborhood of neutrosophic Hough space. The potential peaks are detected on the basis of thresholding on the neutrosophic Hough space, and these peak locations are then used to detect the lines in the image domain. Extensive experiments on noisy and noise-free images are performed in order to show the efficiency of the proposed NHT algorithm. The authors also compare their proposed NHT with the traditional HT and fuzzy HT methods on a variety of images. The obtained results show the efficiency of the proposed NHT for noisy images.

**Florentin Smarandache, Jun Ye, Yanhui Guo**

*Guest Editors*

axioms

Article

# Neutrosophic Quadruple *BCK/BCI*-Algebras

**Young Bae Jun [1], Seok-Zun Song [2,*], Florentin Smarandache [3] and Hashem Bordbar [4]**

1 Department of Mathematics Education, Gyeongsang National University, Jinju 52828, Korea; skywine@gmail.com
2 Department of Mathematics, Jeju National University, Jeju 63243, Korea
3 Mathematics & Science Department, University of New Mexico, 705 Gurley Ave., Gallup, NM 87301, USA; fsmarandache@gmail.com
4 Department of Mathematics, Shahid Beheshti University, Tehran 1983963113, Iran; bordbar.amirh@gmail.com
* Correspondence: szsong@jejunu.ac.kr

Received: 20 April 2018; Accepted: 18 May 2018; Published: 18 June 2018

**Abstract:** The notion of a neutrosophic quadruple $BCK/BCI$-number is considered, and a neutrosophic quadruple $BCK/BCI$-algebra, which consists of neutrosophic quadruple $BCK/BCI$-numbers, is constructed. Several properties are investigated, and a (positive implicative) ideal in a neutrosophic quadruple $BCK$-algebra and a closed ideal in a neutrosophic quadruple $BCI$-algebra are studied. Given subsets $A$ and $B$ of a $BCK/BCI$-algebra, the set $NQ(A,B)$, which consists of neutrosophic quadruple $BCK/BCI$-numbers with a condition, is established. Conditions for the set $NQ(A,B)$ to be a (positive implicative) ideal of a neutrosophic quadruple $BCK$-algebra are provided, and conditions for the set $NQ(A,B)$ to be a (closed) ideal of a neutrosophic quadruple $BCI$-algebra are given. An example to show that the set $\{\tilde{0}\}$ is not a positive implicative ideal in a neutrosophic quadruple $BCK$-algebra is provided, and conditions for the set $\{\tilde{0}\}$ to be a positive implicative ideal in a neutrosophic quadruple $BCK$-algebra are then discussed.

**Keywords:** neutrosophic quadruple $BCK/BCI$-number; neutrosophic quadruple $BCK/BCI$-algebra; neutrosophic quadruple subalgebra; (positive implicative) neutrosophic quadruple ideal

**MSC:** 06F35; 03G25; 08A72

---

## 1. Introduction

The notion of a neutrosophic set was developed by Smarandache [1–3] and is a more general platform that extends the notions of classic sets, (intuitionistic) fuzzy sets, and interval valued (intuitionistic) fuzzy sets. Neutrosophic set theory is applied to a different field (see [4–8]). Neutrosophic algebraic structures in $BCK/BCI$-algebras are discussed in [9–16]. Neutrosophic quadruple algebraic structures and hyperstructures are discussed in [17,18].

In this paper, we will use neutrosophic quadruple numbers based on a set and construct neutrosophic quadruple $BCK/BCI$-algebras. We investigate several properties and consider ideals and positive implicative ideals in neutrosophic quadruple $BCK$-algebra, and closed ideals in neutrosophic quadruple $BCI$-algebra. Given subsets $A$ and $B$ of a neutrosophic quadruple $BCK/BCI$-algebra, we consider sets $NQ(A,B)$, which consist of neutrosophic quadruple $BCK/BCI$-numbers with a condition. We provide conditions for the set $NQ(A,B)$ to be a (positive implicative) ideal of a neutrosophic quadruple $BCK$-algebra and for the set $NQ(A,B)$ to be a (closed) ideal of a neutrosophic quadruple $BCI$-algebra. We give an example to show that the set $\{\tilde{0}\}$ is not a positive implicative ideal in a neutrosophic quadruple $BCK$-algebra, and we then consider conditions for the set $\{\tilde{0}\}$ to be a positive implicative ideal in a neutrosophic quadruple $BCK$-algebra.

## 2. Preliminaries

A $BCK/BCI$-algebra is an important class of logical algebras introduced by Iséki (see [19,20]).

By a $BCI$-algebra, we mean a set $X$ with a special element 0 and a binary operation $*$ that satisfies the following conditions:

(I) $(\forall x, y, z \in X)\ (((x * y) * (x * z)) * (z * y) = 0)$;
(II) $(\forall x, y \in X)\ ((x * (x * y)) * y = 0)$;
(III) $(\forall x \in X)\ (x * x = 0)$;
(IV) $(\forall x, y \in X)\ (x * y = 0,\ y * x = 0\ \Rightarrow\ x = y)$.

If a $BCI$-algebra $X$ satisfies the identity

(V) $(\forall x \in X)\ (0 * x = 0)$,

then $X$ is called a *BCK-algebra*. Any $BCK/BCI$-algebra $X$ satisfies the following conditions:

$$(\forall x \in X)\ (x * 0 = x) \tag{1}$$
$$(\forall x, y, z \in X)\ (x \leq y\ \Rightarrow\ x * z \leq y * z,\ z * y \leq z * x) \tag{2}$$
$$(\forall x, y, z \in X)\ ((x * y) * z = (x * z) * y) \tag{3}$$
$$(\forall x, y, z \in X)\ ((x * z) * (y * z) \leq x * y) \tag{4}$$

where $x \leq y$ if and only if $x * y = 0$. Any $BCI$-algebra $X$ satisfies the following conditions (see [21]):

$$(\forall x, y \in X)(x * (x * (x * y)) = x * y), \tag{5}$$
$$(\forall x, y \in X)(0 * (x * y) = (0 * x) * (0 * y)). \tag{6}$$

A $BCK$-algebra $X$ is said to be *positive implicative* if the following assertion is valid.

$$(\forall x, y, z \in X)\ ((x * z) * (y * z) = (x * y) * z). \tag{7}$$

A nonempty subset $S$ of a $BCK/BCI$-algebra $X$ is called a *subalgebra* of $X$ if $x * y \in S$ for all $x, y \in S$. A subset $I$ of a $BCK/BCI$-algebra $X$ is called an *ideal* of $X$ if it satisfies

$$0 \in I, \tag{8}$$
$$(\forall x \in X)\ (\forall y \in I)\ (x * y \in I\ \Rightarrow\ x \in I). \tag{9}$$

A subset $I$ of a $BCI$-algebra $X$ is called a *closed ideal* (see [21]) of $X$ if it is an ideal of $X$ which satisfies

$$(\forall x \in X)(x \in I\ \Rightarrow\ 0 * x \in I). \tag{10}$$

A subset $I$ of a $BCK$-algebra $X$ is called a *positive implicative ideal* (see [22]) of $X$ if it satisfies (8) and

$$(\forall x, y, z \in X)(((x * y) * z \in I,\ y * z \in I\ \Rightarrow\ x * z \in I). \tag{11}$$

Observe that every positive implicative ideal is an ideal, but the converse is not true (see [22]). Note also that a $BCK$-algebra $X$ is positive implicative if and only if every ideal of $X$ is positive implicative (see [22]).

We refer the reader to the books [21,22] for further information regarding $BCK/BCI$-algebras, and to the site "http://fs.gallup.unm.edu/neutrosophy.htm" for further information regarding neutrosophic set theory.

## 3. Neutrosophic Quadruple BCK/BCI-Algebras

We consider neutrosophic quadruple numbers based on a set instead of real or complex numbers.

**Definition 1.** *Let $X$ be a set. A neutrosophic quadruple X-number is an ordered quadruple $(a, xT, yI, zF)$ where $a, x, y, z \in X$ and $T$, $I$, $F$ have their usual neutrosophic logic meanings.*

The set of all neutrosophic quadruple X-numbers is denoted by $NQ(X)$, that is,

$$NQ(X) := \{(a, xT, yI, zF) \mid a, x, y, z \in X\},$$

and it is called the *neutrosophic quadruple set* based on X. If X is a $BCK/BCI$-algebra, a neutrosophic quadruple X-number is called a *neutrosophic quadruple $BCK/BCI$-number* and we say that $NQ(X)$ is the *neutrosophic quadruple $BCK/BCI$-set*.

Let X be a $BCK/BCI$-algebra. We define a binary operation $\odot$ on $NQ(X)$ by

$$(a, xT, yI, zF) \odot (b, uT, vI, wF) = (a * b, (x * u)T, (y * v)I, (z * w)F)$$

for all $(a, xT, yI, zF)$, $(b, uT, vI, wF) \in NQ(X)$. Given $a_1, a_2, a_3, a_4 \in X$, the neutrosophic quadruple $BCK/BCI$-number $(a_1, a_2T, a_3I, a_4F)$ is denoted by $\tilde{a}$, that is,

$$\tilde{a} = (a_1, a_2T, a_3I, a_4F),$$

and the zero neutrosophic quadruple $BCK/BCI$-number $(0, 0T, 0I, 0F)$ is denoted by $\tilde{0}$, that is,

$$\tilde{0} = (0, 0T, 0I, 0F).$$

We define an order relation "$\ll$" and the equality "$=$" on $NQ(X)$ as follows:

$$\tilde{x} \ll \tilde{y} \Leftrightarrow x_i \leq y_i \text{ for } i = 1, 2, 3, 4$$
$$\tilde{x} = \tilde{y} \Leftrightarrow x_i = y_i \text{ for } i = 1, 2, 3, 4$$

for all $\tilde{x}, \tilde{y} \in NQ(X)$. It is easy to verify that "$\ll$" is an equivalence relation on $NQ(X)$.

**Theorem 1.** *If X is a $BCK/BCI$-algebra, then $(NQ(X); \odot, \tilde{0})$ is a $BCK/BCI$-algebra.*

**Proof.** Let X be a $BCI$-algebra. For any $\tilde{x}, \tilde{y}, \tilde{z} \in NQ(X)$, we have

$$\begin{aligned}
(\tilde{x} \odot \tilde{y}) \odot (\tilde{x} \odot \tilde{z}) &= (x_1 * y_1, (x_2 * y_2)T, (x_3 * y_3)I, (x_4 * y_4)F) \\
&\qquad \odot (x_1 * z_1, (x_2 * z_2)T, (x_3 * z_3)I, (x_4 * z_4)F) \\
&= ((x_1 * y_1) * (x_1 * z_1), ((x_2 * y_2) * (x_2 * z_2))T, \\
&\qquad ((x_3 * y_3) * (x_3 * z_3))I, ((x_4 * y_4) * (x_4 * z_4))T) \\
&\ll (z_1 * y_1, (z_2 * y_2)T, (z_3 * y_3)I, (z_4 * y_4)F) \\
&= \tilde{z} \odot \tilde{y}
\end{aligned}$$

$$\begin{aligned}
\tilde{x} \odot (\tilde{x} \odot \tilde{y}) &= (x_1, x_2T, x_3I, x_4F) \odot (x_1 * y_1, (x_2 * y_2)T, (x_3 * y_3)I, (x_4 * y_4)F) \\
&= (x_1 * (x_1 * y_1), (x_2 * (x_2 * y_2))T, (x_3 * (x_3 * y_3))I, (x_4 * (x_4 * y_4))F) \\
&\ll (y_1, y_2T, y_3I, y_4F) \\
&= \tilde{y}
\end{aligned}$$

$$\begin{aligned}
\tilde{x} \odot \tilde{x} &= (x_1, x_2T, x_3I, x_4F) \odot (x_1, x_2T, x_3I, x_4F) \\
&= (x_1 * x_1, (x_2 * x_2)T, (x_3 * x_3)I, (x_4 * x_4)F) \\
&= (0, 0T, 0I, 0F) = \tilde{0}.
\end{aligned}$$

Assume that $\tilde{x} \odot \tilde{y} = \tilde{0}$ and $\tilde{y} \odot \tilde{x} = \tilde{0}$. Then

$$(x_1 * y_1, (x_2 * y_2)T, (x_3 * y_3)I, (x_4 * y_4)F) = (0, 0T, 0I, 0F)$$

and

$$(y_1 * x_1, (y_2 * x_2)T, (y_3 * x_3)I, (y_4 * x_4)F) = (0, 0T, 0I, 0F).$$

It follows that $x_1 * y_1 = 0 = y_1 * x_1$, $x_2 * y_2 = 0 = y_2 * x_2$, $x_3 * y_3 = 0 = y_3 * x_3$ and $x_4 * y_4 = 0 = y_4 * x_4$. Hence, $x_1 = y_1$, $x_2 = y_2$, $x_3 = y_3$, and $x_4 = y_4$, which implies that

$$\tilde{x} = (x_1, x_2T, x_3I, x_4F) = (y_1, y_2T, y_3I, y_4F) = \tilde{y}.$$

Therefore, we know that $(NQ(X); \odot, \tilde{0})$ is a $BCI$-algebra. We call it the *neutrosophic quadruple $BCI$-algebra*. Moreover, if X is a $BCK$-algebra, then we have

$$\tilde{0} \odot \tilde{x} = (0 * x_1, (0 * x_2)T, (0 * x_3)I, (0 * x_4)F) = (0, 0T, 0I, 0F) = \tilde{0}.$$

Hence, $(NQ(X); \odot, \tilde{0})$ is a $BCK$-algebra. We call it the *neutrosophic quadruple $BCK$-algebra*. □

**Example 1.** *If $X = \{0, a\}$, then the neutrosophic quadruple set $NQ(X)$ is given as follows:*

$$NQ(X) = \{\tilde{0}, \tilde{1}, \tilde{2}, \tilde{3}, \tilde{4}, \tilde{5}, \tilde{6}, \tilde{7}, \tilde{8}, \tilde{9}, \tilde{10}, \tilde{11}, \tilde{12}, \tilde{13}, \tilde{14}, \tilde{15}\}$$

*where*

$\tilde{0} = (0, 0T, 0I, 0F)$, $\tilde{1} = (0, 0T, 0I, aF)$, $\tilde{2} = (0, 0T, aI, 0F)$, $\tilde{3} = (0, 0T, aI, aF)$,
$\tilde{4} = (0, aT, 0I, 0F)$, $\tilde{5} = (0, aT, 0I, aF)$, $\tilde{6} = (0, aT, aI, 0F)$, $\tilde{7} = (0, aT, aI, aF)$,
$\tilde{8} = (a, 0T, 0I, 0F)$, $\tilde{9} = (a, 0T, 0I, aF)$, $\tilde{10} = (a, 0T, aI, 0F)$, $\tilde{11} = (a, 0T, aI, aF)$,
$\tilde{12} = (a, aT, 0I, 0F)$, $\tilde{13} = (a, aT, 0I, aF)$, $\tilde{14} = (a, aT, aI, 0F)$, *and* $\tilde{15} = (a, aT, aI, aF)$.

*Consider a BCK-algebra $X = \{0, a\}$ with the binary operation $*$, which is given in Table 1.*

**Table 1.** Cayley table for the binary operation "$*$".

| $*$ | 0 | $a$ |
|---|---|---|
| 0 | 0 | 0 |
| $a$ | $a$ | 0 |

*Then $(NQ(X), \odot, \tilde{0})$ is a BCK-algebra in which the operation $\odot$ is given by Table 2.*

**Table 2.** Cayley table for the binary operation "$\odot$".

| $\odot$ | $\tilde{0}$ | $\tilde{1}$ | $\tilde{2}$ | $\tilde{3}$ | $\tilde{4}$ | $\tilde{5}$ | $\tilde{6}$ | $\tilde{7}$ | $\tilde{8}$ | $\tilde{9}$ | $\tilde{10}$ | $\tilde{11}$ | $\tilde{12}$ | $\tilde{13}$ | $\tilde{14}$ | $\tilde{15}$ |
|---|---|---|---|---|---|---|---|---|---|---|---|---|---|---|---|---|
| $\tilde{0}$ | $\tilde{0}$ | $\tilde{0}$ | $\tilde{0}$ | $\tilde{0}$ | $\tilde{0}$ | $\tilde{0}$ | $\tilde{0}$ | $\tilde{0}$ | $\tilde{0}$ | $\tilde{0}$ | $\tilde{0}$ | $\tilde{0}$ | $\tilde{0}$ | $\tilde{0}$ | $\tilde{0}$ | $\tilde{0}$ |
| $\tilde{1}$ | $\tilde{1}$ | $\tilde{0}$ | $\tilde{1}$ | $\tilde{0}$ | $\tilde{1}$ | $\tilde{0}$ | $\tilde{1}$ | $\tilde{0}$ | $\tilde{1}$ | $\tilde{0}$ | $\tilde{1}$ | $\tilde{0}$ | $\tilde{1}$ | $\tilde{0}$ | $\tilde{1}$ | $\tilde{0}$ |
| $\tilde{2}$ | $\tilde{2}$ | $\tilde{2}$ | $\tilde{0}$ | $\tilde{0}$ | $\tilde{2}$ | $\tilde{2}$ | $\tilde{0}$ | $\tilde{0}$ | $\tilde{2}$ | $\tilde{2}$ | $\tilde{0}$ | $\tilde{0}$ | $\tilde{2}$ | $\tilde{2}$ | $\tilde{0}$ | $\tilde{0}$ |
| $\tilde{3}$ | $\tilde{3}$ | $\tilde{2}$ | $\tilde{1}$ | $\tilde{0}$ | $\tilde{3}$ | $\tilde{2}$ | $\tilde{1}$ | $\tilde{0}$ | $\tilde{3}$ | $\tilde{2}$ | $\tilde{1}$ | $\tilde{0}$ | $\tilde{3}$ | $\tilde{2}$ | $\tilde{1}$ | $\tilde{0}$ |
| $\tilde{4}$ | $\tilde{4}$ | $\tilde{4}$ | $\tilde{4}$ | $\tilde{4}$ | $\tilde{0}$ | $\tilde{0}$ | $\tilde{0}$ | $\tilde{0}$ | $\tilde{4}$ | $\tilde{4}$ | $\tilde{4}$ | $\tilde{4}$ | $\tilde{0}$ | $\tilde{0}$ | $\tilde{0}$ | $\tilde{0}$ |
| $\tilde{5}$ | $\tilde{5}$ | $\tilde{4}$ | $\tilde{5}$ | $\tilde{4}$ | $\tilde{1}$ | $\tilde{0}$ | $\tilde{1}$ | $\tilde{0}$ | $\tilde{5}$ | $\tilde{4}$ | $\tilde{5}$ | $\tilde{4}$ | $\tilde{1}$ | $\tilde{0}$ | $\tilde{1}$ | $\tilde{0}$ |
| $\tilde{6}$ | $\tilde{6}$ | $\tilde{6}$ | $\tilde{4}$ | $\tilde{4}$ | $\tilde{2}$ | $\tilde{2}$ | $\tilde{0}$ | $\tilde{0}$ | $\tilde{6}$ | $\tilde{6}$ | $\tilde{4}$ | $\tilde{4}$ | $\tilde{2}$ | $\tilde{2}$ | $\tilde{0}$ | $\tilde{0}$ |
| $\tilde{7}$ | $\tilde{7}$ | $\tilde{6}$ | $\tilde{5}$ | $\tilde{4}$ | $\tilde{3}$ | $\tilde{2}$ | $\tilde{1}$ | $\tilde{0}$ | $\tilde{7}$ | $\tilde{6}$ | $\tilde{5}$ | $\tilde{4}$ | $\tilde{3}$ | $\tilde{2}$ | $\tilde{1}$ | $\tilde{0}$ |
| $\tilde{8}$ | $\tilde{8}$ | $\tilde{8}$ | $\tilde{8}$ | $\tilde{8}$ | $\tilde{8}$ | $\tilde{8}$ | $\tilde{8}$ | $\tilde{8}$ | $\tilde{0}$ | $\tilde{0}$ | $\tilde{0}$ | $\tilde{0}$ | $\tilde{0}$ | $\tilde{0}$ | $\tilde{0}$ | $\tilde{0}$ |
| $\tilde{9}$ | $\tilde{9}$ | $\tilde{8}$ | $\tilde{8}$ | $\tilde{8}$ | $\tilde{9}$ | $\tilde{8}$ | $\tilde{9}$ | $\tilde{8}$ | $\tilde{9}$ | $\tilde{0}$ | $\tilde{1}$ | $\tilde{0}$ | $\tilde{1}$ | $\tilde{0}$ | $\tilde{1}$ | $\tilde{0}$ |
| $\tilde{10}$ | $\tilde{10}$ | $\tilde{10}$ | $\tilde{8}$ | $\tilde{8}$ | $\tilde{10}$ | $\tilde{10}$ | $\tilde{8}$ | $\tilde{8}$ | $\tilde{2}$ | $\tilde{2}$ | $\tilde{0}$ | $\tilde{2}$ | $\tilde{2}$ | $\tilde{2}$ | $\tilde{0}$ | $\tilde{0}$ |

**Table 2.** *Cont.*

| $\odot$ | $\tilde{0}$ | $\tilde{1}$ | $\tilde{2}$ | $\tilde{3}$ | $\tilde{4}$ | $\tilde{5}$ | $\tilde{6}$ | $\tilde{7}$ | $\tilde{8}$ | $\tilde{9}$ | $\tilde{10}$ | $\tilde{11}$ | $\tilde{12}$ | $\tilde{13}$ | $\tilde{14}$ | $\tilde{15}$ |
|---|---|---|---|---|---|---|---|---|---|---|---|---|---|---|---|---|
| $\tilde{11}$ | $\tilde{11}$ | $\tilde{10}$ | $\tilde{9}$ | $\tilde{8}$ | $\tilde{11}$ | $\tilde{10}$ | $\tilde{9}$ | $\tilde{8}$ | $\tilde{3}$ | $\tilde{2}$ | $\tilde{1}$ | $\tilde{0}$ | $\tilde{3}$ | $\tilde{2}$ | $\tilde{1}$ | $\tilde{0}$ |
| $\tilde{12}$ | $\tilde{12}$ | $\tilde{12}$ | $\tilde{12}$ | $\tilde{12}$ | $\tilde{8}$ | $\tilde{8}$ | $\tilde{8}$ | $\tilde{8}$ | $\tilde{4}$ | $\tilde{4}$ | $\tilde{4}$ | $\tilde{4}$ | $\tilde{0}$ | $\tilde{0}$ | $\tilde{0}$ | $\tilde{0}$ |
| $\tilde{13}$ | $\tilde{13}$ | $\tilde{12}$ | $\tilde{13}$ | $\tilde{12}$ | $\tilde{9}$ | $\tilde{8}$ | $\tilde{9}$ | $\tilde{8}$ | $\tilde{5}$ | $\tilde{4}$ | $\tilde{5}$ | $\tilde{4}$ | $\tilde{1}$ | $\tilde{0}$ | $\tilde{1}$ | $\tilde{0}$ |
| $\tilde{14}$ | $\tilde{14}$ | $\tilde{14}$ | $\tilde{12}$ | $\tilde{12}$ | $\tilde{10}$ | $\tilde{10}$ | $\tilde{8}$ | $\tilde{8}$ | $\tilde{6}$ | $\tilde{6}$ | $\tilde{4}$ | $\tilde{4}$ | $\tilde{2}$ | $\tilde{2}$ | $\tilde{0}$ | $\tilde{0}$ |
| $\tilde{15}$ | $\tilde{15}$ | $\tilde{14}$ | $\tilde{13}$ | $\tilde{12}$ | $\tilde{11}$ | $\tilde{10}$ | $\tilde{9}$ | $\tilde{8}$ | $\tilde{7}$ | $\tilde{6}$ | $\tilde{5}$ | $\tilde{4}$ | $\tilde{3}$ | $\tilde{2}$ | $\tilde{1}$ | $\tilde{0}$ |

**Theorem 2.** *The neutrosophic quadruple set $NQ(X)$ based on a positive implicative BCK-algebra $X$ is a positive implicative BCK-algebra.*

**Proof.** Let $X$ be a positive implicative $BCK$-algebra. Then $X$ is a $BCK$-algebra, so $(NQ(X); \odot, \tilde{0})$ is a $BCK$-algebra by Theorem 1. Let $\tilde{x}, \tilde{y}, \tilde{z} \in NQ(X)$. Then

$$(x_i * z_i) * (y_i * z_i) = (x_i * y_i) * z_i$$

for all $i = 1,2,3,4$ since $x_i, y_i, z_i \in X$ and $X$ is a positive implicative $BCK$-algebra. Hence, $(\tilde{x} \odot \tilde{z}) \odot (\tilde{y} * \tilde{z}) = (\tilde{x} \odot \tilde{y}) \odot \tilde{z}$; therefore, $NQ(X)$ based on a positive implicative $BCK$-algebra $X$ is a positive implicative $BCK$-algebra. □

**Proposition 1.** *The neutrosophic quadruple set $NQ(X)$ based on a positive implicative BCK-algebra $X$ satisfies the following assertions.*

$$(\forall \tilde{x}, \tilde{y}, \tilde{z} \in NQ(X)) \ (\tilde{x} \odot \tilde{y} \ll \tilde{z} \Rightarrow \tilde{x} \odot \tilde{z} \ll \tilde{y} \odot \tilde{z}) \tag{12}$$

$$(\forall \tilde{x}, \tilde{y} \in NQ(X)) \ (\tilde{x} \odot \tilde{y} \ll \tilde{y} \Rightarrow \tilde{x} \ll \tilde{y}). \tag{13}$$

**Proof.** Let $\tilde{x}, \tilde{y}, \tilde{z} \in NQ(X)$. If $\tilde{x} \odot \tilde{y} \ll \tilde{z}$, then

$$\tilde{0} = (\tilde{x} \odot \tilde{y}) \odot \tilde{z} = (\tilde{x} \odot \tilde{z}) \odot (\tilde{y} \odot \tilde{z}),$$

so $\tilde{x} \odot \tilde{z} \ll \tilde{y} \odot \tilde{z}$. Assume that $\tilde{x} \odot \tilde{y} \ll \tilde{y}$. Using Equation (12) implies that

$$\tilde{x} \odot \tilde{y} \ll \tilde{y} \odot \tilde{y} = \tilde{0},$$

so $\tilde{x} \odot \tilde{y} = \tilde{0}$, i.e., $\tilde{x} \ll \tilde{y}$. □

Let $X$ be a $BCK/BCI$-algebra. Given $a, b \in X$ and subsets $A$ and $B$ of $X$, consider the sets

$$NQ(a, B) := \{(a, aT, yI, zF) \in NQ(X) \mid y, z \in B\}$$

$$NQ(A, b) := \{(a, xT, bI, bF) \in NQ(X) \mid a, x \in A\}$$

$$NQ(A, B) := \{(a, xT, yI, zF) \in NQ(X) \mid a, x \in A; y, z \in B\}$$

$$NQ(A^*, B) := \bigcup_{a \in A} NQ(a, B)$$

$$NQ(A, B^*) := \bigcup_{b \in B} NQ(A, b)$$

and

$$NQ(A \cup B) := NQ(A,0) \cup NQ(0,B).$$

The set $NQ(A,A)$ is denoted by $NQ(A)$.

**Proposition 2.** *Let X be a BCK/BCI-algebra. Given $a, b \in X$ and subsets A and B of X, we have*

(1) *$NQ(A^*, B)$ and $NQ(A, B^*)$ are subsets of $NQ(A,B)$.*
(1) *If $0 \in A \cap B$ then $NQ(A \cup B)$ is a subset of $NQ(A,B)$.*

**Proof.** Straightforward. □

Let $X$ be a $BCK/BCI$-algebra. Given $a, b \in X$ and subalgebras $A$ and $B$ of $X$, $NQ(a,B)$ and $NQ(A,b)$ may not be subalgebras of $NQ(X)$ since

$$(a, aT, x_3I, x_4F) \odot (a, aT, u_3I, v_4F) = (0, 0T, (x_3 * u_3)I, (x_4 * v_4)F) \notin NQ(a,B)$$

and

$$(x_1, x_2T, bI, bF) \odot (u_1, u_2T, bI, bF) = (x_1 * u_1, (x_2 * u_2)T, 0I, 0F) \notin NQ(A,b)$$

for $(a,\ aT,\ x_3I,\ x_4F) \in NQ(a,B)$, $(a,\ aT,\ u_3I,\ v_4F) \in NQ(a,B)$, $(x_1,\ x_2T,\ bI,\ bF) \in NQ(A,b)$, and $(u_1, u_2T, bI, bF) \in NQ(A,b)$.

**Theorem 3.** *If A and B are subalgebras of a BCK/BCI-algebra X, then the set $NQ(A,B)$ is a subalgebra of $NQ(X)$, which is called a neutrosophic quadruple subalgebra.*

**Proof.** Assume that $A$ and $B$ are subalgebras of a $BCK/BCI$-algebra $X$. Let $\tilde{x} = (x_1, x_2T, x_3I, x_4F)$ and $\tilde{y} = (y_1, y_2T, y_3I, y_4F)$ be elements of $NQ(A,B)$. Then $x_1, x_2, y_1, y_2 \in A$ and $x_3, x_4, y_3, y_4 \in B$, which implies that $x_1 * y_1 \in A$, $x_2 * y_2 \in A$, $x_3 * y_3 \in B$, and $x_4 * y_4 \in B$. Hence,

$$\tilde{x} \odot \tilde{y} = (x_1 * y_1, (x_2 * y_2)T, (x_3 * y_3)I, (x_4 * y_4)F) \in NQ(A,B),$$

so $NQ(A,B)$ is a subalgebra of $NQ(X)$. □

**Theorem 4.** *If A and B are ideals of a BCK/BCI-algebra X, then the set $NQ(A,B)$ is an ideal of $NQ(X)$, which is called a neutrosophic quadruple ideal.*

**Proof.** Assume that $A$ and $B$ are ideals of a $BCK/BCI$-algebra $X$. Obviously, $\tilde{0} \in NQ(A,B)$. Let $\tilde{x} = (x_1,\ x_2T,\ x_3I,\ x_4F)$ and $\tilde{y} = (y_1,\ y_2T,\ y_3I,\ y_4F)$ be elements of $NQ(X)$ such that $\tilde{x} \odot \tilde{y} \in NQ(A,B)$ and $\tilde{y} \in NQ(A,B)$. Then

$$\tilde{x} \odot \tilde{y} = (x_1 * y_1, (x_2 * y_2)T, (x_3 * y_3)I, (x_4 * y_4)F) \in NQ(A,B),$$

so $x_1 * y_1 \in A$, $x_2 * y_2 \in A$, $x_3 * y_3 \in B$ and $x_4 * y_4 \in B$. Since $\tilde{y} \in NQ(A,B)$, we have $y_1, y_2 \in A$ and $y_3, y_4 \in B$. Since $A$ and $B$ are ideals of $X$, it follows that $x_1, x_2 \in A$ and $x_3, x_4 \in B$. Hence, $\tilde{x} = (x_1, x_2T, x_3I, x_4F) \in NQ(A,B)$, so $NQ(A,B)$ is an ideal of $NQ(X)$. □

Since every ideal is a subalgebra in a $BCK$-algebra, we have the following corollary.

**Corollary 1.** *If A and B are ideals of a BCK-algebra X, then the set $NQ(A,B)$ is a subalgebra of $NQ(X)$.*

The following example shows that Corollary 1 is not true in a $BCI$-algebra.

**Example 2.** *Consider a BCI-algebra* $(\mathbb{Z}, -, 0)$. *If we take* $A = \mathbb{N}$ *and* $B = \mathbb{Z}$, *then* $NQ(A, B)$ *is an ideal of* $NQ(\mathbb{Z})$. *However, it is not a subalgebra of* $NQ(\mathbb{Z})$ *since*

$$(2, 3T, -5I, 6F) \odot (3, 5T, 6I, -7F) = (-1, -2T, -11I, 13F) \notin NQ(A, B)$$

*for* $(2, 3T, -5I, 6F), (3, 5T, 6I, -7F) \in NQ(A, B)$.

**Theorem 5.** *If A and B are closed ideals of a BCI-algebra X, then the set* $NQ(A, B)$ *is a closed ideal of* $NQ(X)$.

**Proof.** If $A$ and $B$ are closed ideals of a $BCI$-algebra $X$, then the set $NQ(A, B)$ is an ideal of $NQ(X)$ by Theorem 4. Let $\tilde{x} = (x_1, x_2T, x_3I, x_4F) \in NQ(A, B)$. Then

$$\tilde{0} \odot \tilde{x} = (0 * x_1, (0 * x_2)T, (0 * x_3)I, (0 * x_4)F) \in NQ(A, B)$$

since $0 * x_1, 0 * x_2 \in A$ and $0 * x_3, 0 * x_4 \in B$. Therefore, $NQ(A, B)$ is a closed ideal of $NQ(X)$. □

Since every closed ideal of a $BCI$-algebra $X$ is a subalgebra of $X$, we have the following corollary.

**Corollary 2.** *If A and B are closed ideals of a BCI-algebra X, then the set* $NQ(A, B)$ *is a subalgebra of* $NQ(X)$.

In the following example, we know that there exist ideals $A$ and $B$ in a $BCI$-algebra $X$ such that $NQ(A, B)$ is not a closed ideal of $NQ(X)$.

**Example 3.** *Consider BCI-algebras* $(Y, *, 0)$ *and* $(\mathbb{Z}, -, 0)$. *Then* $X = Y \times \mathbb{Z}$ *is a BCI-algebra (see [21]). Let* $A = Y \times \mathbb{N}$ *and* $B = \{0\} \times \mathbb{N}$. *Then A and B are ideals of X, so* $NQ(A, B)$ *is an ideal of* $NQ(X)$ *by Theorem 4. Let* $((0,0), (0,1)T, (0,2)I, (0,3)F) \in NQ(A, B)$. *Then*

$$\begin{aligned}
&((0,0), (0,0)T, (0,0)I, (0,0)F) \odot ((0,0), (0,1)T, (0,2)I, (0,3)F) \\
&= ((0,0), (0,-1)T, (0,-2)I, (0,-3)F) \notin NQ(A, B).
\end{aligned}$$

*Hence,* $NQ(A, B)$ *is not a closed ideal of* $NQ(X)$.

We provide conditions wherethe set $NQ(A, B)$ is a closed ideal of $NQ(X)$.

**Theorem 6.** *Let A and B be ideals of a BCI-algebra X and let*

$$\Gamma := \{\tilde{a} \in NQ(X) \mid (\forall \tilde{x} \in NQ(X))(\tilde{x} \ll \tilde{a} \Rightarrow \tilde{x} = \tilde{a})\}.$$

*Assume that, if* $\Gamma \subseteq NQ(A, B)$, *then* $|\Gamma| < \infty$. *Then* $NQ(A, B)$ *is a closed ideal of* $NQ(X)$.

**Proof.** If $A$ and $B$ are ideals of $X$, then $NQ(A, B)$ is an ideal of $NQ(X)$ by Theorem 4. Let $\tilde{a} = (a_1, a_2T, a_3I, a_4F) \in NQ(A, B)$. For any $n \in \mathbb{N}$, denote $n(\tilde{a}) := \tilde{0} \odot (\tilde{0} \odot \tilde{a})^n$. Then $n(\tilde{a}) \in \Gamma$ and

$$\begin{aligned}
n(\tilde{a}) &= (0 * (0 * a_1)^n, (0 * (0 * a_2)^n)T, (0 * (0 * a_3)^n)I, (0 * (0 * a_4)^n)F) \\
&= (0 * (0 * a_1^n), (0 * (0 * a_2^n))T, (0 * (0 * a_3^n))I, (0 * (0 * a_4^n))F) \\
&= \tilde{0} \odot (\tilde{0} \odot \tilde{a}^n).
\end{aligned}$$

Hence,

$$\begin{aligned}
n(\tilde{a}) \odot \tilde{a}^n &= (\tilde{0} \odot (\tilde{0} \odot \tilde{a}^n)) \odot \tilde{a}^n \\
&= (\tilde{0} \odot \tilde{a}^n) \odot (\tilde{0} \odot \tilde{a}^n) \\
&= \tilde{0} \in NQ(A, B),
\end{aligned}$$

so $n(\tilde{a}) \in NQ(A,B)$, since $\tilde{a} \in NQ(A,B)$, and $NQ(A,B)$ is an ideal of $NQ(X)$. Since $|\Gamma| < \infty$, it follows that $k \in \mathbb{N}$ such that $n(\tilde{a}) = (n+k)(\tilde{a})$, that is, $n(\tilde{a}) = n(\tilde{a}) \odot (\tilde{0} \odot \tilde{a})^k$, and thus

$$\begin{aligned} k(\tilde{a}) &= \tilde{0} \odot (\tilde{0} \odot \tilde{a})^k \\ &= (n(\tilde{a}) \odot (\tilde{0} \odot \tilde{a})^k) \odot n(\tilde{a}) \\ &= n(\tilde{a}) \odot n(\tilde{a}) = \tilde{0}, \end{aligned}$$

i.e., $(k-1)(\tilde{a}) \odot (\tilde{0} \odot \tilde{a}) = \tilde{0}$. Since $\tilde{0} \odot \tilde{a} \in \Gamma$, it follows that $\tilde{0} \odot \tilde{a} = (k-1)(\tilde{a}) \in NQ(A,B)$. Therefore, $NQ(A,B)$ is a closed ideal of $NQ(X)$. □

**Theorem 7.** *Given two elements a and b in a BCI-algebra X, let*

$$A_a := \{x \in X \mid a * x = a\} \text{ and } B_b := \{x \in X \mid b * x = b\}. \tag{14}$$

*Then $NQ(A_a, B_b)$ is a closed ideal of $NQ(X)$.*

**Proof.** Since $a * 0 = a$ and $b * 0 = b$, we have $0 \in A_a \cap B_b$. Thus, $\tilde{0} \in NQ(A_a, B_b)$. If $x \in A_a$ and $y \in B_b$, then

$$0 * x = (a * x) * a = a * a = 0 \text{ and } 0 * y = (b * y) * b = b * b = 0. \tag{15}$$

Let $x, y, c, d \in X$ be such that $x, y * x \in A_a$ and $c, d * c \in B_b$. Then

$$(a * y) * a = 0 * y = (0 * y) * 0 = (0 * y) * (0 * x) = 0 * (y * x) = 0$$

and

$$(b * d) * b = 0 * d = (0 * d) * 0 = (0 * d) * (0 * c) = 0 * (d * c) = 0,$$

that is, $a * y \leq a$ and $b * d \leq b$. On the other hand,

$$a = a * (y * x) = (a * x) * (y * x) \leq a * y$$

and

$$b = b * (d * c) = (b * c) * (d * c) \leq b * d.$$

Thus, $a * y = a$ and $b * d = b$, i.e., $y \in A_a$ and $d \in B_b$. Hence, $A_a$ and $B_b$ are ideals of $X$, and $NQ(A_a, B_b)$ is therefore an ideal of $NQ(X)$ by Theorem 4. Let $\tilde{x} = (x_1, x_2T, x_3I, x_4F) \in NQ(A_a, B_b)$. Then $x_1, x_2 \in A_a$, and $x_3, x_4 \in B_b$. It follows from Equation (15) that $0 * x_1 = 0 \in A_a$, $0 * x_2 = 0 \in A_a$, $0 * x_3 = 0 \in B_b$, and $0 * x_4 = 0 \in B_b$. Hence,

$$\tilde{0} \odot \tilde{x} = (0 * x_1, (0 * x_2)T, (0 * x_3)I, (0 * x_4)F) \in NQ(A_a, B_b).$$

Therefore, $NQ(A_a, B_b)$ is a closed ideal of $NQ(X)$. □

**Proposition 3.** *Let A and B be ideals of a BCK-algebra X. Then*

$$NQ(A) \cap NQ(B) = \{\tilde{0}\} \Leftrightarrow (\forall \tilde{x} \in NQ(A))(\forall \tilde{y} \in NQ(B))(\tilde{x} \odot \tilde{y} = \tilde{x}). \tag{16}$$

**Proof.** Note that $NQ(A)$ and $NQ(B)$ are ideals of $NQ(X)$. Assume that $NQ(A) \cap NQ(B) = \{\tilde{0}\}$. Let

$$\tilde{x} = (x_1, x_2T, x_3I, x_4F) \in NQ(A) \text{ and } \tilde{y} = (y_1, y_2T, y_3I, y_4F) \in NQ(B).$$

Since $\tilde{x} \odot (\tilde{x} \odot \tilde{y}) \ll \tilde{x}$ and $\tilde{x} \odot (\tilde{x} \odot \tilde{y}) \ll \tilde{y}$, it follows that $\tilde{x} \odot (\tilde{x} \odot \tilde{y}) \in NQ(A) \cap NQ(B) = \{\tilde{0}\}$. Obviously, $(\tilde{x} \odot \tilde{y}) \odot \tilde{x} \in \{\tilde{0}\}$. Hence, $\tilde{x} \odot \tilde{y} = \tilde{x}$.

Conversely, suppose that $\tilde{x} \odot \tilde{y} = \tilde{x}$ for all $\tilde{x} \in NQ(A)$ and $\tilde{y} \in NQ(B)$. If $\tilde{z} \in NQ(A) \cap NQ(B)$, then $\tilde{z} \in NQ(A)$ and $\tilde{z} \in NQ(B)$, which is implied from the hypothesis that $\tilde{z} = \tilde{z} \odot \tilde{z} = \tilde{0}$. Hence $NQ(A) \cap NQ(B) = \{\tilde{0}\}$. □

**Theorem 8.** *Let A and B be subsets of a BCK-algebra X such that*

$$(\forall a, b \in A \cap B)(K(a, b) \subseteq A \cap B) \tag{17}$$

*where* $K(a, b) := \{x \in X \mid x * a \leq b\}$. *Then the set* $NQ(A, B)$ *is an ideal of* $NQ(X)$.

**Proof.** If $x \in A \cap B$, then $0 \in K(x, x)$ since $0 * x \leq x$. Hence, $0 \in A \cap B$ by Equation (17), so it is clear that $\tilde{0} \in NQ(A, B)$. Let $\tilde{x} = (x_1, x_2T, x_3I, x_4F)$ and $\tilde{y} = (y_1, y_2T, y_3I, y_4F)$ be elements of $NQ(X)$ such that $\tilde{x} \odot \tilde{y} \in NQ(A, B)$ and $\tilde{y} \in NQ(A, B)$. Then

$$\tilde{x} \odot \tilde{y} = (x_1 * y_1, (x_2 * y_2)T, (x_3 * y_3)I, (x_4 * y_4)F) \in NQ(A, B),$$

so $x_1 * y_1 \in A$, $x_2 * y_2 \in A$, $x_3 * y_3 \in B$, and $x_4 * y_4 \in B$. Using (II), we have $x_1 \in K(x_1 * y_1, y_1) \subseteq A$, $x_2 \in K(x_2 * y_2, y_2) \subseteq A$, $x_3 \in K(x_3 * y_3, y_3) \subseteq B$, and $x_4 \in K(x_4 * y_4, y_4) \subseteq B$. This implies that $\tilde{x} = (x_1, x_2T, x_3I, x_4F) \in NQ(A, B)$. Therefore, $NQ(A, B)$ is an ideal of $NQ(X)$. □

**Corollary 3.** *Let A and B be subsets of a BCK-algebra X such that*

$$(\forall a, x, y \in X)(x, y \in A \cap B,\ (a * x) * y = 0 \Rightarrow a \in A \cap B). \tag{18}$$

*Then the set* $NQ(A, B)$ *is an ideal of* $NQ(X)$.

**Theorem 9.** *Let A and B be nonempty subsets of a BCK-algebra X such that*

$$(\forall a, x, y \in X)(x, y \in A \text{ (or } B),\ a * x \leq y \Rightarrow a \in A \text{ (or } B)). \tag{19}$$

*Then the set* $NQ(A, B)$ *is an ideal of* $NQ(X)$.

**Proof.** Assume that the condition expressed by Equation (19) is valid for nonempty subsets $A$ and $B$ of X. Since $0 * x \leq x$ for any $x \in A$ (or $B$), we have $0 \in A$ (or $B$) by Equation (19). Hence, it is clear that $\tilde{0} \in NQ(A, B)$. Let $\tilde{x} = (x_1, x_2T, x_3I, x_4F)$ and $\tilde{y} = (y_1, y_2T, y_3I, y_4F)$ be elements of $NQ(X)$ such that $\tilde{x} \odot \tilde{y} \in NQ(A, B)$ and $\tilde{y} \in NQ(A, B)$. Then

$$\tilde{x} \odot \tilde{y} = (x_1 * y_1, (x_2 * y_2)T, (x_3 * y_3)I, (x_4 * y_4)F) \in NQ(A, B),$$

so $x_1 * y_1 \in A$, $x_2 * y_2 \in A$, $x_3 * y_3 \in B$, and $x_4 * y_4 \in B$. Note that $x_i * (x_i * y_i) \leq y_i$ for $i = 1, 2, 3, 4$. It follows from Equation (19) that $x_1, x_2 \in A$ and $x_3, x_4 \in B$. Hence,

$$\tilde{x} = (x_1, x_2T, x_3I, x_4F) \in NQ(A, B);$$

therefore, $NQ(A, B)$ is an ideal of $NQ(X)$. □

**Theorem 10.** *If A and B are positive implicative ideals of a BCK-algebra X, then the set* $NQ(A, B)$ *is a positive implicative ideal of* $NQ(X)$*, which is called a positive implicative neutrosophic quadruple ideal.*

**Proof.** Assume that $A$ and $B$ are positive implicative ideals of a $BCK$-algebra $X$. Obviously, $\tilde{0} \in NQ(A,B)$. Let $\tilde{x} = (x_1, x_2T, x_3I, x_4F)$, $\tilde{y} = (y_1, y_2T, y_3I, y_4F)$, and $\tilde{z} = (z_1, z_2T, z_3I, z_4F)$ be elements of $NQ(X)$ such that $(\tilde{x} \odot \tilde{y}) \odot \tilde{z} \in NQ(A,B)$ and $\tilde{y} \odot \tilde{z} \in NQ(A,B)$. Then

$$(\tilde{x} \odot \tilde{y}) \odot \tilde{z} = ((x_1 * y_1) * z_1, ((x_2 * y_2) * z_2)T, \\ ((x_3 * y_3) * z_3)I, ((x_4 * y_4) * z_4)F) \in NQ(A,B),$$

and

$$\tilde{y} \odot \tilde{z} = (y_1 * z_1, (y_2 * z_2)T, (y_3 * z_3)I, (y_4 * z_4)F) \in NQ(A,B),$$

so $(x_1 * y_1) * z_1 \in A$, $(x_2 * y_2) * z_2 \in A$, $(x_3 * y_3) * z_3 \in B$, $(x_4 * y_4) * z_4 \in B$, $y_1 * z_1 \in A$, $y_2 * z_2 \in A$, $y_3 * z_3 \in B$, and $y_4 * z_4 \in B$. Since $A$ and $B$ are positive implicative ideals of $X$, it follows that $x_1 * z_1, x_2 * z_2 \in A$ and $x_3 * z_3, x_4 * z_4 \in B$. Hence,

$$\tilde{x} \odot \tilde{z} = (x_1 * z_1, (x_2 * z_2)T, (x_3 * z_3)I, (x_4 * z_4)F) \in NQ(A,B),$$

so $NQ(A,B)$ is a positive implicative ideal of $NQ(X)$. □

**Theorem 11.** *Let $A$ and $B$ be ideals of a $BCK$-algebra $X$ such that*

$$(\forall x, y, z \in X)((x * y) * z \in A \text{ (or } B) \Rightarrow (x * z) * (y * z) \in A \text{ (or } B)). \tag{20}$$

*Then $NQ(A,B)$ is a positive implicative ideal of $NQ(X)$.*

**Proof.** Since $A$ and $B$ are ideals of $X$, it follows from Theorem 4 that $NQ(A,B)$ is an ideal of $NQ(X)$. Let $\tilde{x} = (x_1, x_2T, x_3I, x_4F)$, $\tilde{y} = (y_1, y_2T, y_3I, y_4F)$, and $\tilde{z} = (z_1, z_2T, z_3I, z_4F)$ be elements of $NQ(X)$ such that $(\tilde{x} \odot \tilde{y}) \odot \tilde{z} \in NQ(A,B)$ and $\tilde{y} \odot \tilde{z} \in NQ(A,B)$. Then

$$(\tilde{x} \odot \tilde{y}) \odot \tilde{z} = ((x_1 * y_1) * z_1, ((x_2 * y_2) * z_2)T, \\ ((x_3 * y_3) * z_3)I, ((x_4 * y_4) * z_4)F) \in NQ(A,B),$$

and

$$\tilde{y} \odot \tilde{z} = (y_1 * z_1, (y_2 * z_2)T, (y_3 * z_3)I, (y_4 * z_4)F) \in NQ(A,B),$$

so $(x_1 * y_1) * z_1 \in A$, $(x_2 * y_2) * z_2 \in A$, $(x_3 * y_3) * z_3 \in B$, $(x_4 * y_4) * z_4 \in B$, $y_1 * z_1 \in A$, $y_2 * z_2 \in A$, $y_3 * z_3 \in B$, and $y_4 * z_4 \in B$. It follows from Equation (20) that $(x_1 * z_1) * (y_1 * z_1) \in A$, $(x_2 * z_2) * (y_2 * z_2) \in A$, $(x_3 * z_3) * (y_3 * z_3) \in B$, and $(x_4 * z_4) * (y_4 * z_4) \in B$. Since $A$ and $B$ are ideals of $X$, we get $x_1 * z_1 \in A$, $x_2 * z_2 \in A$, $x_3 * z_3 \in B$, and $x_4 * z_4 \in B$. Hence,

$$\tilde{x} \odot \tilde{z} = (x_1 * z_1, (x_2 * z_2)T, (x_3 * z_3)I, (x_4 * z_4)F) \in NQ(A,B).$$

Therefore, $NQ(A,B)$ is a positive implicative ideal of $NQ(X)$. □

**Corollary 4.** *Let $A$ and $B$ be ideals of a $BCK$-algebra $X$ such that*

$$(\forall x, y \in X)((x * y) * y \in A \text{ (or } B) \Rightarrow x * y \in A \text{ (or } B)). \tag{21}$$

*Then $NQ(A,B)$ is a positive implicative ideal of $NQ(X)$.*

**Proof.** If the condition expressed in Equation (21) is valid, then the condition expressed in Equation (20) is true. Hence, $NQ(A,B)$ is a positive implicative ideal of $NQ(X)$ by Theorem 11. □

**Theorem 12.** *Let A and B be subsets of a BCK-algebra X such that* $0 \in A \cap B$ *and*

$$((x * y) * y) * z \in A \text{ (or } B), \; z \in A \text{ (or } B) \;\Rightarrow\; x * y \in A \text{ (or } B) \tag{22}$$

*for all* $x, y, z \in X$. *Then* $NQ(A, B)$ *is a positive implicative ideal of* $NQ(X)$.

**Proof.** Since $0 \in A \cap B$, it is clear that $\tilde{0} \in NQ(A, B)$. We first show that

$$(\forall x, y \in X)(x * y \in A \text{ (or } B), \; y \in A \text{ (or } B) \Rightarrow x \in A \text{ (or } B)). \tag{23}$$

Let $x, y \in X$ be such that $x * y \in A$ (or $B$) and $y \in A$ (or $B$). Then

$$((x * 0) * 0) * y = x * y \in A \text{ (or } B)$$

by Equation (1), which, based on Equations (1) and (22), implies that $x = x * 0 \in A$ (or $B$). Let $\tilde{x} = (x_1, x_2T, x_3I, x_4F)$, $\tilde{y} = (y_1, y_2T, y_3I, y_4F)$, and $\tilde{z} = (z_1, z_2T, z_3I, z_4F)$ be elements of $NQ(X)$ such that $(\tilde{x} \odot \tilde{y}) \odot \tilde{z} \in NQ(A, B)$ and $\tilde{y} \odot \tilde{z} \in NQ(A, B)$. Then

$$\begin{aligned}(\tilde{x} \odot \tilde{y}) \odot \tilde{z} = (&(x_1 * y_1) * z_1, ((x_2 * y_2) * z_2)T, \\ &((x_3 * y_3) * z_3)I, ((x_4 * y_4) * z_4)F) \in NQ(A, B),\end{aligned}$$

and

$$\tilde{y} \odot \tilde{z} = (y_1 * z_1, (y_2 * z_2)T, (y_3 * z_3)I, (y_4 * z_4)F) \in NQ(A, B),$$

so $(x_1 * y_1) * z_1 \in A$, $(x_2 * y_2) * z_2 \in A$, $(x_3 * y_3) * z_3 \in B$, $(x_4 * y_4) * z_4 \in B$, $y_1 * z_1 \in A$, $y_2 * z_2 \in A$, $y_3 * z_3 \in B$, and $y_4 * z_4 \in B$. Note that

$$(((x_i * z_i) * z_i) * (y_i * z_i)) * ((x_i * y_i) * z_i) = 0 \in A \text{ (or } B)$$

for $i = 1, 2, 3, 4$. Since $(x_i * y_i) * z_i \in A$ for $i = 1, 2$ and $(x_j * y_j) * z_j \in B$ for $j = 3, 4$, it follows from Equation (23) that $((x_i * z_i) * z_i) * (y_i * z_i) \in A$ for $i = 1, 2$, and $((x_j * z_j) * z_j) * (y_j * z_j) \in B$ for $j = 3, 4$. Moreover, since $y_i * z_i \in A$ for $i = 1, 2$, and $y_j * z_j \in B$ for $j = 3, 4$, we have $x_1 * z_1 \in A$, $x_2 * z_2 \in A$, $x_3 * z_3 \in B$, and $x_4 * z_4 \in B$ by Equation (22). Hence,

$$\tilde{x} \odot \tilde{z} = (x_1 * z_1, (x_2 * z_2)T, (x_3 * z_3)I, (x_4 * z_4)F) \in NQ(A, B).$$

Therefore, $NQ(A, B)$ is a positive implicative ideal of $NQ(X)$. □

**Theorem 13.** *Let A and B be subsets of a BCK-algebra X such that* $NQ(A, B)$ *is a positive implicative ideal of* $NQ(X)$. *Then the set*

$$\Omega_{\tilde{a}} := \{\tilde{x} \in NQ(X) \mid \tilde{x} \odot \tilde{a} \in NQ(A, B)\} \tag{24}$$

*is an ideal of* $NQ(X)$ *for any* $\tilde{a} \in NQ(X)$.

**Proof.** Obviously, $\tilde{0} \in \Omega_{\tilde{a}}$. Let $\tilde{x}$, $\tilde{y} \in NQ(X)$ be such that $\tilde{x} \odot \tilde{y} \in \Omega_{\tilde{a}}$ and $\tilde{y} \in \Omega_{\tilde{a}}$. Then $(\tilde{x} \odot \tilde{y}) \odot \tilde{a} \in NQ(A, B)$ and $\tilde{y} \odot \tilde{a} \in NQ(A, B)$. Since $NQ(A, B)$ is a positive implicative ideal of $NQ(X)$, it follows from Equation (11) that $\tilde{x} \odot \tilde{a} \in NQ(A, B)$ and therefore that $\tilde{x} \in \Omega_{\tilde{a}}$. Hence, $\Omega_{\tilde{a}}$ is an ideal of $NQ(X)$. □

Combining Theorems 12 and 13, we have the following corollary.

**Corollary 5.** *If A and B are subsets of a BCK-algebra X satisfying $0 \in A \cap B$ and the condition expressed in Equation (22), then the set $\Omega_{\tilde{a}}$ in Equation (24) is an ideal of $NQ(X)$ for all $\tilde{a} \in NQ(X)$.*

**Theorem 14.** *For any subsets A and B of a BCK-algebra X, if the set $\Omega_{\tilde{a}}$ in Equation (24) is an ideal of $NQ(X)$ for all $\tilde{a} \in NQ(X)$, then $NQ(A,B)$ is a positive implicative ideal of $NQ(X)$.*

**Proof.** Since $\tilde{0} \in \Omega_{\tilde{a}}$, we have $\tilde{0} = \tilde{0} \odot \tilde{a} \in NQ(A,B)$. Let $\tilde{x}$, $\tilde{y}$, $\tilde{z} \in NQ(X)$ be such that $(\tilde{x} \odot \tilde{y}) \odot \tilde{z} \in NQ(A,B)$ and $\tilde{y} \odot \tilde{z} \in NQ(A,B)$. Then $\tilde{x} \odot \tilde{y} \in \Omega_{\tilde{z}}$ and $\tilde{y} \in \Omega_{\tilde{z}}$. Since $\Omega_{\tilde{z}}$ is an ideal of $NQ(X)$, it follows that $\tilde{x} \in \Omega_{\tilde{z}}$. Hence, $\tilde{x} \odot \tilde{z} \in NQ(A,B)$. Therefore, $NQ(A,B)$ is a positive implicative ideal of $NQ(X)$. □

**Theorem 15.** *For any ideals A and B of a BCK-algebra X and for any $\tilde{a} \in NQ(X)$, if the set $\Omega_{\tilde{a}}$ in Equation (24) is an ideal of $NQ(X)$, then $NQ(X)$ is a positive implicative BCK-algebra.*

**Proof.** Let $\Omega$ be any ideal of $NQ(X)$. For any $\tilde{x}, \tilde{y}, \tilde{z} \in NQ(X)$, assume that $(\tilde{x} \odot \tilde{y}) \odot \tilde{z} \in \Omega$ and $\tilde{y} \odot \tilde{z} \in \Omega$. Then $\tilde{x} \odot \tilde{y} \in \Omega_{\tilde{z}}$ and $\tilde{y} \in \Omega_{\tilde{z}}$. Since $\Omega_{\tilde{z}}$ is an ideal of $NQ(X)$, it follows that $\tilde{x} \in \Omega_{\tilde{z}}$. Hence, $\tilde{x} \odot \tilde{z} \in \Omega$, which shows that $\Omega$ is a positive implicative ideal of $NQ(X)$. Therefore, $NQ(X)$ is a positive implicative BCK-algebra. □

In general, the set $\{\tilde{0}\}$ is an ideal of any neutrosophic quadruple BCK-algebra $NQ(X)$, but it is not a positive implicative ideal of $NQ(X)$ as seen in the following example.

**Example 4.** *Consider a BCK-algebra $X = \{0,1,2\}$ with the binary operation $*$, which is given in Table 3.*

**Table 3.** Cayley table for the binary operation "$*$".

| $*$ | **0** | **1** | **2** |
|---|---|---|---|
| 0 | 0 | 0 | 0 |
| 1 | 1 | 0 | 0 |
| 2 | 2 | 1 | 0 |

*Then the neutrosophic quadruple BCK-algebra $NQ(X)$ has 81 elements. If we take $\tilde{a} = (2,2T,2I,2F)$ and $\tilde{b} = (1,1T,1I,1F)$ in $NQ(X)$, then*

$$\begin{aligned}(\tilde{a} \odot \tilde{b}) \odot \tilde{b} &= ((2*1)*1, ((2*1)*1)T, ((2*1)*1)I, ((2*1)*1)F) \\ &= (1*1, (1*1)T, (1*1)I, (1*1)F) = (0,0T,0I,0F) = \tilde{0},\end{aligned}$$

*and $\tilde{b} \odot \tilde{b} = \tilde{0}$. However,*

$$\tilde{a} \odot \tilde{b} = (2*1, (2*1)T, (2*1)I, (2*1)F) = (1,1T,1I,1F) \neq \tilde{0}.$$

*Hence, $\{\tilde{0}\}$ is not a positive implicative ideal of $NQ(X)$.*

We now provide conditions for the set $\{\tilde{0}\}$ to be a positive implicative ideal in the neutrosophic quadruple BCK-algebra.

**Theorem 16.** *Let $NQ(X)$ be a neutrosophic quadruple BCK-algebra. If the set*

$$\Omega(\tilde{a}) := \{\tilde{x} \in NQ(X) \mid \tilde{x} \ll \tilde{a}\} \tag{25}$$

*is an ideal of $NQ(X)$ for all $\tilde{a} \in NQ(X)$, then $\{\tilde{0}\}$ is a positive implicative ideal of $NQ(X)$.*

**Proof.** We first show that

$$(\forall \tilde{x}, \tilde{y} \in NQ(X))((\tilde{x} \odot \tilde{y}) \odot \tilde{y} = \tilde{0} \Rightarrow \tilde{x} \odot \tilde{y} = \tilde{0}). \tag{26}$$

Assume that $(\tilde{x} \odot \tilde{y}) \odot \tilde{y} = \tilde{0}$ for all $\tilde{x}, \tilde{y} \in NQ(X)$. Then $\tilde{x} \odot \tilde{y} \ll \tilde{y}$, so $\tilde{x} \odot \tilde{y} \in \Omega(\tilde{y})$. Since $\tilde{y} \in \Omega(\tilde{y})$ and $\Omega(\tilde{y})$ is an ideal of $NQ(X)$, we have $\tilde{x} \in \Omega(\tilde{y})$. Thus, $\tilde{x} \ll \tilde{y}$, that is, $\tilde{x} \odot \tilde{y} = \tilde{0}$. Let $\tilde{u} := (\tilde{x} \odot \tilde{y}) \odot \tilde{y}$. Then

$$((\tilde{x} \odot \tilde{u}) \odot \tilde{y}) \odot \tilde{y} = ((\tilde{x} \odot \tilde{y}) \odot \tilde{y}) \odot \tilde{u} = \tilde{0},$$

which implies, based on Equations (3) and (26), that

$$(\tilde{x} \odot \tilde{y}) \odot ((\tilde{x} \odot \tilde{y}) \odot \tilde{y}) = (\tilde{x} \odot \tilde{y}) \odot \tilde{u} = (\tilde{x} \odot \tilde{u}) \odot \tilde{y} = \tilde{0},$$

that is, $\tilde{x} \odot \tilde{y} \ll (\tilde{x} \odot \tilde{y}) \odot \tilde{y}$. Since $(\tilde{x} \odot \tilde{y}) \odot \tilde{y} \ll \tilde{x} \odot \tilde{y}$, it follows that

$$(\tilde{x} \odot \tilde{y}) \odot \tilde{y} = \tilde{x} \odot \tilde{y}. \tag{27}$$

If we put $\tilde{y} = \tilde{x} \odot (\tilde{y} \odot (\tilde{y} \odot \tilde{x}))$ in Equation (27), then

$$\begin{aligned}
\tilde{x} \odot (\tilde{x} \odot (\tilde{y} \odot (\tilde{y} \odot \tilde{x}))) &= (\tilde{x} \odot (\tilde{x} \odot (\tilde{y} \odot (\tilde{y} \odot \tilde{x})))) \odot (\tilde{x} \odot (\tilde{y} \odot (\tilde{y} \odot \tilde{x}))) \\
&\ll (\tilde{y} \odot (\tilde{y} \odot \tilde{x})) \odot (\tilde{x} \odot (\tilde{y} \odot (\tilde{y} \odot \tilde{x}))) \\
&\ll (\tilde{y} \odot (\tilde{y} \odot \tilde{x})) \odot (\tilde{x} \odot \tilde{y}) \\
&= (\tilde{y} \odot (\tilde{x} \odot \tilde{y})) \odot (\tilde{y} \odot \tilde{x}) \\
&= ((\tilde{y} \odot (\tilde{x} \odot \tilde{y})) \odot (\tilde{y} \odot \tilde{x})) \odot (\tilde{y} \odot \tilde{x}) \\
&\ll (\tilde{x} \odot (\tilde{x} \odot \tilde{y})) \odot (\tilde{y} \odot \tilde{x}).
\end{aligned}$$

On the other hand,

$$\begin{aligned}
&((\tilde{x} \odot (\tilde{x} \odot \tilde{y})) \odot (\tilde{y} \odot \tilde{x})) \odot (\tilde{x} \odot (\tilde{x} \odot (\tilde{y} \odot (\tilde{y} \odot \tilde{x})))) \\
&= ((\tilde{x} \odot (\tilde{x} \odot (\tilde{x} \odot (\tilde{y} \odot (\tilde{y} \odot \tilde{x}))))) \odot (\tilde{x} \odot \tilde{y})) \odot (\tilde{y} \odot \tilde{x})) \\
&= ((\tilde{x} \odot (\tilde{y} \odot (\tilde{y} \odot \tilde{x}))) \odot (\tilde{x} \odot \tilde{y})) \odot (\tilde{y} \odot \tilde{x})) \\
&\ll (\tilde{y} \odot (\tilde{y} \odot (\tilde{y} \odot \tilde{x}))) \odot (\tilde{y} \odot \tilde{x})) = \tilde{0},
\end{aligned}$$

so $((\tilde{x} \odot (\tilde{x} \odot \tilde{y})) \odot (\tilde{y} \odot \tilde{x})) \odot (\tilde{x} \odot (\tilde{x} \odot (\tilde{y} \odot (\tilde{y} \odot \tilde{x})))) = \tilde{0}$, that is,

$$((\tilde{x} \odot (\tilde{x} \odot \tilde{y})) \odot (\tilde{y} \odot \tilde{x})) \ll \tilde{x} \odot (\tilde{x} \odot (\tilde{y} \odot (\tilde{y} \odot \tilde{x}))).$$

Hence,

$$\tilde{x} \odot (\tilde{x} \odot (\tilde{y} \odot (\tilde{y} \odot \tilde{x}))) = ((\tilde{x} \odot (\tilde{x} \odot \tilde{y})) \odot (\tilde{y} \odot \tilde{x})). \tag{28}$$

If we use $\tilde{y} \odot \tilde{x}$ instead of $\tilde{x}$ in Equation (28), then

$$\begin{aligned}
\tilde{y} \odot \tilde{x} &= (\tilde{y} \odot \tilde{x}) \odot \tilde{0} \\
&= (\tilde{y} \odot \tilde{x}) \odot ((\tilde{y} \odot \tilde{x}) \odot (\tilde{y} \odot (\tilde{y} \odot (\tilde{y} \odot \tilde{x})))) \\
&= ((\tilde{y} \odot \tilde{x}) \odot ((\tilde{y} \odot \tilde{x}) \odot \tilde{y})) \odot (\tilde{y} \odot (\tilde{y} \odot \tilde{x})) \\
&= (\tilde{y} \odot \tilde{x}) \odot (\tilde{y} \odot (\tilde{y} \odot \tilde{x})),
\end{aligned}$$

which, by taking $\tilde{x} = \tilde{y} \odot \tilde{x}$, implies that

$$\begin{aligned}\tilde{y} \odot (\tilde{y} \odot \tilde{x}) &= (\tilde{y} \odot (\tilde{y} \odot \tilde{x})) \odot (\tilde{y} \odot (\tilde{y} \odot (\tilde{y} \odot \tilde{x}))) \\ &= (\tilde{y} \odot (\tilde{y} \odot \tilde{x})) \odot (\tilde{y} \odot \tilde{x}).\end{aligned}$$

It follows that

$$\begin{aligned}(\tilde{y} \odot (\tilde{y} \odot \tilde{x})) \odot (\tilde{x} \odot \tilde{y}) &= ((\tilde{y} \odot (\tilde{y} \odot \tilde{x})) \odot (\tilde{y} \odot \tilde{x})) \odot (\tilde{x} \odot \tilde{y}) \\ &\ll (\tilde{x} \odot (\tilde{y} \odot \tilde{x})) \odot (\tilde{x} \odot \tilde{y}) \\ &= (\tilde{x} \odot (\tilde{x} \odot \tilde{y})) \odot (\tilde{y} \odot \tilde{x}),\end{aligned}$$

so,

$$\begin{aligned}\tilde{y} \odot \tilde{x} &= (\tilde{y} \odot (\tilde{y} \odot (\tilde{y} \odot \tilde{x}))) \odot \tilde{0} \\ &= (\tilde{y} \odot (\tilde{y} \odot (\tilde{y} \odot \tilde{x}))) \odot ((\tilde{y} \odot \tilde{x}) \odot \tilde{y}) \\ &\ll ((\tilde{y} \odot \tilde{x}) \odot ((\tilde{y} \odot \tilde{x}) \odot \tilde{y})) \odot (\tilde{y} \odot (\tilde{y} \odot \tilde{x})) \\ &= (\tilde{y} \odot \tilde{x}) \odot (\tilde{y} \odot (\tilde{y} \odot \tilde{x})) \\ &\ll (\tilde{y} \odot \tilde{x}) \odot \tilde{x}.\end{aligned}$$

Since $(\tilde{y} \odot \tilde{x}) \odot \tilde{x} \ll \tilde{y} \odot \tilde{x}$, it follows that

$$(\tilde{y} \odot \tilde{x}) \odot \tilde{x} = \tilde{y} \odot \tilde{x}. \tag{29}$$

Based on Equation (29), it follows that

$$\begin{aligned}&((\tilde{x} \odot \tilde{z}) * (\tilde{y} \odot \tilde{z})) \odot ((\tilde{x} \odot \tilde{y}) \odot \tilde{z}) \\ &= (((\tilde{x} \odot \tilde{z}) \odot \tilde{z}) \odot (\tilde{y} \odot \tilde{z})) \odot ((\tilde{x} \odot \tilde{y}) \odot \tilde{z}) \\ &\ll ((\tilde{x} \odot \tilde{z}) \odot \tilde{y}) \odot ((\tilde{x} \odot \tilde{y}) \odot \tilde{z}) \\ &= \tilde{0},\end{aligned}$$

that is, $(\tilde{x} \odot \tilde{z}) * (\tilde{y} \odot \tilde{z}) \ll (\tilde{x} \odot \tilde{y}) \odot \tilde{z}$. Note that

$$\begin{aligned}&((\tilde{x} \odot \tilde{y}) \odot \tilde{z}) \odot ((x \odot \tilde{z}) \odot (\tilde{y} \odot \tilde{z})) \\ &= ((\tilde{x} \odot \tilde{y}) \odot \tilde{z}) \odot ((x \odot (\tilde{y} \odot \tilde{z})) \odot \tilde{z}) \\ &\ll (\tilde{x} \odot \tilde{y}) \odot (\tilde{x} \odot (\tilde{y} \odot \tilde{z})) \\ &\ll (\tilde{y} \odot \tilde{z}) \odot \tilde{y} = \tilde{0},\end{aligned}$$

which shows that $(\tilde{x} \odot \tilde{y}) \odot \tilde{z} \ll (\tilde{x} \odot \tilde{z}) \odot (\tilde{y} \odot \tilde{z})$. Hence, $(\tilde{x} \odot \tilde{y}) \odot \tilde{z} = (\tilde{x} \odot \tilde{z}) \odot (\tilde{y} \odot \tilde{z})$. Therefore, $NQ(X)$ is a positive implicative, so $\{\tilde{0}\}$ is a positive implicative ideal of $NQ(X)$. □

## 4. Conclusions

We have considered a neutrosophic quadruple $BCK/BCI$-number on a set and established neutrosophic quadruple $BCK/BCI$-algebras, which consist of neutrosophic quadruple $BCK/BCI$-numbers. We have investigated several properties and considered ideal theory in a neutrosophic quadruple $BCK$-algebra and a closed ideal in a neutrosophic quadruple $BCI$-algebra. Using subsets $A$ and $B$ of a neutrosophic quadruple $BCK/BCI$-algebra, we have considered sets $NQ(A, B)$, which consist of neutrosophic quadruple $BCK/BCI$-numbers with a condition. We have provided conditions for the set $NQ(A, B)$ to be a (positive implicative) ideal of a neutrosophic quadruple $BCK$-algebra, and the set $NQ(A, B)$ to be a (closed) ideal of a neutrosophic quadruple $BCI$-algebra. We have provided an example to show that the set $\{\tilde{0}\}$ is not a positive implicative ideal in a neutrosophic quadruple $BCK$-algebra, and we have considered conditions for the set $\{\tilde{0}\}$ to be a positive implicative ideal in a neutrosophic quadruple $BCK$-algebra.

**Author Contributions:** Y.B.J. and S.-Z.S. initiated the main idea of this work and wrote the paper. F.S. and H.B. provided examples and checked the content. All authors conceived and designed the new definitions and results, and have read and approved the final manuscript for submission.

**Acknowledgments:** The authors wish to thank the anonymous reviewers for their valuable suggestions. The second author, Seok-Zun Song, was supported under the framework of international cooperation program managed by the National Research Foundation of Korea (2017K2A9A1A01092970).

**Conflicts of Interest:** The authors declare no conflict of interest.

## References

1. Smarandache, F. Neutrosophy, Neutrosophic Probability, Set, and Logic, ProQuest Information & Learning, Ann Arbor, Michigan, USA, p. 105, 1998. Available online: http://fs.gallup.unm.edu/eBook-neutrosophics6.pdf (accessed on 1 September 2007).
2. Smarandache, F. *A Unifying Field in Logics: Neutrosophic Logic. Neutrosophy, Neutrosophic Set, Neutrosophic Probability*; American Reserch Press: Rehoboth, NM, USA, 1999.
3. Smarandache, F. Neutrosophic set—A generalization of the intuitionistic fuzzy set. *Int. J. Pure Appl. Math.* **2005**, *24*, 287–297.
4. Garg, H. Linguistic single-valued neutrosophic prioritized aggregation operators and their applications to multiple-attribute group decision-making. *J. Ambient Intell. Humaniz. Comput.* **2018**, in press. [CrossRef]
5. Garg, H. Non-linear programming method for multi-criteria decision making problems under interval neutrosophic set environment. *Appl. Intell.* **2017**, in press. [CrossRef]
6. Garg, H. Some New Biparametric Distance Measures on Single-Valued Neutrosophic Sets with Applications to Pattern Recognition and Medical Diagnosis. *Information* **2017**, *8*, 162. [CrossRef]
7. Garg, H. Novel single-valued neutrosophic aggregated operators under Frank norm operation and its application to decision-making process. *Int. J. Uncertain. Quantif.* **2016**, *6*, 361–375.
8. Garg, H.; Garg, N. On single-valued neutrosophic entropy of order $\alpha$. *Neutrosophic Sets Syst.* **2016**, *14*, 21–28.
9. Saeid, A.B.; Jun, Y.B. Neutrosophic subalgebras of $BCK/BCI$-algebras based on neutrosophic points. *Ann. Fuzzy Math. Inform.* **2017**, *14*, 87–97.
10. Jun, Y.B. Neutrosophic subalgebras of several types in $BCK/BCI$-algebras. *Ann. Fuzzy Math. Inform.* **2017**, *14*, 75–86.
11. Jun, Y.B.; Kim, S.J.; Smarandache, F. Interval neutrosophic sets with applications in $BCK/BCI$-algebra. *Axioms* **2018**, *7*, 23. [CrossRef]
12. Jun, Y.B.; Smarandache, F.; Bordbar, H. Neutrosophic $\mathcal{N}$-structures applied to $BCK/BCI$-algebras. *Information* **2017**, *8*, 128. [CrossRef]
13. Jun, Y.B.; Smarandache, F.; Song, S.Z.; Khan, M. Neutrosophic positive implicative $\mathcal{N}$-ideals in $BCK/BCI$-algebras. *Axioms* **2018**, *7*, 3. [CrossRef]
14. Khan, M.; Anis, S.; Smarandache, F.; Jun, Y.B. Neutrosophic $\mathcal{N}$-structures and their applications in semigroups. *Ann. Fuzzy Math. Inform.* **2017**, *14*, 583–598.
15. Öztürk, M.A.; Jun, Y.B. Neutrosophic ideals in $BCK/BCI$-algebras based on neutrosophic points. *J. Inter. Math. Virtual Inst.* **2018**, *8*, 1–17.
16. Song, S.Z.; Smarandache, F.; Jun, Y.B. Neutrosophic commutative $\mathcal{N}$-ideals in $BCK$-algebras. *Information* **2017**, *8*, 130. [CrossRef]
17. Agboola, A.A.A.; Davvaz, B.; Smarandache, F. Neutrosophic quadruple algebraic hyperstructures. *Ann. Fuzzy Math. Inform.* **2017**, *14*, 29–42.
18. Akinleye, S.A.; Smarandache, F.; Agboola, A.A.A. On neutrosophic quadruple algebraic structures. *Neutrosophic Sets Syst.* **2016**, *12*, 122–126.
19. Iséki, K. On $BCI$-algebras. *Math. Semin. Notes* **1980**, *8*, 125–130.

20. Iséki, K.; Tanaka, S. An introduction to the theory of *BCK*-algebras. *Math. Jpn.* **1978**, *23*, 1–26.
21. Huang, Y. *BCI-Algebra*; Science Press: Beijing, China, 2006.
22. Meng, J.; Jun, Y.B. *BCK-Algebras*; Kyungmoonsa Co.: Seoul, Korea, 1994.

axioms

MDPI

Article

# Decision-Making with Bipolar Neutrosophic TOPSIS and Bipolar Neutrosophic ELECTRE-I

**Muhammad Akram [1,*], Shumaiza [1] and Florentin Smarandache [2]**

1 Department of Mathematics, University of the Punjab, New Campus, Lahore 54590, Pakistan; shumaiza00@gmail.com
2 Mathematics and Science Department, University of New Mexico, Gallup, NM 87301, USA; fsmarandache@gmail.com
* Correspondence: m.akram@pucit.edu.pk

Received: 12 April 2018; Accepted: 11 May 2018; Published: 15 May 2018

**Abstract:** Technique for the order of preference by similarity to ideal solution (TOPSIS) and elimination and choice translating reality (ELECTRE) are widely used methods to solve multi-criteria decision making problems. In this research article, we present bipolar neutrosophic TOPSIS method and bipolar neutrosophic ELECTRE-I method to solve such problems. We use the revised closeness degree to rank the alternatives in our bipolar neutrosophic TOPSIS method. We describe bipolar neutrosophic TOPSIS method and bipolar neutrosophic ELECTRE-I method by flow charts. We solve numerical examples by proposed methods. We also give a comparison of these methods.

**Keywords:** neutrosophic sets; bipolar neutrosophic TOPSIS; bipolar neutrosophic ELECTRE-I; normalized Euclidean distance

## 1. Introduction

The theory of fuzzy sets was introduced by Zadeh [1]. Fuzzy set theory allows objects to be members of the set with a degree of membership, which can take any value within the unit closed interval $[0,1]$. Smarandache [2] originally introduced neutrosophy, a branch of philosophy which examines the origin, nature, and scope of neutralities, as well as their connections with different intellectual spectra. To apply neutrosophic set in real-life problems more conveniently, Smarandache [2] and Wang et al. [3] defined single-valued neutrosophic sets which takes the value from the subset of $[0,1]$. Thus, a single-valued neutrosophic set is an instance of neutrosophic set, and can be used feasibly to deal with real-world problems, especially in decision support. Deli et al. [4] dealt with bipolar neutrosophic sets, which is an extension of bipolar fuzzy sets [5].

Multi-criteria decision making (MCDM) is a process to make an ideal choice that has the highest degree of achievement from a set of alternatives that are characterized in terms of multiple conflicting criteria. Hwang and Yoon [6] developed the TOPSIS method, which is one of the most favorable and effective MCDM methods to solve MCDM problems. In classical MCDM methods, the attribute values and weights are determined precisely. To deal with problems consisting of incomplete and vague information, in 2000 Chen [7] conferred the fuzzy version of TOPSIS method for the first time. Chung and Chu [8] presented fuzzy TOPSIS method under group decision for facility location selection problem. Hadi et al. [9] proposed the fuzzy inferior ratio method for multiple attribute decision making problems. Joshi and Kumar [10] discussed the TOPSIS method based on intuitionistic fuzzy entropy and distance measure for multi criteria decision making. A comparative study of multiple criteria decision making methods under stochastic inputs is described by Kolios et al. [11]. Akram et al. [12–14] considered decision support systems based on bipolar fuzzy graphs. Applications of bipolar fuzzy sets to graphs have been discussed in [15,16]. Faizi et al. [17] presented group decision making for hesitant fuzzy sets based on characteristic objects method. Recently, Alghamdi et al. [18] have studied multi-criteria decision-making methods in bipolar fuzzy environment. Dey et al. [19] considered TOPSIS method for solving the decision making problem under bipolar neutrosophic environment.

On the other hand, the ELECTRE is one of the useful MCDM methods. This outranking method was proposed by Benayoun et al. [20], which was later referred to as ELECTRE-I method. Different versions of ELECTRE method have been developed as ELECTRE-I, II, III, IV and TRI. Hatami-Marbini and Tavana [21] extended the ELECTRE-I method and gave an alternative fuzzy outranking method to deal with uncertain and linguistic information. Aytac et al. [22] considered fuzzy ELECTRE-I method for evaluating catering firm alternatives. Wu and Chen [23] proposed the multi-criteria analysis approach ELECTRE based on intuitionistic fuzzy sets. In this research article, we present bipolar neutrosophic TOPSIS method and bipolar neutrosophic ELECTRE-I method to solve MCDM problems. We use the revised closeness degree to rank the alternatives in our bipolar neutrosophic TOPSIS method. We describe bipolar neutrosophic TOPSIS method and bipolar neutrosophic ELECTRE-I method by flow charts. We solve numerical examples by proposed methods. We also give a comparison of these methods. For other notions and applications that are not mentioned in this paper, the readers are referred to [24–29].

## 2. Bipolar Neutrosophic TOPSIS Method

**Definition 1.** *Ref. [4] Let C be a nonempty set. A bipolar neutrosophic set (BNS) $\widetilde{B}$ on C is defined as follows*

$$\widetilde{B} = \{c, \langle T^+_{\widetilde{B}}(c), I^+_{\widetilde{B}}(c), F^+_{\widetilde{B}}(c), T^-_{\widetilde{B}}(c), I^-_{\widetilde{B}}(c), F^-_{\widetilde{B}}(c)\rangle \mid c \in C\},$$

*where, $T^+_{\widetilde{B}}(c), I^+_{\widetilde{B}}(c), F^+_{\widetilde{B}}(c) : C \to [0,1]$ and $T^-_{\widetilde{B}}(c), I^-_{\widetilde{B}}(c), F^-_{\widetilde{B}}(c) : C \to [-1,0]$.*

We now describe our proposed bipolar neutrosophic TOPSIS method.

Let $S = \{S_1, S_2, \cdots, S_m\}$ be a set of $m$ favorable alternatives and let $T = \{T_1, T_2, \cdots, T_n\}$ be a set of $n$ attributes. Let $W = [w_1 \ w_2 \ \cdots \ w_n]^T$ be the weight vector such that $0 \leq w_j \leq 1$ and $\sum\limits_{j=1}^{n} w_j = 1$. Suppose that the rating value of each alternative $S_i, (i = 1, 2, \cdots, m)$ with respect to the attributes $T_j, (j = 1, 2, \cdots, n)$ is given by decision maker in the form of bipolar neutrosophic sets (BNSs). The steps of bipolar neutrosophic TOPSIS method are described as follows:

(i) Each value of alternative is estimated with respect to $n$ criteria. The value of each alternative under each criterion is given in the form of BNSs and they can be expressed in the decision matrix as

$$K = [k_{ij}]_{m \times n} = \begin{bmatrix} k_{11} & k_{12} & \ldots & k_{1n} \\ k_{21} & k_{22} & \ldots & k_{2n} \\ . & . & \ldots & . \\ . & . & \ldots & . \\ k_{m1} & k_{m2} & \ldots & k_{mn} \end{bmatrix}.$$

Each entry $k_{ij} =< T^+_{ij}, I^+_{ij}, F^+_{ij}, T^-_{ij}, I^-_{ij}, F^-_{ij} >$, where, $T^+_{ij}, I^+_{ij}$ and $F^+_{ij}$ represent the degree of positive truth, indeterminacy and falsity membership, respectively, whereas, $T^-_{ij}, I^-_{ij}$ and $F^-_{ij}$ represent the degree of negative truth, indeterminacy and falsity membership, respectively, such that $T^+_{ij}, I^+_{ij}, F^+_{ij} \in [0,1]$, $T^-_{ij}, I^-_{ij}, F^-_{ij} \in [-1,0]$ and $0 \leq T^+_{ij} + I^+_{ij} + F^+_{ij} - T^-_{ij} - I^-_{ij} - F^-_{ij} \leq 6$, $i = 1, 2, 3, ..., m; j = 1, 2, 3, ..., n$.

(ii) Suppose that the weights of the criteria are not equally assigned and they are totally unknown to the decision maker. We use the maximizing deviation method [30] to determine the unknown weights of the criteria. Therefore, the weight of the attribute $T_j$ is given as

$$w_j = \frac{\sum\limits_{i=1}^{m}\sum\limits_{l=1}^{m}|k_{ij}-k_{lj}|}{\sqrt{\sum\limits_{j=1}^{n}\left(\sum\limits_{i=1}^{m}\sum\limits_{l=1}^{m}|k_{ij}-k_{lj}|\right)^2}},$$

and the normalized weight of the attribute $T_j$ is given as

$$w_j^* = \frac{\sum\limits_{i=1}^{m}\sum\limits_{l=1}^{m}|k_{ij}-k_{lj}|}{\sum\limits_{j=1}^{n}\left(\sum\limits_{i=1}^{m}\sum\limits_{l=1}^{m}|k_{ij}-k_{lj}|\right)}.$$

(iii) The accumulated weighted bipolar neutrosophic decision matrix is computed by multiplying the weights of the attributes to aggregated decision matrix as follows:

$$K \otimes W = [k_{ij}^{w_j}]_{m\times n} = \begin{bmatrix} k_{11}^{w_1} & k_{12}^{w_2} & \dots & k_{1n}^{w_n} \\ k_{21}^{w_1} & k_{22}^{w_2} & \dots & k_{2n}^{w_n} \\ . & . & \dots & . \\ . & . & \dots & . \\ k_{m1}^{w_1} & k_{m2}^{w_2} & \dots & k_{mn}^{w_n} \end{bmatrix}.$$

where

$$\begin{aligned} k_{ij}^{w_j} &=< T_{ij}^{w_j+}, I_{ij}^{w_j+}, F_{ij}^{w_j+}, T_{ij}^{w_j-}, I_{ij}^{w_j-}, F_{ij}^{w_j-} > \\ &=< 1-(1-T_{ij}^{+})^{w_j}, (I_{ij}^{+})^{w_j}, (F_{ij}^{+})^{w_j}, -(-T_{ij}^{-})^{w_j}, -(-I_{ij}^{-})^{w_j}, -(1-(1-(-F_{ij}^{-}))^{w_j}) >, \end{aligned}$$

(iv) Two types of attributes, benefit type attributes and cost type attributes, are mostly applicable in real life decision making. The bipolar neutrosophic relative positive ideal solution (BNRPIS) and bipolar neutrosophic relative negative ideal solution (BNRNIS) for both type of attributes are defined as follows:

$$\begin{aligned} BNRPIS = \Big(&\langle {}^{+}T_1^{w_1+}, {}^{+}I_1^{w_1+}, {}^{+}F_1^{w_1+}, {}^{+}T_1^{w_1-}, {}^{+}I_1^{w_1-}, {}^{+}F_1^{w_1-}\rangle, \langle {}^{+}T_2^{w_2+}, {}^{+}I_2^{w_2+}, {}^{+}F_2^{w_2+}, {}^{+}T_2^{w_2-}, \\ &{}^{+}I_2^{w_2-}, {}^{+}F_2^{w_2-}\rangle, \dots, \langle {}^{+}T_n^{w_n+}, {}^{+}I_n^{w_n+}, {}^{+}F_n^{w_n+}, {}^{+}T_n^{w_n-}, {}^{+}I_n^{w_n-}, {}^{+}F_n^{w_n-}\rangle\Big), \\ BNRNIS = \Big(&\langle {}^{-}T_1^{w_1+}, {}^{-}I_1^{w_1+}, {}^{-}F_1^{w_1+}, {}^{-}T_1^{w_1-}, {}^{-}I_1^{w_1-}, {}^{-}F_1^{w_1-}\rangle, \langle {}^{-}T_2^{w_2+}, {}^{-}I_2^{w_2+}, {}^{-}F_2^{w_2+}, {}^{-}T_2^{w_2-}, \\ &{}^{-}I_2^{w_2-}, {}^{-}F_2^{w_2-}\rangle, \dots, \langle {}^{-}T_n^{w_n+}, {}^{-}I_n^{w_n+}, {}^{-}F_n^{w_n+}, {}^{-}T_n^{w_n-}, {}^{-}I_n^{w_n-}, {}^{-}F_n^{w_n-}\rangle\Big), \end{aligned}$$

such that, for benefit type criteria, $j = 1, 2, ..., n$

$$\langle {}^{+}T_j^{w_j+}, {}^{+}I_j^{w_j+}, {}^{+}F_j^{w_j+}, {}^{+}T_j^{w_j-}, {}^{+}I_j^{w_j-}, {}^{+}F_j^{w_j-} \rangle = \langle \max(T_{ij}^{w_j+}), \min(I_{ij}^{w_j+}), \min(F_{ij}^{w_j+}), \min(T_{ij}^{w_j-}), \max(I_{ij}^{w_j-}), \max(F_{ij}^{w_j-}) \rangle,$$

$$\langle {}^{-}T_j^{w_j+}, {}^{-}I_j^{w_j+}, {}^{-}F_j^{w_j+}, {}^{-}T_j^{w_j-}, {}^{-}I_j^{w_j-}, {}^{-}F_j^{w_j-} \rangle = \langle \min(T_{ij}^{w_j+}), \max(I_{ij}^{w_j+}), \max(F_{ij}^{w_j+}), \max(T_{ij}^{w_j-}), \min(I_{ij}^{w_j-}), \min(F_{ij}^{w_j-}) \rangle.$$

Similarly, for cost type criteria, $j = 1, 2, ..., n$

$$\langle {}^{+}T_j^{w_j+}, {}^{+}I_j^{w_j+}, {}^{+}F_j^{w_j+}, {}^{+}T_j^{w_j-}, {}^{+}I_j^{w_j-}, {}^{+}F_j^{w_j-} \rangle = \langle \min(T_{ij}^{w_j+}), \max(I_{ij}^{w_j+}), \max(F_{ij}^{w_j+}), \max(T_{ij}^{w_j-}), \min(I_{ij}^{w_j-}), \min(F_{ij}^{w_j-}) \rangle,$$

$$\langle {}^{-}T_j^{w_j+}, {}^{-}I_j^{w_j+}, {}^{-}F_j^{w_j+}, {}^{-}T_j^{w_j-}, {}^{-}I_j^{w_j-}, {}^{-}F_j^{w_j-} \rangle = \langle \max(T_{ij}^{w_j+}), \min(I_{ij}^{w_j+}), \min(F_{ij}^{w_j+}), \min(T_{ij}^{w_j-}), \max(I_{ij}^{w_j-}), \max(F_{ij}^{w_j-}) \rangle.$$

(v) The normalized Euclidean distance of each alternative $\langle T_{ij}^{w_j+}, I_{ij}^{w_j+}, F_{ij}^{w_j+}, T_{ij}^{w_j-}, I_{ij}^{w_j-}, F_{ij}^{w_j-} \rangle$ from the BNRPIS $\langle {}^{+}T_j^{w_j+}, {}^{+}I_j^{w_j+}, {}^{+}F_j^{w_j+}, {}^{+}T_j^{w_j-}, {}^{+}I_j^{w_j-}, {}^{+}F_j^{w_j-} \rangle$ can be calculated as

$$d_N(S_i, BNRPIS) = \sqrt{\frac{1}{6n} \sum_{j=1}^{n} \left\{ \begin{array}{c} (T_{ij}^{w_j+} - {}^{+}T_j^{w_j+})^2 + (I_{ij}^{w_j+} - {}^{+}I_j^{w_j+})^2 + (F_{ij}^{w_j+} - {}^{+}F_j^{w_j+})^2 + \\ (T_{ij}^{w_j-} - {}^{+}T_j^{w_j-})^2 + (I_{ij}^{w_j-} - {}^{+}I_j^{w_j-})^2 + (F_{ij}^{w_j-} - {}^{+}F_j^{w_j-})^2 \end{array} \right\}},$$

and the normalized Euclidean distance of each alternative $\langle T_{ij}^{w_j+}, I_{ij}^{w_j+}, F_{ij}^{w_j+}, T_{ij}^{w_j-}, I_{ij}^{w_j-}, F_{ij}^{w_j-} \rangle$ from the BNRNIS $\langle {}^{-}T_j^{w_j+}, {}^{-}I_j^{w_j+}, {}^{-}F_j^{w_j+}, {}^{-}T_j^{w_j-}, {}^{-}I_j^{w_j-}, {}^{-}F_j^{w_j-} \rangle$ can be calculated as

$$d_N(S_i, BNRNIS) = \sqrt{\frac{1}{6n} \sum_{j=1}^{n} \left\{ \begin{array}{c} (T_{ij}^{w_j+} - {}^{-}T_j^{w_j+})^2 + (I_{ij}^{w_j+} - {}^{-}I_j^{w_j+})^2 + (F_{ij}^{w_j+} - {}^{-}F_j^{w_j+})^2 + \\ (T_{ij}^{w_j-} - {}^{-}T_j^{w_j-})^2 + (I_{ij}^{w_j-} - {}^{-}I_j^{w_j-})^2 + (F_{ij}^{w_j-} - {}^{-}F_j^{w_j-})^2 \end{array} \right\}}.$$

(vi) Revised closeness degree of each alternative to BNRPIS represented as $\rho_i$ and it is calculated using formula

$$\rho(S_i) = \frac{d_N(S_i, BNRNIS)}{\max\{d_N(S_i, BNRNIS)\}} - \frac{d_N(S_i, BNRPIS)}{\min\{d_N(S_i, BNRPIS)\}}, \quad i = 1, 2, ..., m.$$

(vii) By using the revised closeness degrees, the inferior ratio to each alternative is determined as follows:

$$IR(i) = \frac{\rho(S_i)}{\min\limits_{1 \le i \le m} (\rho(S_i))}.$$

It is clear that each value of $IR(i)$ lies in the closed unit interval [0,1].

(viii) The alternatives are ranked according to the ascending order of inferior ratio values and the best alternative with minimum choice value is chosen.

Geometric representation of the procedure of our proposed bipolar neutrosophic TOPSIS method is shown in Figure 1.

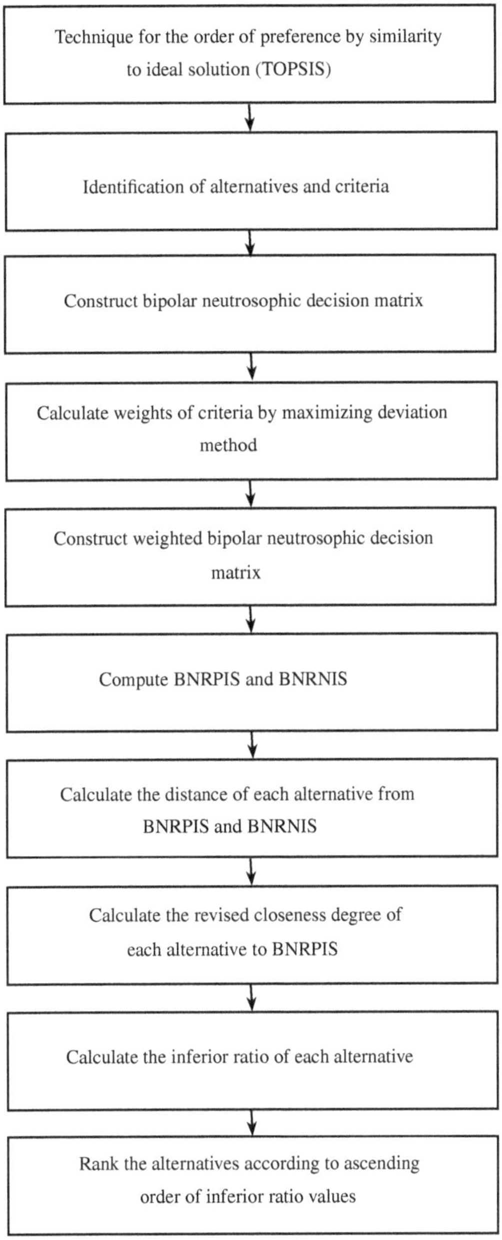
Technique for the order of preference by similarity to ideal solution (TOPSIS)
Identification of alternatives and criteria
Construct bipolar neutrosophic decision matrix
Calculate weights of criteria by maximizing deviation method
Construct weighted bipolar neutrosophic decision matrix
Compute BNRPIS and BNRNIS
Calculate the distance of each alternative from BNRPIS and BNRNIS
Calculate the revised closeness degree of each alternative to BNRPIS
Calculate the inferior ratio of each alternative
Rank the alternatives according to ascending order of inferior ratio values

**Figure 1.** Flow chart of bipolar neutrosophic TOPSIS.

## 3. Applications

In this section, we apply bipolar neutrosophic TOPSIS method to solve real life problems: the best electronic commerce web site, heart surgeon and employee were chosen.

### 3.1. *Electronic Commerce Web Site*

Electronic Commerce (*e*-commerce, for short) is a process of trading the services and goods through electronic networks like computer structures as well as the internet. In recent times *e*-commerce has become a very fascinating and convenient choice for both the businesses and customers. Many companies are interested in advancing their online stores rather than the brick and mortar buildings, because of the appealing requirements of customers for online purchasing. Suppose that a person wants to launch his own online store for selling his products. He will choose the *e*-commerce web site that has comparatively better ratings and that is most popular among internet users. After initial screening four web sites, $S_1$ = Shopify, $S_2$ = $3d$ Cart, $S_3$ = BigCommerce and $S_4$ = Shopsite, are considered. Four attributes, $T_1$ = Customer satisfaction, $T_2$ = Comparative prices, $T_3$ = On-time delivery and $T_4$ = Digital marketing, are designed to choose the best alternative.

**Step 1.** The decision matrix in the form of bipolar neutrosophic information is given as in Table 1:

**Table 1.** Bipolar neutrosophic decision matrix.

| $S\setminus T$ | $T_1$ | $T_2$ | $T_3$ | $T_4$ |
|---|---|---|---|---|
| $S_1$ | $(0.4, 0.2, 0.5, -0.6, -0.4, -0.4)$ | $(0.5, 0.3, 0.3, -0.7, -0.2, -0.4)$ | $(0.2, 0.7, 0.5, -0.4, -0.4, -0.3)$ | $(0.4, 0.6, 0.5, -0.3, -0.7, -0.4)$ |
| $S_2$ | $(0.3, 0.6, 0.1, -0.5, -0.7, -0.5)$ | $(0.2, 0.6, 0.1, -0.5, -0.3, -0.7)$ | $(0.4, 0.2, 0.5, -0.6, -0.3, -0.1)$ | $(0.2, 0.7, 0.5, -0.5, -0.3, -0.2)$ |
| $S_3$ | $(0.3, 0.5, 0.2, -0.4, -0.3, -0.7)$ | $(0.4, 0.5, 0.2, -0.3, -0.8, -0.5)$ | $(0.9, 0.5, 0.7, -0.3, -0.4, -0.3)$ | $(0.3, 0.7, 0.6, -0.5, -0.5, -0.4)$ |
| $S_4$ | $(0.6, 0.7, 0.5, -0.2, -0.1, -0.3)$ | $(0.8, 0.4, 0.6, -0.1, -0.3, -0.4)$ | $(0.6, 0.3, 0.6, -0.1, -0.4, -0.2)$ | $(0.8, 0.3, 0.2, -0.1, -0.3, -0.1)$ |

**Step 2.** The normalized weights of the criteria are calculated by using maximizing deviation method as given below:

$$w_1 = 0.2567, w_2 = 0.2776, w_3 = 0.2179, w_4 = 0.2478, \text{where} \sum_{j=1}^{4} w_j = 1.$$

**Step 3.** The weighted bipolar neutrosophic decision matrix is constructed by multiplying the weights to decision matrix as given in Table 2:

**Table 2.** Weighted bipolar neutrosophic decision matrix.

| $S\setminus T$ | $T_1$ | $T_2$ | $T_3$ | $T_4$ |
|---|---|---|---|---|
| $S_1$ | $(0.123, 0.662, 0.837, -0.877, -0.79, -0.123)$ | $(0.175, 0.716, 0.716, -0.906, -0.64, -0.132)$ | $(0.047, 0.925, 0.86, -0.819, -0.819, -0.075)$ | $(0.119, 0.881, 0.842, -0.742, -0.915, -0.119)$ |
| $S_2$ | $(0.087, 0.877, 0.554, -0.837, -0.913, -0.163)$ | $(0.06, 0.868, 0.528, -0.825, -0.716, -0.284)$ | $(0.105, 0.704, 0.86, -0.895, -0.769, -0.023)$ | $(0.054, 0.915, 0.842, -0.842, -0.742, -0.054)$ |
| $S_3$ | $(0.087, 0.837, 0.662, -0.79, -0.734, -0.266)$ | $(0.132, 0.825, 0.64, -0.716, -0.94, -0.175)$ | $(0.395, 0.86, 0.925, -0.769, -0.819, -0.075)$ | $(0.085, 0.915, 0.881, -0.842, -0.842, -0.119)$ |
| $S_4$ | $(0.21, 0.913, 0.837, -0.662, -0.554, -0.087)$ | $(0.36, 0775, 0.868, -0.528, -0.716, -0.132)$ | $(0.181, 0.769, 0.895, -0.605, -0.819, -0.047)$ | $(0.329, 0.742, 0.671, -0.565, -0.742, -0.026)$ |

**Step 4.** The BNRPIS and BNRNIS are given by

$$\begin{aligned} BNRPIS =< & (0.21, 0.662, 0.554, -0.877, -0.554, -0.087), \\ & (0.06, 0.868, 0.868, -0.528, -0.94, -0.284), \\ & (0.395, 0.704, 0.86, -0.895, -0.769, -0.023), \\ & (0.329, 0.742, 0.671, -0.842, -0.742, -0.062) >; \end{aligned}$$

$$\begin{aligned} BNRNIS =< & (0.087, 0.913, 0.837, -0.662, -0.913, -0.266), \\ & (0.36, 0.716, 0.528, -0.906, -0.64, -0.132), \\ & (0.047, 0.925, 0.925, -0.605, -0.819, -0.075), \\ & (0.054, 0.915, 0.881, -0.565, -0.915, -0.119) > . \end{aligned}$$

**Step 5.** The normalized Euclidean distances of each alternative from the BNRPISs and the BNRNISs are given as follows:

$$\begin{aligned} d_N(S_1, BNRPIS) &= 0.1805, & d_N(S_1, BNRNIS) &= 0.1125, \\ d_N(S_2, BNRPIS) &= 0.1672, & d_N(S_2, BNRNIS) &= 0.1485, \\ d_N(S_3, BNRPIS) &= 0.135, & d_N(S_3, BNRNIS) &= 0.1478, \\ d_N(S_4, BNRPIS) &= 0.155, & d_N(S_4, BNRNIS) &= 0.1678. \end{aligned}$$

**Step 6.** The revised closeness degree of each alternative is given as

$$\rho(S_1) = -0.667, \rho(S_2) = -0.354, \rho(S_3) = -0.119, \rho(S_4) = -0.148.$$

**Step 7.** The inferior ratio to each alternative is given as

$$IR(1) = 1, IR(2) = 0.52, IR(3) = 0.18, IR(4) = 0.22.$$

**Step 8.** Ordering the web stores according to ascending order of alternatives, we obtain: $S_3 < S_4 < S_2 < S_1$. Therefore, the person will choose the BigCommerce for opening a web store.

### 3.2. Heart Surgeon

Suppose that a heart patient wants to select a best cardiac surgeon for heart surgery. After initial screening, five surgeons are considered for further evaluation. These surgeons represent the alternatives and are denoted by $S_1, S_2, S_3, S_4$, and $S_5$ in our MCDM problem. Suppose that he concentrates on four characteristics, $T_1$ = Availability of medical equipment, $T_2$ = Surgeon reputation, $T_3$ = Expenditure and $T_4$ = Suitability of time, in order to select the best surgeon. These characteristics represent the criteria for this MCDM problem.

**Step 1.** The decision matrix in the form of bipolar neutrosophic information is given as in Table 3:

**Table 3.** Bipolar neutrosophic decision matrix.

| $S \setminus T$ | $T_1$ | $T_2$ | $T_3$ | $T_4$ |
|---|---|---|---|---|
| $S_1$ | (0.6, 0.5, 0.3, −0.5, −0.7, −0.4) | (0.5, 0.7, 0.4, −0.6, −0.4, −0.5) | (0.3, 0.5, 0.5, −0.7, −0.3, −0.4) | (0.5, 0.3, 0.6, −0.4, −0.7, −0.5) |
| $S_2$ | (0.9, 0.3, 0.2, −0.3, −0.6, −0.5) | (0.7, 0.4, 0.2, −0.4, −0.5, −0.7) | (0.4, 0.7, 0.6, −0.6, −0.3, −0.3) | (0.8, 0.3, 0.2, −0.2, −0.5, −0.7) |
| $S_3$ | (0.4, 0.6, 0.6, −0.7, −0.4, −0.3) | (0.5, 0.3, 0.6, −0.6, −0.4, −0.4) | (0.7, 0.5, 0.3, −0.4, −0.4, −0.6) | (0.4, 0.6, 0.7, −0.5, −0.4, −0.4) |
| $S_4$ | (0.8, 0.5, 0.3, −0.3, −0.4, −0.5) | (0.6, 0.4, 0.3, −0.5, −0.7, −0.8) | (0.4, 0.5, 0.7, −0.5, −0.4, −0.2) | (0.5, 0.4, 0.6, −0.6, −0.7, −0.3) |
| $S_5$ | (0.6, 0.4, 0.6, −0.4, −0.7, −0.3) | (0.4, 0.7, 0.6, −0.7, −0.5, −0.6) | (0.6, 0.3, 0.5, −0.3, −0.7, −0.4) | (0.5, 0.7, 0.4, −0.3, −0.6, −0.5) |

**Step 2.** The normalized weights of the criteria are calculated by using maximizing deviation method as given below:

$$w_1 = 0.2480, w_2 = 0.2424, w_3 = 0.2480, w_4 = 0.2616, \text{where} \sum_{j=1}^{4} w_j = 1.$$

**Step 3.** The weighted bipolar neutrosophic decision matrix is constructed by multiplying the weights to decision matrix as given in Table 4:

**Table 4.** Weighted bipolar neutrosophic decision matrix.

| $S\setminus T$ | $T_1$ | $T_2$ | $T_3$ | $T_4$ |
|---|---|---|---|---|
| $S_1$ | (0.203, 0.842, 0.742, −0.842, −0.915, −0.119) | (0.155, 0.917, 0.801, −0.884, −0.801, −0.155) | (0.085, 0.842, 0.842, −0.915, −0.742, −0.119) | (0.166, 0.730, 0.875, −0.787, −0.911, −0.166) |
| $S_2$ | (0.435, 0.742, 0.671, −0.742, −0.881, −0.158) | (0.253, 0.801, 0.677, −0.801, −0.845, −0.253) | (0.119, 0.915, 0.881, −0.881, −0.742, −0.085) | (0.344, 0.730, 0.656, −0.656, −0.834, −0.270) |
| $S_3$ | (0.119, 0.881, 0.881, −0.915, −0.797, −0.085) | (0.155, 0.747, 0.884, −0.884, −0.801, −0.116) | (0.258, 0.842, 0.742, −0.797, −0.797, −0.203) | (0.125, 0.875, 0.911, −0.834, −0.787, −0.125) |
| $S_4$ | (0.329, 0.842, 0.742, −0.742, −0.797, −0.158) | (0.199, 0.801, 0.747, −0.845, −0.917, −0.323) | (0.119, 0.842, 0.915, −0.842, −0.797, −0.054) | (0.166, 0.787, 0.875, −0.875, −0.911, −0.089) |
| $S_5$ | (0.203, 0.797, 0.881, −0.797, −0.915, −0.085) | (0.116, 0.917, 0.884, −0.917, −0.845, −0.199) | (0.203, 0.742, 0.842, −0.742, −0.915, −0.119) | (0.166, 0.911, 0.787, −0.730, −0.875, −0.166) |

**Step 4.** The BNRPIS and BNRNIS are given by

$$\begin{aligned} BNRPIS =< & (0.435, 0.742, 0.671, -0.915, -0.797, -0.085), \\ & (0.253, 0.747, 0.677, -0.917, -0.801, -0.116), \\ & (0.085, 0.915, 0.915, -0.742, -0.915, -0.203), \\ & (0.344, 0.730, 0.656, -0.875, -0.787, -0.089) >; \end{aligned}$$

$$\begin{aligned} BNRNIS =< & (0.119, 0.881, 0.881, -0.742, -0.915, -0.158), \\ & (0.116, 0.917, 0.884, -0.801, -0.917, -0.323), \\ & (0.258, 0.742, 0.742, -0.915, -0.742, -0.054), \\ & (0.125, 0.911, 0.911, -0.656, -0.911, -0.270) > . \end{aligned}$$

**Step 5.** The normalized Euclidean distances of each alternative from the BNRPISs and the BNRNISs are given as follows:

$$\begin{aligned} d_N(S_1, BNRPIS) &= 0.1176, & d_N(S_1, BNRNIS) &= 0.0945, \\ d_N(S_2, BNRPIS) &= 0.0974, & d_N(S_2, BNRNIS) &= 0.1402, \\ d_N(S_3, BNRPIS) &= 0.1348, & d_N(S_3, BNRNIS) &= 0.1043, \\ d_N(S_4, BNRPIS) &= 0.1089, & d_N(S_4, BNRNIS) &= 0.1093, \\ d_N(S_5, BNRPIS) &= 0.1292, & d_N(S_5, BNRNIS) &= 0.0837. \end{aligned}$$

**Step 6.** The revised closeness degree of each alternative is given as

$$\rho(S_1) = -0.553, \rho(S_2) = 0, \rho(S_3) = -0.64, \rho(S_4) = -0.338, \rho(S_5) = -0.729$$

**Step 7.** The inferior ratio to each alternative is given as

$$IR(1) = 0.73, IR(2) = 0, IR(3) = 0.88, IR(4) = 0.46, IR(5) = 1.$$

**Step 8.** Ordering the alternatives in ascending order, we obtain: $S_2 < S_4 < S_1 < S_3 < S_5$. Therefore, $S_2$ is best among all other alternatives.

### 3.3. Employee (Marketing Manager)

Process of employee selection has an analytical importance for any kind of business. According to firm hiring requirements and the job position, this process may vary from a very simple process to a complicated procedure. Suppose that a company wants to hire an employee for the post of marketing manager. After initial screening, four candidates are considered as alternatives and denoted by $S_1, S_2, S_3$ and $S_4$ in our MCDM problem. The requirements for this post, $T_1$ = Confidence, $T_2$ = Qualification, $T_3$ = Leading skills and $T_4$ = Communication skills, are considered as criteria in order to select the most relevant candidate.

**Step 1.** The decision matrix in the form of bipolar neutrosophic information is given as in Table 5:

**Table 5.** Bipolar neutrosophic decision matrix.

| $S \diagdown T$ | $T_1$ | $T_2$ | $T_3$ | $T_4$ |
|---|---|---|---|---|
| $S_1$ | $(0.8, 0.5, 0.3, -0.3, -0.6, -0.5)$ | $(0.7, 0.3, 0.2, -0.3, -0.5, -0.4)$ | $(0.5, 0.4, 0.6, -0.5, -0.3, -0.4)$ | $(0.9, 0.3, 0.2, -0.3, -0.4, -0.2)$ |
| $S_2$ | $(0.5, 0.7, 0.6 -0.4, -0.2, -0.4)$ | $(0.4, 0.7, 0.5, -0.6, -0.2, -0.3)$ | $(0.6, 0.8, 0.5, -0.3, -0.5, -0.7)$ | $(0.5, 0.3, 0.6, -0.6, -0.4, -0.3)$ |
| $S_3$ | $(0.4, 0.6, 0.8, -0.7, -0.3, -0.4)$ | $(0.6, 0.3, 0.5, -0.2, -0.4, -0.6)$ | $(0.3, 0.5, 0.7, -0.8, -0.4, -0.2)$ | $(0.5, 0.7, 0.4, -0.6, -0.3, -0.5)$ |
| $S_4$ | $(0.7, 0.3, 0.5, -0.4, -0.2, -0.5)$ | $(0.5, 0.4, 0.6, -0.4, -0.5, -0.3)$ | $(0.6, 0.4, 0.3, -0.3, -0.5, -0.7)$ | $(0.4, 0.5, 0.7, -0.6, -0.5, -0.3)$ |

**Step 2.** The normalized weights of the criteria are calculated by using maximizing deviation method as given below:

$$w_1 = 0.25, w_2 = 0.2361, w_3 = 0.2708, w_4 = 0.2431, \text{where } \sum_{j=1}^{4} w_j = 1.$$

**Step 3.** The weighted bipolar neutrosophic decision matrix is constructed by multiplying the weights to decision matrix as given in Table 6:

**Table 6.** Weighted bipolar neutrosophic decision matrix.

| $S \diagdown T$ | $T_1$ | $T_2$ | $T_3$ | $T_4$ |
|---|---|---|---|---|
| $S_1$ | $(0.3313, 0.8409, 0.7401, -0.7401, -0.8801, -0.1591)$ | $(0.2474, 0.7526, 0.6839, -0.7526, -0.8490, -0.1136)$ | $(0.1711, 0.7803, 0.8708, -0.8289, -0.7218, -0.1292)$ | $(0.4287, 0.7463, 0.6762, -0.7463, -0.8003, -0.0528)$ |
| $S_2$ | $(0.1591, 0.9147, 0.8801, -0.7953, -0.6687, -0.1199)$ | $(0.1136, 0.9192, 0.8490, -0.8864, -0.6839, -0.0808)$ | $(0.2197, 0.9414, 0.8289, -0.7218, -0.8289, -0.2782)$ | $(0.1551, 0.7463, 0.8832, -0.8832, -0.8003, -0.0831)$ |
| $S_3$ | $(0.1199, 0.8801, 0.9457, -0.9147, -0.7401, -0.1199)$ | $(0.1945, 0.7526, 0.8490, -0.6839, -0.8055, -0.1945)$ | $(0.0921, 0.8289, 0.9079, -0.9414, -0.7803, -0.0586)$ | $(0.1551, 0.9169, 0.8003, -0.8832, -0.7463, -0.1551)$ |
| $S_4$ | $(0.2599, 0.7401, 0.8409, -0.7953, -0.6687, -0.1591)$ | $(0.1510, 0.8055, 0.8864, -0.8055, -0.8490, -0.0808)$ | $(0.2197, 0.7803, 0.7218, -0.7218, -0.8289, -0.2782)$ | $(0.1168, 0.8449, 0.9169, -0.8832, -0.8449, -0.0831)$ |

**Step 4.** The BNRPIS and BNRNIS are given by

$$\begin{aligned} BNRPIS = < & (0.3313, 0.7401, 0.7401, -0.9147, -0.6687, -0.1199), \\ & (0.2474, 0.7526, 0.6839, -0.8864, -0.6839, -0.0808), \\ & (0.2197, 0.7803, 0.7218, -0.9414, -0.7218, -0.0586), \\ & (0.4287, 0.7463, 0.6762, -0.8832, -0.7463, -0.0528) >; \end{aligned}$$

$$\begin{aligned} BNRNIS =< &(0.1199, 0.9147, 0.9457, -0.7401, -0.8801, -0.1591), \\ &(0.1136, 0.9192, 0.8864, -0.6839, -0.8490, -0.1945), \\ &(0.0921, 0.9414, 0.9079, -0.7218, -0.8289, -0.2782), \\ &(0.1168, 0.9169, 0.9169, -0.7463, -0.8449, -0.1551) > . \end{aligned}$$

**Step 5.** The normalized Euclidean distances of each alternative from the BNRPISs and the BNRNISs are given as follows:

$$\begin{aligned} d_N(S_1, BNRPIS) &= 0.0906, \quad d_N(S_1, BNRNIS) = 0.1393, \\ d_N(S_2, BNRPIS) &= 0.1344, \quad d_N(S_2, BNRNIS) = 0.0953, \\ d_N(S_3, BNRPIS) &= 0.1286, \quad d_N(S_3, BNRNIS) = 0.1011, \\ d_N(S_4, BNRPIS) &= 0.1293, \quad d_N(S_4, BNRNIS) = 0.0999. \end{aligned}$$

**Step 6.** The revised closeness degree of each alternative is given as

$$\rho(S_1) = 0, \rho(S_2) = -0.799, \rho(S_3) = -0.694, \rho(S_4) = -0.780.$$

**Step 7.** The inferior ratio to each alternative is given as

$$IR(1) = 0, IR(2) = 1, IR(3) = 0.87, IR(4) = 0.98.$$

**Step 8.** Ordering the alternatives in ascending order, we obtain: $S_1 < S_3 < S_4 < S_2$. Therefore, the company will select the candidate $S_1$ for this post.

## 4. Bipolar Neutrosophic ELECTRE-I Method

In this section, we propose bipolar neutrosophic ELECTRE-I method to solve MCDM problems. Consider a set of alternatives, denoted by $S = \{S_1, S_2, S_3, \cdots, S_m\}$ and the set of criteria, denoted by $T = \{T_1, T_2, T_3, \cdots, T_n\}$ which are used to evaluate the alternatives.

(i–iii) As in the section of bipolar neutrosophic TOPSIS, the rating values of alternatives with respect to the criteria are expressed in the form of matrix $[k_{ij}]_{m\times n}$. The weights $w_j$ of the criteria $T_j$ are evaluated by maximizing deviation method and the weighted bipolar neutrosophic decision matrix $[k_{ij}^{w_j}]_{m\times n}$ is constructed.

(iv) The bipolar neutrosophic concordance sets $E_{xy}$ and bipolar neutrosophic discordance sets $F_{xy}$ are defined as follows:

$$\begin{aligned} E_{xy} &= \{1 \leq j \leq n \mid \rho_{xj} \geq \rho_{yj}\}, \ x \neq y, \ x, y = 1, 2, \cdots, m, \\ F_{xy} &= \{1 \leq j \leq n \mid \rho_{xj} \leq \rho_{yj}\}, \ x \neq y, \ x, y = 1, 2, \cdots, m, \end{aligned}$$

where, $\rho_{ij} = T_{ij}^{+} + I_{ij}^{+} + F_{ij}^{+} + T_{ij}^{-} + I_{ij}^{-} + F_{ij}^{-}$, $i = 1, 2, \cdots, m$, $j = 1, 2, \cdots, n$.

(v) The bipolar neutrosophic concordance matrix $E$ is constructed as follows:

$$E = \begin{bmatrix} - & e_{12} & \cdot & \cdot & \cdot & e_{1m} \\ e_{21} & - & \cdot & \cdot & \cdot & e_{2m} \\ \cdot & & & & & \\ \cdot & & & & & \\ \cdot & & & & & \\ e_{m1} & e_{m2} & \cdot & \cdot & \cdot & - \end{bmatrix},$$

where, the bipolar neutrosophic concordance indices $e_{xy}$'s are determined as

$$e_{xy} = \sum_{j \in E_{xy}} w_j.$$

(vi) The bipolar neutrosophic discordance matrix $F$ is constructed as follows:

$$F = \begin{bmatrix} - & f_{12} & \cdot & \cdot & \cdot & f_{1m} \\ f_{21} & - & \cdot & \cdot & \cdot & f_{2m} \\ \cdot & & & & & \\ \cdot & & & & & \\ \cdot & & & & & \\ f_{m1} & f_{m2} & \cdot & \cdot & \cdot & - \end{bmatrix},$$

where, the bipolar neutrosophic discordance indices $f_{xy}$'s are determined as

$$f_{xy} = \frac{\max\limits_{j \in F_{xy}} \sqrt{\frac{1}{6n} \left\{ \begin{array}{c} (T_{xj}^{w_j+} - T_{yj}^{w_j+})^2 + (I_{xj}^{w_j+} - I_{yj}^{w_j+})^2 + (F_{xj}^{w_j+} - F_{yj}^{w_j+})^2 + \\ (T_{xj}^{w_j-} - T_{yj}^{w_j-})^2 + (I_{xj}^{w_j-} - I_{yj}^{w_j-})^2 + (F_{xj}^{w_j-} - F_{yj}^{w_j-})^2 \end{array} \right\}}}{\max\limits_{j} \sqrt{\frac{1}{6n} \left\{ \begin{array}{c} (T_{xj}^{w_j+} - T_{yj}^{w_j+})^2 + (I_{xj}^{w_j+} - I_{yj}^{w_j+})^2 + (F_{xj}^{w_j+} - F_{yj}^{w_j+})^2 + \\ (T_{xj}^{w_j-} - T_{yj}^{w_j-})^2 + (I_{xj}^{w_j-} - I_{yj}^{w_j-})^2 + (F_{xj}^{w_j-} - F_{yj}^{w_j-})^2 \end{array} \right\}}}.$$

(vii) Concordance and discordance levels are computed to rank the alternatives. The bipolar neutrosophic concordance level $\hat{e}$ is defined as the average value of the bipolar neutrosophic concordance indices as

$$\hat{e} = \frac{1}{m(m-1)} \sum_{\substack{x=1, \\ x \neq y}}^{m} \sum_{\substack{y=1, \\ y \neq x}}^{m} e_{xy},$$

similarly, the bipolar neutrosophic discordance level $\hat{f}$ is defined as the average value of the bipolar neutrosophic discordance indices as

$$\hat{f} = \frac{1}{m(m-1)} \sum_{\substack{x=1, \\ x \neq y}}^{m} \sum_{\substack{y=1, \\ y \neq x}}^{m} f_{xy}.$$

(viii) The bipolar neutrosophic concordance dominance matrix $\phi$ on the basis of $\hat{e}$ is determined as follows:

$$\phi = \begin{bmatrix} - & \phi_{12} & \cdot & \cdot & \cdot & \phi_{1m} \\ \phi_{21} & - & \cdot & \cdot & \cdot & \phi_{2m} \\ \cdot & & & & & \\ \cdot & & & & & \\ \cdot & & & & & \\ \phi_{m1} & \phi_{m2} & \cdot & \cdot & \cdot & - \end{bmatrix},$$

where, $\phi_{xy}$ is defined as

$$\phi_{xy} = \begin{cases} 1, & \text{if } e_{xy} \geq \hat{e}, \\ 0, & \text{if } e_{xy} < \hat{e}. \end{cases}$$

(ix) The bipolar neutrosophic discordance dominance matrix $\psi$ on the basis of $\hat{f}$ is determined as follows:

$$\psi = \begin{bmatrix} - & \psi_{12} & \cdot & \cdot & \cdot & \psi_{1m} \\ \psi_{21} & - & \cdot & \cdot & \cdot & \psi_{2m} \\ \cdot & & & & & \\ \cdot & & & & & \\ \cdot & & & & & \\ \psi_{m1} & \psi_{m2} & \cdot & \cdot & \cdot & - \end{bmatrix},$$

where, $\psi_{xy}$ is defined as

$$\psi_{xy} = \begin{cases} 1, & \text{if } f_{xy} \leq \hat{f}, \\ 0, & \text{if } f_{xy} > \hat{f}. \end{cases}$$

(x) Consequently, the bipolar neutrosophic aggregated dominance matrix $\pi$ is evaluated by multiplying the corresponding entries of $\phi$ and $\psi$, that is

$$\pi = \begin{bmatrix} - & \pi_{12} & \cdot & \cdot & \cdot & \pi_{1m} \\ \pi_{21} & - & \cdot & \cdot & \cdot & \pi_{2m} \\ \cdot & & & & & \\ \cdot & & & & & \\ \cdot & & & & & \\ \pi_{m1} & \pi_{m2} & \cdot & \cdot & \cdot & - \end{bmatrix},$$

where, $\pi_{xy}$ is defined as

$$\pi_{xy} = \phi_{xy}\psi_{xy}.$$

(xi) Finally, the alternatives are ranked according to the outranking values $\pi_{xy}'s$. That is, for each pair of alternatives $S_x$ and $S_y$, an arrow from $S_x$ to $S_y$ exists if and only if $\pi_{xy} = 1$. As a result, we have three possible cases:

(a) There exits a unique arrow from $S_x$ into $S_y$.
(b) There exist two possible arrows between $S_x$ and $S_y$.
(c) There is no arrow between $S_x$ and $S_y$.

For case a, we decide that $S_x$ is preferred to $S_y$. For the second case, $S_x$ and $S_y$ are indifferent, whereas, $S_x$ and $S_y$ are incomparable in case c.

Geometric representation of proposed bipolar neutrosophic ELECTRE-I method is shown in Figure 2.

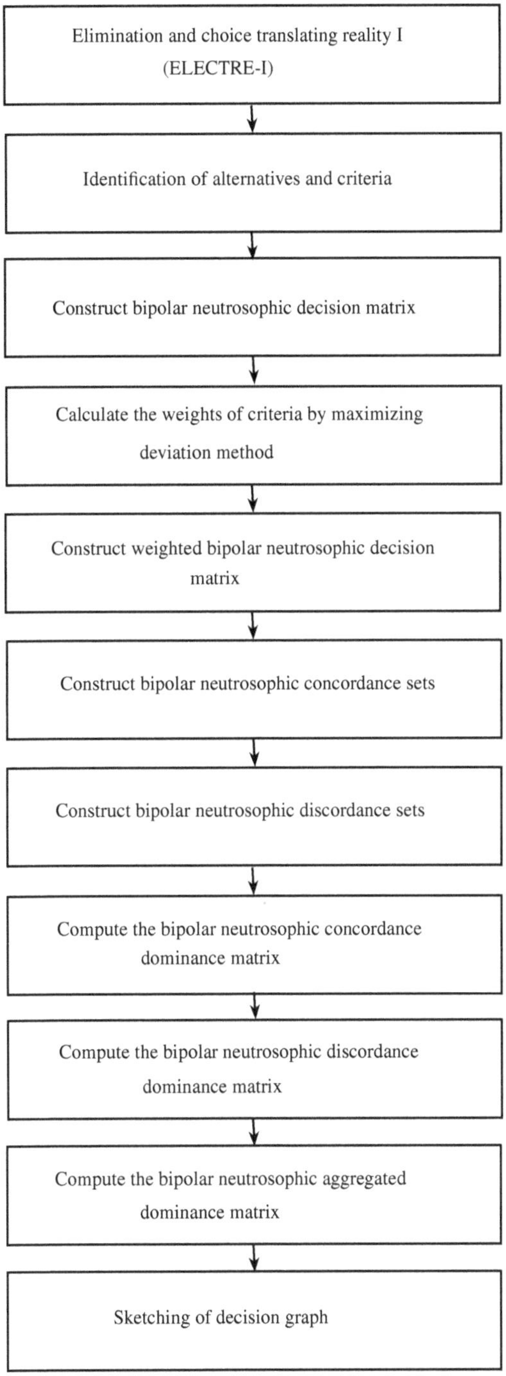
Elimination and choice translating reality I (ELECTRE-I)
Identification of alternatives and criteria
Construct bipolar neutrosophic decision matrix
Calculate the weights of criteria by maximizing deviation method
Construct weighted bipolar neutrosophic decision matrix
Construct bipolar neutrosophic concordance sets
Construct bipolar neutrosophic discordance sets
Compute the bipolar neutrosophic concordance dominance matrix
Compute the bipolar neutrosophic discordance dominance matrix
Compute the bipolar neutrosophic aggregated dominance matrix
Sketching of decision graph

**Figure 2.** Flow chart of bipolar neutrosophic ELECTRE-I.

*Numerical Example*

In Section 3, MCDM problems are presented using the bipolar neutrosophic TOPSIS method. In this section, we apply our proposed bipolar neutrosophic ELECTRE-I method to select the "electronic commerce web site" to compare these two MCDM methods. Steps (1–3) have already been done in Section 3.1. So we move on to Step 4.

**Step 4.** The bipolar neutrosophic concordance sets $E_{xy}$'s are given as in Table 7:

**Table 7.** Bipolar neutrosophic concordance sets.

| $E_{xy} \setminus y$ | 1 | 2 | 3 | 4 |
|---|---|---|---|---|
| $E_{1y}$ | - | {1, 2, 3} | {1, 2} | { } |
| $E_{2y}$ | {4} | - | {4} | { } |
| $E_{3y}$ | {3, 4} | {1, 2, 3} | - | {3} |
| $E_{4y}$ | {1, 2, 3, 4} | {1, 2, 3, 4} | {1, 2, 4} | - |

**Step 5.** The bipolar neutrosophic discordance sets $F_{xy}$'s are given as in Table 8.

**Table 8.** Bipolar neutrosophic discordance sets.

| $F_{xy} \setminus y$ | 1 | 2 | 3 | 4 |
|---|---|---|---|---|
| $F_{1y}$ | - | {4} | {3, 4} | {1, 2, 3, 4} |
| $F_{2y}$ | {1, 2, 3} | - | {1, 2, 3} | {1, 2, 3, 4} |
| $F_{3y}$ | {1, 2} | {4} | - | {1, 2, 4} |
| $F_{4y}$ | { } | { } | {3} | - |

**Step 6.** The bipolar neutrosophic concordance matrix $E$ is computed as follows

$$E = \begin{bmatrix} - & 0.7522 & 0.5343 & 0 \\ 0.2478 & - & 0.2478 & 0 \\ 0.4657 & 0.7522 & - & 0.2179 \\ 1 & 1 & 0.7821 & - \end{bmatrix}$$

**Step 7.** The bipolar neutrosophic discordance matrix $F$ is computed as follows

$$F = \begin{bmatrix} - & 0.5826 & 0.9464 & 1 \\ 1 & - & 1 & 1 \\ 1 & 0.3534 & - & 1 \\ 0 & 0 & 0.6009 & - \end{bmatrix}$$

**Step 8.** The bipolar neutrosophic concordance level is $\hat{e} = 0.5003$ and bipolar neutrosophic discordance level is $\hat{f} = 0.7069$. The bipolar neutrosophic concordance dominance matrix $\phi$ and bipolar neutrosophic discordance dominance matrix $\psi$ are as follows

$$\phi = \begin{bmatrix} - & 1 & 1 & 0 \\ 0 & - & 0 & 0 \\ 0 & 1 & - & 0 \\ 1 & 1 & 1 & - \end{bmatrix}, \psi = \begin{bmatrix} - & 1 & 0 & 0 \\ 0 & - & 0 & 0 \\ 0 & 1 & - & 0 \\ 0 & 0 & 0 & - \end{bmatrix}.$$

**Step 9.** The bipolar neutrosophic aggregated dominance matrix $\pi$ is computed as

$$\pi = \begin{bmatrix} - & 1 & 0 & 0 \\ 0 & - & 0 & 0 \\ 0 & 1 & - & 0 \\ 0 & 0 & 0 & - \end{bmatrix}.$$

According to nonzero values of $\pi_{xy}$, we get the alternatives in the following sequence:

$$S_1 \rightarrow S_2 \leftarrow S_3$$

Therefore, the most favorable alternatives are $S_3$ and $S_1$.

## 5. Comparison of Bipolar Neutrosophic TOPSIS and Bipolar Neutrosophic ELECTRE-I

TOPSIS and ELECTRE-I are the most commonly used MCDM methods to solve decision making problems, in which the best possible alternative is selected among others. The main idea of the TOPSIS method is that the chosen alternative has the shortest distance from positive ideal solution and the greatest distance from negative ideal solution, whereas the ELECTRE-I method is based on the binary comparison of alternatives. The proposed MCDM methods TOPSIS and ELECTRE-I are based on bipolar neutrosophic information. In the bipolar neutrosophic TOPSIS method, the normalized Euclidean distance is used to compute the revised closeness coefficient of alternatives to BNRPIS and BNRNIS. Alternatives are ranked in increasing order on the basis of inferior ratio values. Bipolar neutrosophic TOPSIS is an effective method because it has a simple process and is able to deal with any number of alternatives and criteria. Throughout history, one drawback of the TOPSIS method is that more rank reversals are created by increasing the number of alternatives. The proposed bipolar neutrosophic ELECTRE-I is an outranking relation theory that compares all pairs of alternatives and figures out which alternatives are preferred to the others by systematically comparing them for each criterion. The connection between different alternatives shows the bipolar neutrosophic concordance and bipolar neutrosophic discordance behavior of alternatives. The bipolar neutrosophic TOPSIS method gives only one possible alternative but the bipolar neutrosophic ELECTRE-I method sometimes provides a set of alternatives as a final selection to consider the MCDM problem. Despite all of the above comparisons, it is difficult to determine which method is most convenient, because both methods have their own importance and can be used according to the choice of the decision maker.

## 6. Conclusions

A single-valued neutrosophic set as an instance of a neutrosophic set provides an additional possibility to represent imprecise, uncertainty, inconsistent and incomplete information which exist in real situations. Single valued neutrosophic models are more flexible and practical than fuzzy and intuitionistic fuzzy models.We have presented the procedure, technique and implication of TOPSIS and ELECTRE-I methods under bipolar neutrosophic environment. The rating values of alternatives with respect to attributes are expressed in the form of BNSs. The unknown weights of the attributes are calculated by maximizing the deviation method to construct the weighted decision matrix. The normalized Euclidean distance is used to calculate the distance of each alternative from BNRPIS and BNRNIS. Revised closeness degrees are computed and then the inferior ratio method is used to rank the alternatives in bipolar neutrosophic TOPSIS. The concordance and discordance matrices are evaluated to rank the alternatives in bipolar neutrosophic ELECTRE-I. We have also presented some examples to explain these methods.

**Author Contributions:** M.A. and S. conceived and designed the experiments; F.S. analyzed the data; S. wrote the paper.

**Acknowledgments:** The authors are very thankful to the editor and referees for their valuable comments and suggestions for improving the paper.

**Conflicts of Interest:** The authors declare no conflicts of interest.

## References


1. Zadeh, L.A. Fuzzy sets. *Inf. Control* **1965**, *8*, 338–353. [CrossRef]
2. Smarandache, F. *A Unifying Field of Logics, Neutrosophy: Neutrosophic Probability, Set and Logic*; American Research Press: Rehoboth, DE, USA, 1998.
3. Wang, H.; Smarandache, F.; Zhang, Y.; Sunderraman, R. Single valued neutrosophic sets. In *Multi-Space and Multi-Structure*; Infinite Study: New Delhi, India, 2010; Volume 4, pp. 410–413.
4. Deli, M.; Ali, F. Smarandache, Bipolar neutrosophic sets and their application based on multi-criteria decision making problems. In Proceedings of the 2015 International Conference on Advanced Mechatronic Systems, Beiging, China, 20–24 August 2015; pp. 249–254.
5. Zhang, W.R. Bipolar fuzzy sets and relations: A computational framework for cognitive modeling and multiagent decision analysis. In Processing of the IEEE Conference, Fuzzy Information Processing Society Biannual Conference, San Antonio, TX, USA, 81–21 December 1994; pp. 305–309.
6. Hwang, C.L.; Yoon, K. *Multi Attribute Decision Making: Methods and Applications*; Springer: New York, NY, USA, 1981.
7. Chen, C.-T. Extensions of the TOPSIS for group decision-making under fuzzy environment. *Fuzzy Sets Syst.* **2000**, *114*, 1–9. [CrossRef]
8. Chu, T.C. Facility location selection using fuzzy TOPSIS under group decisions. *Int. J. Uncertain. Fuzziness Knowl.-Based Syst.* **2002**, *10*, 687–701. [CrossRef]
9. Hadi-Vencheh, A.; Mirjaberi, M. Fuzzy inferior ratio method for multiple attribute decision making problems. *Inf. Sci.* **2014**, *277*, 263–272. [CrossRef]
10. Joshi, D.; Kumar, S. Intuitionistic fuzzy entropy and distance measure based TOPSIS method for multi-criteria decision making. *Egypt. Inform. J.* **2014**, *15*, 97–104. [CrossRef]
11. Kolios, A.; Mytilinou, V.; Lozano-Minguez, E.; Salonitis, K. A comparative study of multiple-criteria decision-making methods under stochastic inputs. *Energies* **2016**, *9*, 566. [CrossRef]
12. Akram, M.; Alshehri, N.O.; Davvaz, B.; Ashraf, A. Bipolar fuzzy digraphs in decision support systems. *J. Multiple-Valued Log. Soft Comput.* **2016**, *27*, 531–551.
13. Akram, M.; Feng, F.; Saeid, A.B.; Fotea, V. A new multiple criteria decision-making method based on bipolar fuzzy soft graphs. *Iran. J. Fuzzy Syst.* **2017**. [CrossRef]
14. Akram, M.; Waseem, N. Novel applications of bipolar fuzzy graphs to decision making problems. *J. Appl. Math. Comput.* **2018**, *56*, 73–91. [CrossRef]
15. Sarwar, M.; Akram, M. Certain algorithms for computing strength of competition in bipolar fuzzy graphs. *Int. J. Uncertain. Fuzziness Knowl.-Based Syst.* **2017**, *25*, 877–896. [CrossRef]
16. Sarwar, M.; Akram, M. Novel concepts of bipolar fuzzy competition graphs. *J. Appl. Math. Comput.* **2017**, *54*, 511–547. [CrossRef]
17. Faizi, S.; Salabun, W.; Rashid, T.; Watróbski, J.; Zafar, S. Group decision-making for hesitant fuzzy sets based on characteristic objects method. *Symmetry* **2017**, *9*, 136. [CrossRef]
18. Alghamdi, M.A.; Alshehri, N.O.; Akram, M. Multi-criteria decision-making methods in bipolar fuzzy environment. *Int. J. Fuzzy Syst.* **2018**. [CrossRef]
19. Dey, P.P.; Pramanik, S.; Giri, B.C. TOPSIS for solving multi-attribute decision making problems under bipolar neutrosophic environment. In *New Trends in Neutrosophic Theory and Applications*; Pons Editions: Brussels, Belgium, 2016; pp. 65–77.
20. Benayoun, R.; Roy, B.; Sussman, N. *Manual de Reference du Programme Electre*; Note de Synthese et Formation, No. 25; Direction Scientific SEMA: Paris, France, 1966.
21. Hatami-Marbini, A.; Tavana, M. An extension of the ELECTRE method for group decision making under a fuzzy environment. *Omega* **2011**, *39*, 373–386. [CrossRef]
22. Aytaç, E.; Isiķ, A.T.; Kundaki, N. Fuzzy ELECTRE I method for evaluating catering firm alternatives. *Ege Acad. Rev.* **2011**, *11*, 125–134.
23. Wu, M.-C.; Chen, T.-Y. The ELECTRE multicriteria analysis approach based on Attanssov's intuitionistic fuzzy sets. *Expert Syst. Appl.* **2011**, *38*, 12318–12327. [CrossRef]
24. Chu, T-C. Selecting plant location via a fuzzy TOPSIS approach. *Int. J. Adv. Manuf. Technol.* **2002**, *20*, 859–864.
25. Guarini, M.R.; Battisti, F.; Chiovitti, A. A Methodology for the selection of multi-criteria decision analysis methods in real estate and land management processes. *Sustainability* **2018**, *10*, 507. [CrossRef]

26. Hatefi, S.M.; Tamošaitiene, J. Construction projects assessment based on the sustainable development criteria by an integrated fuzzy AHP and improved GRA model. *Sustainability* **2018**, *10*, 991. [CrossRef]
27. Huang, Y.P.; Basanta, H.; Kuo, H.C.; Huang, A. Health symptom checking system for elderly people using fuzzy analytic hierarchy process. *Appl. Syst. Innov.* **2018**, *1*, 10. [CrossRef]
28. Salabun, W.; Piegat, A. Comparative analysis of MCDM methods for the assessment of mortality in patients with acute coronary syndrome. *Artif. Intell. Rev.* **2017**, *48*, 557–571. [CrossRef]
29. Triantaphyllou, E. *Multi-Criteria Decision Making Methods, Multi-Criteria Decision Making Methods: A Comparative Study*; Springer: Boston, MA, USA, 2000; pp. 5–21.
30. Yang, Y.M. Using the method of maximizing deviations to make decision for multi-indices. *Syst. Eng. Electron.* **1998**, *7*, 24–31.

axiomes

MDPI

Article

# Interval Neutrosophic Sets with Applications in *BCK*/*BCI*-Algebra

**Young Bae Jun [1,*], Seon Jeong Kim [2] and Florentin Smarandache [3]**

1 Department of Mathematics Education Gyeongsang National University, Jinju 52828, Korea
2 Department of Mathematics, Natural Science of College, Gyeongsang National University, Jinju 52828, Korea; skim@gnu.ac.kr
3 Mathematics & Science Department, University of New Mexico, 705 Gurley Ave., Gallup, NM 87301, USA; fsmarandache@gmail.com
* Correspondence: skywine@gmail.com

Received: 27 February 2018; Accepted: 06 April 2018; Published: 9 April 2018

**Abstract:** For $i, j, k, l, m, n \in \{1, 2, 3, 4\}$, the notion of $(T(i,j), I(k,l), F(m,n))$-interval neutrosophic subalgebra in $BCK/BCI$-algebra is introduced, and their properties and relations are investigated. The notion of interval neutrosophic length of an interval neutrosophic set is also introduced, and related properties are investigated.

**Keywords:** interval neutrosophic set; interval neutrosophic subalgebra; interval neutrosophic length

**MSC:** 06F35, 03G25, 03B52

## 1. Introduction

Intuitionistic fuzzy set, which is introduced by Atanassov [1], is a generalization of Zadeh's fuzzy sets [2], and consider both truth-membership and falsity-membership. Since the sum of degree true, indeterminacy and false is one in intuitionistic fuzzy sets, incomplete information is handled in intuitionistic fuzzy sets. On the other hand, neutrosophic sets can handle the indeterminate information and inconsistent information that exist commonly in belief systems in a neutrosophic set since indeterminacy is quantified explicitly and truth-membership, indeterminacy-membership and falsity-membership are independent, which is mentioned in [3]. As a formal framework that generalizes the concept of the classic set, fuzzy set, interval valued fuzzy set, intuitionistic fuzzy set, interval valued intuitionistic fuzzy set and paraconsistent set, etc., the neutrosophic set is developed by Smarandache [4,5], which is applied to various parts, including algebra, topology, control theory, decision-making problems, medicines and in many real-life problems. The concept of interval neutrosophic sets is presented by Wang et al. [6], and it is more precise and more flexible than the single-valued neutrosophic set. The interval neutrosophic set can represent uncertain, imprecise, incomplete and inconsistent information, which exists in the real world. $BCK$-algebra is introduced by Imai and Iséki [7], and it has been applied to several branches of mathematics, such as group theory, functional analysis, probability theory and topology, etc. As a generalization of $BCK$-algebra, Iséki introduced the notion of $BCI$-algebra (see [8]).

In this article, we discuss interval neutrosophic sets in $BCK/BCI$-algebra. We introduce the notion of $(T(i,j), I(k,l), F(m,n))$-interval neutrosophic subalgebra in $BCK/BCI$-algebra for $i, j, k, l, m, n \in \{1, 2, 3, 4\}$, and investigate their properties and relations. We also introduce the notion of interval neutrosophic length of an interval neutrosophic set, and investigate related properties.

## 2. Preliminaries

By a *BCI-algebra*, we mean a system $X := (X, *, 0) \in K(\tau)$ in which the following axioms hold:

(I) $((x * y) * (x * z)) * (z * y) = 0,$

(II) $(x * (x * y)) * y = 0,$

(III) $x * x = 0,$

(IV) $x * y = y * x = 0 \Rightarrow x = y$

for all $x, y, z \in X$. If a $BCI$-algebra $X$ satisfies $0 * x = 0$ for all $x \in X$, then we say that $X$ is *BCK-algebra*.

A non-empty subset $S$ of a $BCK/BCI$-algebra $X$ is called a *subalgebra* of $X$ if $x * y \in S$ for all $x, y \in S$.

The collection of all $BCK$-algebra and all $BCI$-algebra are denoted by $\mathcal{B}_K(X)$ and $\mathcal{B}_I(X)$, respectively. In addition, $\mathcal{B}(X) := \mathcal{B}_K(X) \cup \mathcal{B}_I(X)$.

We refer the reader to the books [9,10] for further information regarding $BCK/BCI$-algebra.

By a *fuzzy structure* over a nonempty set $X$, we mean an ordered pair $(X, \rho)$ of $X$ and a fuzzy set $\rho$ on $X$.

**Definition 1** ([11]). *For any $(X, *, 0) \in \mathcal{B}(X)$, a fuzzy structure $(X, \mu)$ over $(X, *, 0)$ is called a*

- *fuzzy subalgebra of $(X, *, 0)$ with type* 1 *(briefly,* 1*-fuzzy subalgebra of $(X, *, 0)$) if*

$$(\forall x, y \in X) \left(\mu(x * y) \geq \min\{\mu(x), \mu(y)\}\right), \tag{1}$$

- *fuzzy subalgebra of $(X, *, 0)$ with type* 2 *(briefly,* 2*-fuzzy subalgebra of $(X, *, 0)$) if*

$$(\forall x, y \in X) \left(\mu(x * y) \leq \min\{\mu(x), \mu(y)\}\right), \tag{2}$$

- *fuzzy subalgebra of $(X, *, 0)$ with type* 3 *(briefly,* 3*-fuzzy subalgebra of $(X, *, 0)$) if*

$$(\forall x, y \in X) \left(\mu(x * y) \geq \max\{\mu(x), \mu(y)\}\right), \tag{3}$$

- *fuzzy subalgebra of $(X, *, 0)$ with type* 4 *(briefly,* 4*-fuzzy subalgebra of $(X, *, 0)$) if*

$$(\forall x, y \in X) \left(\mu(x * y) \leq \max\{\mu(x), \mu(y)\}\right). \tag{4}$$

Let $X$ be a non-empty set. A neutrosophic set (NS) in $X$ (see [4]) is a structure of the form:

$$A := \{\langle x; A_T(x), A_I(x), A_F(x)\rangle \mid x \in X\},$$

where $A_T : X \to [0,1]$ is a truth membership function, $A_I : X \to [0,1]$ is an indeterminate membership function, and $A_F : X \to [0,1]$ is a false membership function.

An interval neutrosophic set (INS) $A$ in $X$ is characterized by truth-membership function $T_A$, indeterminacy membership function $I_A$ and falsity-membership function $F_A$. For each point $x$ in $X$, $T_A(x), I_A(x), F_A(x) \in [0,1]$ (see [3,6]).

## 3. Interval Neutrosophic Subalgebra

In what follows, let $(X, *, 0) \in \mathcal{B}(X)$ and $\mathcal{P}^*([0,1])$ be the family of all subintervals of $[0,1]$ unless otherwise specified.

**Definition 2** ([3,6]). *An interval neutrosophic set in a nonempty set $X$ is a structure of the form:*

$$\mathcal{I} := \{\langle x, \mathcal{I}[T](x), \mathcal{I}[I](x), \mathcal{I}[F](x)\rangle \mid x \in X\},$$

*where*

$$\mathcal{I}[T] : X \to \mathcal{P}^*([0,1]),$$

*which is called interval truth-membership function,*

$$\mathcal{I}[I] : X \to \mathcal{P}^*([0,1]),$$

*which is called interval indeterminacy-membership function, and*

$$\mathcal{I}[F] : X \to \mathcal{P}^*([0,1]),$$

*which is called interval falsity-membership function.*

For the sake of simplicity, we will use the notation $\mathcal{I} := (\mathcal{I}[T], \mathcal{I}[I], \mathcal{I}[F])$ for the interval neutrosophic set

$$\mathcal{I} := \{\langle x, \mathcal{I}[T](x), \mathcal{I}[I](x), \mathcal{I}[F](x)\rangle \mid x \in X\}.$$

Given an interval neutrosophic set $\mathcal{I} := (\mathcal{I}[T], \mathcal{I}[I], \mathcal{I}[F])$ in $X$, we consider the following functions:

$$\begin{aligned}
\mathcal{I}[T]_{\inf} &: X \to [0,1],\ x \mapsto \inf\{\mathcal{I}[T](x)\},\\
\mathcal{I}[I]_{\inf} &: X \to [0,1],\ x \mapsto \inf\{\mathcal{I}[I](x)\},\\
\mathcal{I}[F]_{\inf} &: X \to [0,1],\ x \mapsto \inf\{\mathcal{I}[F](x)\},
\end{aligned}$$

and

$$\begin{aligned}
\mathcal{I}[T]_{\sup} &: X \to [0,1],\ x \mapsto \sup\{\mathcal{I}[T](x)\},\\
\mathcal{I}[I]_{\sup} &: X \to [0,1],\ x \mapsto \sup\{\mathcal{I}[I](x)\},\\
\mathcal{I}[F]_{\sup} &: X \to [0,1],\ x \mapsto \sup\{\mathcal{I}[F](x)\}.
\end{aligned}$$

**Definition 3.** *For any $i, j, k, l, m, n \in \{1,2,3,4\}$, an interval neutrosophic set $\mathcal{I} := (\mathcal{I}[T], \mathcal{I}[I], \mathcal{I}[F])$ in $X$ is called a $(T(i,j), I(k,l), F(m,n))$-interval neutrosophic subalgebra of $X$ if the following assertions are valid.*

(1) *$(X, \mathcal{I}[T]_{\inf})$ is an i-fuzzy subalgebra of $(X, *, 0)$ and $(X, \mathcal{I}[T]_{\sup})$ is a j-fuzzy subalgebra of $(X, *, 0)$,*
(2) *$(X, \mathcal{I}[I]_{\inf})$ is a k-fuzzy subalgebra of $(X, *, 0)$ and $(X, \mathcal{I}[I]_{\sup})$ is an l-fuzzy subalgebra of $(X, *, 0)$,*
(3) *$(X, \mathcal{I}[F]_{\inf})$ is an m-fuzzy subalgebra of $(X, *, 0)$ and $(X, \mathcal{I}[F]_{\sup})$ is an n-fuzzy subalgebra of $(X, *, 0)$.*

**Example 1.** *Consider a BCK-algebra $X = \{0,1,2,3\}$ with the binary operation $*$, which is given in Table 1 (see [10]).*

**Table 1.** Cayley table for the binary operation "$*$".

| $*$ | 0 | 1 | 2 | 3 |
|---|---|---|---|---|
| 0 | 0 | 0 | 0 | 0 |
| 1 | 1 | 0 | 0 | 1 |
| 2 | 2 | 1 | 0 | 2 |
| 3 | 3 | 3 | 3 | 0 |

*(1) Let $\mathcal{I} := (\mathcal{I}[T], \mathcal{I}[I], \mathcal{I}[F])$ be an interval neutrosophic set in $(X, *, 0)$ for which $\mathcal{I}[T]$, $\mathcal{I}[I]$ and $\mathcal{I}[F]$ are given as follows:*

$$\mathcal{I}[T] : X \to \mathcal{P}^*([0,1])\ \ x \mapsto \begin{cases} [0.4, 0.5) & \text{if } x = 0,\\ (0.3, 0.5] & \text{if } x = 1,\\ [0.2, 0.6) & \text{if } x = 2,\\ [0.1, 0.7] & \text{if } x = 3, \end{cases}$$

$$\mathcal{I}[I] : X \rightarrow \mathcal{P}^*([0,1]) \;\; x \mapsto \begin{cases} [0.5, 0.8) & \text{if } x = 0, \\ (0.2, 0.7) & \text{if } x = 1, \\ [0.5, 0.6] & \text{if } x = 2, \\ [0.4, 0.8) & \text{if } x = 3, \end{cases}$$

*and*

$$\mathcal{I}[F] : X \rightarrow \mathcal{P}^*([0,1]) \;\; x \mapsto \begin{cases} [0.4, 0.5) & \text{if } x = 0, \\ (0.2, 0.9) & \text{if } x = 1, \\ [0.1, 0.6] & \text{if } x = 2, \\ (0.4, 0.7] & \text{if } x = 3. \end{cases}$$

*It is routine to verify that* $\mathcal{I} := (\mathcal{I}[T], \mathcal{I}[I], \mathcal{I}[F])$ *is a* $(T(1,4), I(1,4), F(1,4))$*-interval neutrosophic subalgebra of* $(X, *, 0)$.

*(2) Let* $\mathcal{I} := (\mathcal{I}[T], \mathcal{I}[I], \mathcal{I}[F])$ *be an interval neutrosophic set in* $(X, *, 0)$ *for which* $\mathcal{I}[T]$, $\mathcal{I}[I]$ *and* $\mathcal{I}[F]$ *are given as follows:*

$$\mathcal{I}[T] : X \rightarrow \mathcal{P}^*([0,1]) \;\; x \mapsto \begin{cases} [0.1, 0.4) & \text{if } x = 0, \\ (0.3, 0.5) & \text{if } x = 1, \\ [0.2, 0.7] & \text{if } x = 2, \\ [0.4, 0.6) & \text{if } x = 3, \end{cases}$$

$$\mathcal{I}[I] : X \rightarrow \mathcal{P}^*([0,1]) \;\; x \mapsto \begin{cases} (0.2, 0.5) & \text{if } x = 0, \\ [0.5, 0.8] & \text{if } x = 1, \\ (0.4, 0.5] & \text{if } x = 2, \\ [0.2, 0.6] & \text{if } x = 3, \end{cases}$$

*and*

$$\mathcal{I}[F] : X \rightarrow \mathcal{P}^*([0,1]) \;\; x \mapsto \begin{cases} [0.3, 0.4) & \text{if } x = 0, \\ (0.4, 0.7) & \text{if } x = 1, \\ (0.6, 0.8) & \text{if } x = 2, \\ [0.4, 0.6] & \text{if } x = 3. \end{cases}$$

*By routine calculations, we know that* $\mathcal{I} := (\mathcal{I}[T], \mathcal{I}[I], \mathcal{I}[F])$ *is a* $(T(4,4), I(4,4), F(4,4))$*-interval neutrosophic subalgebra of* $(X, *, 0)$.

**Example 2.** *Consider a BCI-algebra* $X = \{0, a, b, c\}$ *with the binary operation* $*$*, which is given in Table 2 (see [10]).*

**Table 2.** Cayley table for the binary operation "$*$".

| $*$ | 0 | $a$ | $b$ | $c$ |
|---|---|---|---|---|
| 0 | 0 | $a$ | $b$ | $c$ |
| $a$ | $a$ | 0 | $c$ | $b$ |
| $b$ | $b$ | $c$ | 0 | $a$ |
| $c$ | $c$ | $b$ | $a$ | 0 |

*Let* $\mathcal{I} := (\mathcal{I}[T], \mathcal{I}[I], \mathcal{I}[F])$ *be an interval neutrosophic set in* $(X, *, 0)$ *for which* $\mathcal{I}[T]$, $\mathcal{I}[I]$ *and* $\mathcal{I}[F]$ *are given as follows:*

$$\mathcal{I}[T] : X \rightarrow \mathcal{P}^*([0,1]) \;\; x \mapsto \begin{cases} [0.3, 0.9) & \text{if } x = 0, \\ (0.7, 0.9) & \text{if } x = a, \\ [0.7, 0.8) & \text{if } x = b, \\ (0.5, 0.8] & \text{if } x = c, \end{cases}$$

$$\mathcal{I}[I] : X \to \mathcal{P}^*([0,1]) \ x \mapsto \begin{cases} [0.2, 0.65) & \text{if } x = 0, \\ [0.5, 0.55] & \text{if } x = a, \\ (0.6, 0.65) & \text{if } x = b, \\ [0.5, 0.55) & \text{if } x = c, \end{cases}$$

*and*

$$\mathcal{I}[F] : X \to \mathcal{P}^*([0,1]) \ x \mapsto \begin{cases} (0.3, 0.6) & \text{if } x = 0, \\ [0.4, 0.6] & \text{if } x = a, \\ (0.4, 0.5] & \text{if } x = b, \\ [0.3, 0.5) & \text{if } x = c. \end{cases}$$

*Routine calculations show that* $\mathcal{I} := (\mathcal{I}[T], \mathcal{I}[I], \mathcal{I}[F])$ *is a* $(T(4,1), I(4,1), F(4,1))$*-interval neutrosophic subalgebra of* $(X, *, 0)$*. However, it is not a* $(T(2,1), I(2,1), F(2,1))$*-interval neutrosophic subalgebra of* $(X, *, 0)$ *since*

$$\mathcal{I}[T]_{\inf}(c * a) = \mathcal{I}[T]_{\inf}(b) = 0.7 > 0.5 = \min\{\mathcal{I}[T]_{\inf}(c), \mathcal{I}[T]_{\inf}(a)\}$$

*and/or*

$$\mathcal{I}[I]_{\inf}(a * c) = \mathcal{I}[I]_{\inf}(b) = 0.6 > 0.5 = \min\{\mathcal{I}[I]_{\inf}(a), \mathcal{I}[I]_{\inf}(c)\}.$$

*In addition, it is not a* $(T(4,3), I(4,3), F(4,3))$*-interval neutrosophic subalgebra of* $(X, *, 0)$ *since*

$$\mathcal{I}[T]_{\sup}(a * b) = \mathcal{I}[T]_{\sup}(c) = 0.8 < 0.9 = \max\{\mathcal{I}[T]_{\inf}(a), \mathcal{I}[T]_{\inf}(c)\}$$

*and/or*

$$\mathcal{I}[F]_{\sup}(a * b) = \mathcal{I}[F]_{\sup}(c) = 0.5 < 0.6 = \max\{\mathcal{I}[F]_{\inf}(a), \mathcal{I}[F]_{\inf}(c)\}.$$

Let $\mathcal{I} := (\mathcal{I}[T], \mathcal{I}[I], \mathcal{I}[F])$ be an interval neutrosophic set in X. We consider the following sets:

$$\begin{aligned} U(\mathcal{I}[T]_{\inf}; \alpha_I) &:= \{x \in X \mid \mathcal{I}[T]_{\inf}(x) \geq \alpha_I\}, \\ L(\mathcal{I}[T]_{\sup}; \alpha_S) &:= \{x \in X \mid \mathcal{I}[T]_{\sup}(x) \leq \alpha_S\}, \\ U(\mathcal{I}[I]_{\inf}; \beta_I) &:= \{x \in X \mid \mathcal{I}[I]_{\inf}(x) \geq \beta_I\}, \\ L(\mathcal{I}[I]_{\sup}; \beta_S) &:= \{x \in X \mid \mathcal{I}[I]_{\sup}(x) \leq \beta_S\}, \end{aligned}$$

and

$$\begin{aligned} U(\mathcal{I}[F]_{\inf}; \gamma_I) &:= \{x \in X \mid \mathcal{I}[F]_{\inf}(x) \geq \gamma_I\}, \\ L(\mathcal{I}[F]_{\sup}; \gamma_S) &:= \{x \in X \mid \mathcal{I}[F]_{\sup}(x) \leq \gamma_S\}, \end{aligned}$$

where $\alpha_I$, $\alpha_S$, $\beta_I$, $\beta_S$, $\gamma_I$ and $\gamma_S$ are numbers in $[0,1]$.

**Theorem 1.** *If an interval neutrosophic set* $\mathcal{I} := (\mathcal{I}[T], \mathcal{I}[I], \mathcal{I}[F])$ *in X is a* $(T(i,4), I(i,4), F(i,4))$*-interval neutrosophic subalgebra of* $(X, *, 0)$ *for* $i \in \{1,3\}$*, then* $U(\mathcal{I}[T]_{\inf}; \alpha_I)$*,* $L(\mathcal{I}[T]_{\sup}; \alpha_S)$*,* $U(\mathcal{I}[I]_{\inf}; \beta_I)$*,* $L(\mathcal{I}[I]_{\sup}; \beta_S)$*,* $U(\mathcal{I}[F]_{\inf}; \gamma_I)$ *and* $L(\mathcal{I}[F]_{\sup}; \gamma_S)$ *are either empty or subalgebra of* $(X, *, 0)$ *for all* $\alpha_I$*,* $\alpha_S$*,* $\beta_I$*,* $\beta_S$*,* $\gamma_I$*,* $\gamma_S \in [0,1]$*.*

**Proof.** Assume that $\mathcal{I} := (\mathcal{I}[T], \mathcal{I}[I], \mathcal{I}[F])$ is a $(T(1,4), I(1,4), F(1,4))$-interval neutrosophic subalgebra of $(X, *, 0)$. Then, $(X, \mathcal{I}[T]_{\inf})$, $(X, \mathcal{I}[I]_{\inf})$ and $(X, \mathcal{I}[F]_{\inf})$ are 1-fuzzy subalgebra of X; and $(X, \mathcal{I}[T]_{\sup})$, $(X, \mathcal{I}[I]_{\sup})$ and $(X, \mathcal{I}[F]_{\sup})$ are 4-fuzzy subalgebra of X. Let $\alpha_I$, $\alpha_S \in [0,1]$ be such that $U(\mathcal{I}[T]_{\inf}; \alpha_I)$ and $L(\mathcal{I}[T]_{\sup}; \alpha_S)$ are nonempty. For any $x, y \in X$, if $x, y \in U(\mathcal{I}[T]_{\inf}; \alpha_I)$, then $\mathcal{I}[T]_{\inf}(x) \geq \alpha_I$ and $\mathcal{I}[T]_{\inf}(y) \geq \alpha_I$, and so

$$\mathcal{I}[T]_{\inf}(x * y) \geq \min\{\mathcal{I}[T]_{\inf}(x), \mathcal{I}[T]_{\inf}(y)\} \geq \alpha_I,$$

that is, $x * y \in U(\mathcal{I}[T]_{\inf}; \alpha_I)$. If $x, y \in L(\mathcal{I}[T]_{\sup}; \alpha_S)$, then $\mathcal{I}[T]_{\sup}(x) \leq \alpha_S$ and $\mathcal{I}[T]_{\sup}(y) \leq \alpha_S$, which imply that

$$\mathcal{I}[T]_{\sup}(x * y) \leq \max\{\mathcal{I}[T]_{\sup}(x), \mathcal{I}[T]_{\sup}(y)\} \leq \alpha_S,$$

that is, $x * y \in L(\mathcal{I}[T]_{\sup}; \alpha_S)$. Hence, $U(\mathcal{I}[T]_{\inf}; \alpha_I)$ and $L(\mathcal{I}[T]_{\sup}; \alpha_S)$ are subalgebra of $(X, *, 0)$ for all $\alpha_I$, $\alpha_S \in [0,1]$. Similarly, we can prove that $U(\mathcal{I}[I]_{\inf}; \beta_I)$, $L(\mathcal{I}[I]_{\sup}; \beta_S)$, $U(\mathcal{I}[F]_{\inf}; \gamma_I)$ and $L(\mathcal{I}[F]_{\sup}; \gamma_S)$ are either empty or subalgebra of $(X, *, 0)$ for all $\beta_I$, $\beta_S$, $\gamma_I$, $\gamma_S \in [0,1]$. Suppose that $\mathcal{I} := (\mathcal{I}[T], \mathcal{I}[I], \mathcal{I}[F])$ is a $(T(3,4), I(3,4), F(3,4))$-interval neutrosophic subalgebra of $(X, *, 0)$. Then, $(X, \mathcal{I}[T]_{\inf})$, $(X, \mathcal{I}[I]_{\inf})$ and $(X, \mathcal{I}[F]_{\inf})$ are 3-fuzzy subalgebra of $X$; and $(X, \mathcal{I}[T]_{\sup})$, $(X, \mathcal{I}[I]_{\sup})$ and $(X, \mathcal{I}[F]_{\sup})$ are 4-fuzzy subalgebra of X. Let $\beta_I$ and $\beta_S \in [0,1]$ be such that $U(\mathcal{I}[I]_{\inf}; \beta_I)$ and $L(\mathcal{I}[I]_{\sup}; \beta_S)$ are nonempty. Let $x, y \in U(\mathcal{I}[I]_{\inf}; \beta_I)$. Then, $\mathcal{I}[I]_{\inf}(x) \geq \beta_I$ and $\mathcal{I}[I]_{\inf}(y) \geq \beta_I$. It follows that

$$\mathcal{I}[I]_{\inf}(x * y) \geq \max\{\mathcal{I}[I]_{\inf}(x), \mathcal{I}[I]_{\inf}(y)\} \geq \beta_I$$

and so $x * y \in U(\mathcal{I}[I]_{\inf}; \beta_I)$. Thus, $U(\mathcal{I}[I]_{\inf}; \beta_I)$ is a subalgebra of $(X, *, 0)$. If $x, y \in L(\mathcal{I}[I]_{\inf}; \beta_S)$, then $\mathcal{I}[I]_{\inf}(x) \leq \beta_S$ and $\mathcal{I}[I]_{\inf}(y) \leq \beta_S$. Hence,

$$\mathcal{I}[I]_{\inf}(x * y) \leq \max\{\mathcal{I}[I]_{\inf}(x), \mathcal{I}[I]_{\inf}(y)\} \leq \beta_S,$$

and so $x * y \in L(\mathcal{I}[I]_{\inf}; \beta_S)$. Thus, $L(\mathcal{I}[I]_{\inf}; \beta_S)$ is a subalgebra of $(X, *, 0)$. Similarly, we can show that $U(\mathcal{I}[T]_{\inf}; \alpha_I)$, $L(\mathcal{I}[T]_{\sup}; \alpha_S)$, $U(\mathcal{I}[F]_{\inf}; \gamma_I)$ and $L(\mathcal{I}[F]_{\sup}; \gamma_S)$ are either empty or subalgebra of $(X, *, 0)$ for all $\alpha_I$, $\alpha_S$, $\gamma_I$, $\gamma_S \in [0,1]$. □

Since every 2-fuzzy subalgebra is a 4-fuzzy subalgebra, we have the following corollary.

**Corollary 1.** *If an interval neutrosophic set $\mathcal{I} := (\mathcal{I}[T], \mathcal{I}[I], \mathcal{I}[F])$ in X is a $(T(i,2), I(i,2), F(i,2))$-interval neutrosophic subalgebra of $(X, *, 0)$ for $i \in \{1,3\}$, then $U(\mathcal{I}[T]_{\inf}; \alpha_I)$, $L(\mathcal{I}[T]_{\sup}; \alpha_S)$, $U(\mathcal{I}[I]_{\inf}; \beta_I)$, $L(\mathcal{I}[I]_{\sup}; \beta_S)$, $U(\mathcal{I}[F]_{\inf}; \gamma_I)$ and $L(\mathcal{I}[F]_{\sup}; \gamma_S)$ are either empty or subalgebra of $(X, *, 0)$ for all $\alpha_I$, $\alpha_S$, $\beta_I$, $\beta_S$, $\gamma_I$, $\gamma_S \in [0,1]$.*

By a similar way to the proof of Theorem 1, we have the following theorems.

**Theorem 2.** *If an interval neutrosophic set $\mathcal{I} := (\mathcal{I}[T], \mathcal{I}[I], \mathcal{I}[F])$ in X is a $(T(i,4), I(i,4), F(i,4))$-interval neutrosophic subalgebra of $(X, *, 0)$ for $i \in \{2,4\}$, then $L(\mathcal{I}[T]_{\inf}; \alpha_I)$, $L(\mathcal{I}[T]_{\sup}; \alpha_S)$, $L(\mathcal{I}[I]_{\inf}; \beta_I)$, $L(\mathcal{I}[I]_{\sup}; \beta_S)$, $L(\mathcal{I}[F]_{\inf}; \gamma_I)$ and $L(\mathcal{I}[F]_{\sup}; \gamma_S)$ are either empty or subalgebra of $(X, *, 0)$ for all $\alpha_I$, $\alpha_S$, $\beta_I$, $\beta_S$, $\gamma_I$, $\gamma_S \in [0,1]$.*

**Corollary 2.** *If an interval neutrosophic set $\mathcal{I} := (\mathcal{I}[T], \mathcal{I}[I], \mathcal{I}[F])$ in X is a $(T(i,2), I(i,2), F(i,2))$-interval neutrosophic subalgebra of $(X, *, 0)$ for $i \in \{2,4\}$, then $L(\mathcal{I}[T]_{\inf}; \alpha_I)$, $L(\mathcal{I}[T]_{\sup}; \alpha_S)$, $L(\mathcal{I}[I]_{\inf}; \beta_I)$, $L(\mathcal{I}[I]_{\sup}; \beta_S)$, $L(\mathcal{I}[F]_{\inf}; \gamma_I)$ and $L(\mathcal{I}[F]_{\sup}; \gamma_S)$ are either empty or subalgebra of $(X, *, 0)$ for all $\alpha_I$, $\alpha_S$, $\beta_I$, $\beta_S$, $\gamma_I$, $\gamma_S \in [0,1]$.*

**Theorem 3.** *If an interval neutrosophic set $\mathcal{I} := (\mathcal{I}[T], \mathcal{I}[I], \mathcal{I}[F])$ in X is a $(T(k,1), I(k,1), F(k,1))$-interval neutrosophic subalgebra of $(X, *, 0)$ for $k \in \{1,3\}$, then $U(\mathcal{I}[T]_{\inf}; \alpha_I)$, $U(\mathcal{I}[T]_{\sup}; \alpha_S)$, $U(\mathcal{I}[I]_{\inf}; \beta_I)$, $U(\mathcal{I}[I]_{\sup}; \beta_S)$, $U(\mathcal{I}[F]_{\inf}; \gamma_I)$ and $U(\mathcal{I}[F]_{\sup}; \gamma_S)$ are either empty or subalgebra of $(X, *, 0)$ for all $\alpha_I$, $\alpha_S$, $\beta_I$, $\beta_S$, $\gamma_I$, $\gamma_S \in [0,1]$.*

**Corollary 3.** *If an interval neutrosophic set $\mathcal{I} := (\mathcal{I}[T], \mathcal{I}[I], \mathcal{I}[F])$ in X is a $(T(k,3), I(k,3), F(k,3))$-interval neutrosophic subalgebra of $(X, *, 0)$ for $k \in \{1,3\}$, then $U(\mathcal{I}[T]_{\inf}; \alpha_I)$, $U(\mathcal{I}[T]_{\sup}; \alpha_S)$, $U(\mathcal{I}[I]_{\inf}; \beta_I)$, $U(\mathcal{I}[I]_{\sup}; \beta_S)$, $U(\mathcal{I}[F]_{\inf}; \gamma_I)$ and $U(\mathcal{I}[F]_{\sup}; \gamma_S)$ are either empty or subalgebra of $(X, *, 0)$ for all $\alpha_I$, $\alpha_S$, $\beta_I$, $\beta_S$, $\gamma_I$, $\gamma_S \in [0,1]$.*

**Theorem 4.** *If an interval neutrosophic set* $\mathcal{I} := (\mathcal{I}[T], \mathcal{I}[I], \mathcal{I}[F])$ *in X is a* $(T(k,1), I(k,1), F(k,1))$*-interval neutrosophic subalgebra of* $(X,*,0)$ *for* $k \in \{2,4\}$*, then* $L(\mathcal{I}[T]_{\inf}; \alpha_I)$*,* $U(\mathcal{I}[T]_{\sup}; \alpha_S)$*,* $L(\mathcal{I}[I]_{\inf}; \beta_I)$*,* $U(\mathcal{I}[I]_{\sup}; \beta_S)$*,* $L(\mathcal{I}[F]_{\inf}; \gamma_I)$ *and* $U(\mathcal{I}[F]_{\sup}; \gamma_S)$ *are either empty or subalgebra of* $(X,*,0)$ *for all* $\alpha_I$*,* $\alpha_S$*,* $\beta_I$*,* $\beta_S$*,* $\gamma_I$*,* $\gamma_S \in [0,1]$.

**Corollary 4.** *If an interval neutrosophic set* $\mathcal{I} := (\mathcal{I}[T], \mathcal{I}[I], \mathcal{I}[F])$ *in X is a* $(T(k,3), I(k,3), F(k,3))$*-interval neutrosophic subalgebra of* $(X,*,0)$ *for* $k \in \{2,4\}$*, then* $L(\mathcal{I}[T]_{\inf}; \alpha_I)$*,* $U(\mathcal{I}[T]_{\sup}; \alpha_S)$*,* $L(\mathcal{I}[I]_{\inf}; \beta_I)$*,* $U(\mathcal{I}[I]_{\sup}; \beta_S)$*,* $L(\mathcal{I}[F]_{\inf}; \gamma_I)$ *and* $U(\mathcal{I}[F]_{\sup}; \gamma_S)$ *are either empty or subalgebra of* $(X,*,0)$ *for all* $\alpha_I$*,* $\alpha_S$*,* $\beta_I$*,* $\beta_S$*,* $\gamma_I$*,* $\gamma_S \in [0,1]$.

**Theorem 5.** *Let* $\mathcal{I} := (\mathcal{I}[T], \mathcal{I}[I], \mathcal{I}[F])$ *be an interval neutrosophic set in X in which* $U(\mathcal{I}[T]_{\inf}; \alpha_I)$*,* $L(\mathcal{I}[T]_{\sup}; \alpha_S)$*,* $U(\mathcal{I}[I]_{\inf}; \beta_I)$*,* $L(\mathcal{I}[I]_{\sup}; \beta_S)$*,* $U(\mathcal{I}[F]_{\inf}; \gamma_I)$ *and* $L(\mathcal{I}[F]_{\sup}; \gamma_S)$ *are nonempty subalgebra of* $(X,*,0)$ *for all* $\alpha_I$*,* $\alpha_S$*,* $\beta_I$*,* $\beta_S$*,* $\gamma_I$*,* $\gamma_S \in [0,1]$*. Then,* $\mathcal{I} := (\mathcal{I}[T], \mathcal{I}[I], \mathcal{I}[F])$ *is a* $(T(1,4), I(1,4), F(1,4))$*-interval neutrosophic subalgebra of* $(X,*,0)$.

**Proof.** Suppose that $(X, \mathcal{I}[T]_{\inf})$ is not a 1-fuzzy subalgebra of $(X,*,0)$. Then, there exists $x, y \in X$ such that

$$\mathcal{I}[T]_{\inf}(x*y) < \min\{\mathcal{I}[T]_{\inf}(x), \mathcal{I}[T]_{\inf}(y)\}.$$

If we take $\alpha_I = \min\{\mathcal{I}[T]_{\inf}(x), \mathcal{I}[T]_{\inf}(y)\}$, then $x, y \in U(\mathcal{I}[T]_{\inf}; \alpha_I)$, but $x*y \notin U(\mathcal{I}[T]_{\inf}; \alpha_I)$. This is a contradiction, and so $(X, \mathcal{I}[T]_{\inf})$ is a 1-fuzzy subalgebra of $(X,*,0)$. If $(X, \mathcal{I}[T]_{\sup})$ is not a 4-fuzzy subalgebra of $(X,*,0)$, then

$$\mathcal{I}[T]_{\sup}(a*b) > \max\{\mathcal{I}[T]_{\sup}(a), \mathcal{I}[T]_{\sup}(b)\}$$

for some $a, b \in X$, and so $a, b \in L(\mathcal{I}[T]_{\sup}; \alpha_S)$ and $a*b \notin L(\mathcal{I}[T]_{\sup}; \alpha_S)$ by taking

$$\alpha_S := \max\{\mathcal{I}[T]_{\sup}(a), \mathcal{I}[T]_{\sup}(b)\}.$$

This is a contradiction, and therefore $(X, \mathcal{I}[T]_{\sup})$ is a 4-fuzzy subalgebra of $(X,*,0)$. Similarly, we can verify that $(X, \mathcal{I}[I]_{\inf})$ is a 1-fuzzy subalgebra of $(X,*,0)$ and $(X, \mathcal{I}[I]_{\sup})$ is a 4-fuzzy subalgebra of $(X,*,0)$; and $(X, \mathcal{I}[F]_{\inf})$ is a 1-fuzzy subalgebra of $(X,*,0)$ and $(X, \mathcal{I}[F]_{\sup})$ is a 4-fuzzy subalgebra of $(X,*,0)$. Consequently, $\mathcal{I} := (\mathcal{I}[T], \mathcal{I}[I], \mathcal{I}[F])$ is a $(T(1,4), I(1,4), F(1,4))$-interval neutrosophic subalgebra of $(X,*,0)$. □

Using the similar method to the proof of Theorem 5, we get the following theorems.

**Theorem 6.** *Let* $\mathcal{I} := (\mathcal{I}[T], \mathcal{I}[I], \mathcal{I}[F])$ *be an interval neutrosophic set in X in which* $L(\mathcal{I}[T]_{\inf}; \alpha_I)$*,* $U(\mathcal{I}[T]_{\sup}; \alpha_S)$*,* $L(\mathcal{I}[I]_{\inf}; \beta_I)$*,* $U(\mathcal{I}[I]_{\sup}; \beta_S)$*,* $L(\mathcal{I}[F]_{\inf}; \gamma_I)$ *and* $U(\mathcal{I}[F]_{\sup}; \gamma_S)$ *are nonempty subalgebra of* $(X,*,0)$ *for all* $\alpha_I$*,* $\alpha_S$*,* $\beta_I$*,* $\beta_S$*,* $\gamma_I$*,* $\gamma_S \in [0,1]$*. Then,* $\mathcal{I} := (\mathcal{I}[T], \mathcal{I}[I], \mathcal{I}[F])$ *is a* $(T(4,1), I(4,1), F(4,1))$*-interval neutrosophic subalgebra of* $(X,*,0)$.

**Theorem 7.** *Let* $\mathcal{I} := (\mathcal{I}[T], \mathcal{I}[I], \mathcal{I}[F])$ *be an interval neutrosophic set in X in which* $L(\mathcal{I}[T]_{\inf}; \alpha_I)$*,* $L(\mathcal{I}[T]_{\sup}; \alpha_S)$*,* $L(\mathcal{I}[I]_{\inf}; \beta_I)$*,* $L(\mathcal{I}[I]_{\sup}; \beta_S)$*,* $L(\mathcal{I}[F]_{\inf}; \gamma_I)$ *and* $L(\mathcal{I}[F]_{\sup}; \gamma_S)$ *are nonempty subalgebra of* $(X,*,0)$ *for all* $\alpha_I$*,* $\alpha_S$*,* $\beta_I$*,* $\beta_S$*,* $\gamma_I$*,* $\gamma_S \in [0,1]$*. Then,* $\mathcal{I} := (\mathcal{I}[T], \mathcal{I}[I], \mathcal{I}[F])$ *is a* $(T(4,4), I(4,4), F(4,4))$*-interval neutrosophic subalgebra of* $(X,*,0)$.

**Theorem 8.** *Let* $\mathcal{I} := (\mathcal{I}[T], \mathcal{I}[I], \mathcal{I}[F])$ *be an interval neutrosophic set in X in which* $U(\mathcal{I}[T]_{\inf}; \alpha_I)$*,* $U(\mathcal{I}[T]_{\sup}; \alpha_S)$*,* $U(\mathcal{I}[I]_{\inf}; \beta_I)$*,* $U(\mathcal{I}[I]_{\sup}; \beta_S)$*,* $U(\mathcal{I}[F]_{\inf}; \gamma_I)$ *and* $U(\mathcal{I}[F]_{\sup}; \gamma_S)$ *are nonempty subalgebra of* $(X,*,0)$ *for all* $\alpha_I$*,* $\alpha_S$*,* $\beta_I$*,* $\beta_S$*,* $\gamma_I$*,* $\gamma_S \in [0,1]$*. Then,* $\mathcal{I} := (\mathcal{I}[T], \mathcal{I}[I], \mathcal{I}[F])$ *is a* $(T(1,1), I(1,1), F(1,1))$*-interval neutrosophic subalgebra of* $(X,*,0)$.

## 4. Interval Neutrosophic Lengths

**Definition 4.** *Given an interval neutrosophic set $\mathcal{I} := (\mathcal{I}[T], \mathcal{I}[I], \mathcal{I}[F])$ in X, we define the interval neutrosophic length of $\mathcal{I}$ as an ordered triple $\mathcal{I}_\ell := (\mathcal{I}[T]_\ell, \mathcal{I}[I]_\ell, \mathcal{I}[F]_\ell)$ where*

$$\mathcal{I}[T]_\ell : X \to [0,1],\ x \mapsto \mathcal{I}[T]_{\sup}(x) - \mathcal{I}[T]_{\inf}(x),$$
$$\mathcal{I}[I]_\ell : X \to [0,1],\ x \mapsto \mathcal{I}[I]_{\sup}(x) - \mathcal{I}[I]_{\inf}(x),$$

*and*

$$\mathcal{I}[F]_\ell : X \to [0,1],\ x \mapsto \mathcal{I}[F]_{\sup}(x) - \mathcal{I}[F]_{\inf}(x),$$

*which are called interval neutrosophic T-length, interval neutrosophic I-length and interval neutrosophic F-length of $\mathcal{I}$, respectively.*

**Example 3.** *Consider the interval neutrosophic set $\mathcal{I} := (\mathcal{I}[T], \mathcal{I}[I], \mathcal{I}[F])$ in X, which is given in Example 2. Then, the interval neutrosophic length of $\mathcal{I}$ is given by Table 3.*

**Table 3.** Interval neutrosophic length of $\mathcal{I}$.

| X | $\mathcal{I}[T]_\ell$ | $\mathcal{I}[I]_\ell$ | $\mathcal{I}[F]_\ell$ |
|---|---|---|---|
| 0 | 0.6 | 0.45 | 0.3 |
| $a$ | 0.2 | 0.05 | 0.2 |
| $b$ | 0.1 | 0.05 | 0.1 |
| $c$ | 0.3 | 0.05 | 0.2 |

**Theorem 9.** *If an interval neutrosophic set $\mathcal{I} := (\mathcal{I}[T], \mathcal{I}[I], \mathcal{I}[F])$ in X is a $(T(i,3), I(i,3), F(i,3))$-interval neutrosophic subalgebra of $(X, *, 0)$ for $i \in \{2,4\}$, then $(X, \mathcal{I}[T]_\ell)$, $(X, \mathcal{I}[I]_\ell)$ and $(X, \mathcal{I}[F]_\ell)$ are 3-fuzzy subalgebra of $(X, *, 0)$.*

**Proof.** Assume that $\mathcal{I} := (\mathcal{I}[T], \mathcal{I}[I], \mathcal{I}[F])$ is a $(T(2,3), I(2,3), F(2,3))$-interval neutrosophic subalgebra of $(X, *, 0)$. Then, $(X, \mathcal{I}[T]_{\inf})$, $(X, \mathcal{I}[I]_{\inf})$ and $(X, \mathcal{I}[F]_{\inf})$ are 2-fuzzy subalgebra of X, and $(X, \mathcal{I}[T]_{\sup})$, $(X, \mathcal{I}[I]_{\sup})$ and $(X, \mathcal{I}[F]_{\sup})$ are 3-fuzzy subalgebra of X. Thus,

$$\mathcal{I}[T]_{\inf}(x*y) \le \min\{\mathcal{I}[T]_{\inf}(x), \mathcal{I}[T]_{\inf}(y)\},$$
$$\mathcal{I}[I]_{\inf}(x*y) \le \min\{\mathcal{I}[I]_{\inf}(x), \mathcal{I}[I]_{\inf}(y)\},$$
$$\mathcal{I}[F]_{\inf}(x*y) \le \min\{\mathcal{I}[F]_{\inf}(x), \mathcal{I}[F]_{\inf}(y)\},$$

and

$$\mathcal{I}[T]_{\sup}(x*y) \ge \max\{\mathcal{I}[T]_{\sup}(x), \mathcal{I}[T]_{\sup}(y)\},$$
$$\mathcal{I}[I]_{\sup}(x*y) \ge \max\{\mathcal{I}[I]_{\sup}(x), \mathcal{I}[I]_{\sup}(y)\},$$
$$\mathcal{I}[F]_{\sup}(x*y) \ge \max\{\mathcal{I}[F]_{\sup}(x), \mathcal{I}[F]_{\sup}(y)\},$$

for all $x, y \in X$. It follows that

$$\mathcal{I}[T]_\ell(x*y) = \mathcal{I}[T]_{\sup}(x*y) - \mathcal{I}[T]_{\inf}(x*y) \ge \mathcal{I}[T]_{\sup}(x) - \mathcal{I}[T]_{\inf}(x) = \mathcal{I}[T]_\ell(x),$$
$$\mathcal{I}[T]_\ell(x*y) = \mathcal{I}[T]_{\sup}(x*y) - \mathcal{I}[T]_{\inf}(x*y) \ge \mathcal{I}[T]_{\sup}(y) - \mathcal{I}[T]_{\inf}(y) = \mathcal{I}[T]_\ell(y),$$
$$\mathcal{I}[I]_\ell(x*y) = \mathcal{I}[I]_{\sup}(x*y) - \mathcal{I}[I]_{\inf}(x*y) \ge \mathcal{I}[I]_{\sup}(x) - \mathcal{I}[I]_{\inf}(x) = \mathcal{I}[I]_\ell(x),$$
$$\mathcal{I}[I]_\ell(x*y) = \mathcal{I}[I]_{\sup}(x*y) - \mathcal{I}[I]_{\inf}(x*y) \ge \mathcal{I}[I]_{\sup}(y) - \mathcal{I}[I]_{\inf}(y) = \mathcal{I}[I]_\ell(y),$$

and

$$\mathcal{I}[F]_\ell(x*y) = \mathcal{I}[F]_{\sup}(x*y) - \mathcal{I}[F]_{\inf}(x*y) \geq \mathcal{I}[F]_{\sup}(x) - \mathcal{I}[F]_{\inf}(x) = \mathcal{I}[F]_\ell(x),$$
$$\mathcal{I}[F]_\ell(x*y) = \mathcal{I}[F]_{\sup}(x*y) - \mathcal{I}[F]_{\inf}(x*y) \geq \mathcal{I}[F]_{\sup}(y) - \mathcal{I}[F]_{\inf}(y) = \mathcal{I}[F]_\ell(y).$$

Hence,

$$\mathcal{I}[T]_\ell(x*y) \geq \max\{\mathcal{I}[T]_\ell(x), \mathcal{I}[T]_\ell(y)\},$$
$$\mathcal{I}[I]_\ell(x*y) \geq \max\{\mathcal{I}[I]_\ell(x), \mathcal{I}[I]_\ell(y)\},$$

and

$$\mathcal{I}[F]_\ell(x*y) \geq \max\{\mathcal{I}[F]_\ell(x), \mathcal{I}[F]_\ell(y)\},$$

for all $x, y \in X$. Therefore, $(X, \mathcal{I}[T]_\ell)$, $(X, \mathcal{I}[I]_\ell)$ and $(X, \mathcal{I}[F]_\ell)$ are 3-fuzzy subalgebra of $(X, *, 0)$.

Suppose that $\mathcal{I} := (\mathcal{I}[T], \mathcal{I}[I], \mathcal{I}[F])$ is a $(T(4,3), I(4,3), F(4,3))$-interval neutrosophic subalgebra of $(X, *, 0)$. Then, $(X, \mathcal{I}[T]_{\inf})$, $(X, \mathcal{I}[I]_{\inf})$ and $(X, \mathcal{I}[F]_{\inf})$ are 4-fuzzy subalgebra of $X$, and $(X, \mathcal{I}[T]_{\sup})$, $(X, \mathcal{I}[I]_{\sup})$ and $(X, \mathcal{I}[F]_{\sup})$ are 3-fuzzy subalgebra of X. Hence,

$$\begin{aligned} &\mathcal{I}[T]_{\inf}(x*y) \leq \max\{\mathcal{I}[T]_{\inf}(x), \mathcal{I}[T]_{\inf}(y)\}, \\ &\mathcal{I}[I]_{\inf}(x*y) \leq \max\{\mathcal{I}[I]_{\inf}(x), \mathcal{I}[I]_{\inf}(y)\}, \\ &\mathcal{I}[F]_{\inf}(x*y) \leq \max\{\mathcal{I}[F]_{\inf}(x), \mathcal{I}[F]_{\inf}(y)\}, \end{aligned} \tag{5}$$

and

$$\mathcal{I}[T]_{\sup}(x*y) \geq \max\{\mathcal{I}[T]_{\sup}(x), \mathcal{I}[T]_{\sup}(y)\},$$
$$\mathcal{I}[I]_{\sup}(x*y) \geq \max\{\mathcal{I}[I]_{\sup}(x), \mathcal{I}[I]_{\sup}(y)\},$$
$$\mathcal{I}[F]_{\sup}(x*y) \geq \max\{\mathcal{I}[F]_{\sup}(x), \mathcal{I}[F]_{\sup}(y)\},$$

for all $x, y \in X$. Label (5) implies that

$$\mathcal{I}[T]_{\inf}(x*y) \leq \mathcal{I}[T]_{\inf}(x) \text{ or } \mathcal{I}[T]_{\inf}(x*y) \leq \mathcal{I}[T]_{\inf}(y),$$
$$\mathcal{I}[I]_{\inf}(x*y) \leq \mathcal{I}[I]_{\inf}(x) \text{ or } \mathcal{I}[I]_{\inf}(x*y) \leq \mathcal{I}[I]_{\inf}(y),$$
$$\mathcal{I}[F]_{\inf}(x*y) \leq \mathcal{I}[F]_{\inf}(x) \text{ or } \mathcal{I}[F]_{\inf}(x*y) \leq \mathcal{I}[F]_{\inf}(y).$$

If $\mathcal{I}[T]_{\inf}(x*y) \leq \mathcal{I}[T]_{\inf}(x)$, then

$$\mathcal{I}[T]_\ell(x*y) = \mathcal{I}[T]_{\sup}(x*y) - \mathcal{I}[T]_{\inf}(x*y) \geq \mathcal{I}[T]_{\sup}(x) - \mathcal{I}[T]_{\inf}(x) = \mathcal{I}[T]_\ell(x).$$

If $\mathcal{I}[T]_{\inf}(x*y) \leq \mathcal{I}[T]_{\inf}(y)$, then

$$\mathcal{I}[T]_\ell(x*y) = \mathcal{I}[T]_{\sup}(x*y) - \mathcal{I}[T]_{\inf}(x*y) \geq \mathcal{I}[T]_{\sup}(y) - \mathcal{I}[T]_{\inf}(y) = \mathcal{I}[T]_\ell(y).$$

It follows that $\mathcal{I}[T]_\ell(x*y) \geq \max\{\mathcal{I}[T]_\ell(x), \mathcal{I}[T]_\ell(y)\}$. Therefore, $(X, \mathcal{I}[T]_\ell)$ is a 3-fuzzy subalgebra of $(X, *, 0)$. Similarly, we can show that $(X, \mathcal{I}[I]_\ell)$ and $(X, \mathcal{I}[F]_\ell)$ are 3-fuzzy subalgebra of $(X, *, 0)$. □

**Corollary 5.** *If an interval neutrosophic set* $\mathcal{I} := (\mathcal{I}[T], \mathcal{I}[I], \mathcal{I}[F])$ *in X is a* $(T(i,3), I(i,3), F(i,3))$*-interval neutrosophic subalgebra of* $(X, *, 0)$ *for* $i \in \{2,4\}$*, then* $(X, \mathcal{I}[T]_\ell)$*,* $(X, \mathcal{I}[I]_\ell)$ *and* $(X, \mathcal{I}[F]_\ell)$ *are 1-fuzzy subalgebra of* $(X, *, 0)$*.*

**Theorem 10.** *If an interval neutrosophic set* $\mathcal{I} := (\mathcal{I}[T], \mathcal{I}[I], \mathcal{I}[F])$ *in* $X$ *is a* $(T(3,4), I(3,4), F(3,4))$*-interval neutrosophic subalgebra of* $(X,*,0)$*, then* $(X, \mathcal{I}[T]_\ell)$*,* $(X, \mathcal{I}[I]_\ell)$ *and* $(X, \mathcal{I}[F]_\ell)$ *are* 4*-fuzzy subalgebra of* $(X,*,0)$*.*

**Proof.** Let $\mathcal{I} := (\mathcal{I}[T], \mathcal{I}[I], \mathcal{I}[F])$ be a $(T(3,4), I(3,4), F(3,4))$-interval neutrosophic subalgebra of $(X,*,0)$. Then, $(X, \mathcal{I}[T]_{\inf})$, $(X, \mathcal{I}[I]_{\inf})$ and $(X, \mathcal{I}[F]_{\inf})$ are 3-fuzzy subalgebra of $X$, and $(X, \mathcal{I}[T]_{\sup})$, $(X, \mathcal{I}[I]_{\sup})$ and $(X, \mathcal{I}[F]_{\sup})$ are 4-fuzzy subalgebra of $X$. Thus,

$$\begin{aligned}
\mathcal{I}[T]_{\inf}(x*y) &\geq \max\{\mathcal{I}[T]_{\inf}(x), \mathcal{I}[T]_{\inf}(y)\},\\
\mathcal{I}[I]_{\inf}(x*y) &\geq \max\{\mathcal{I}[I]_{\inf}(x), \mathcal{I}[I]_{\inf}(y)\},\\
\mathcal{I}[F]_{\inf}(x*y) &\geq \max\{\mathcal{I}[F]_{\inf}(x), \mathcal{I}[F]_{\inf}(y)\},
\end{aligned}$$

and

$$\begin{aligned}
\mathcal{I}[T]_{\sup}(x*y) &\leq \max\{\mathcal{I}[T]_{\sup}(x), \mathcal{I}[T]_{\sup}(y)\},\\
\mathcal{I}[I]_{\sup}(x*y) &\leq \max\{\mathcal{I}[I]_{\sup}(x), \mathcal{I}[I]_{\sup}(y)\},\\
\mathcal{I}[F]_{\sup}(x*y) &\leq \max\{\mathcal{I}[F]_{\sup}(x), \mathcal{I}[F]_{\sup}(y)\},
\end{aligned} \tag{6}$$

for all $x, y \in X$. It follows from Label (6) that

$$\begin{aligned}
&\mathcal{I}[T]_{\sup}(x*y) \leq \mathcal{I}[T]_{\sup}(x) \text{ or } \mathcal{I}[T]_{\sup}(x*y) \leq \mathcal{I}[T]_{\sup}(y),\\
&\mathcal{I}[I]_{\sup}(x*y) \leq \mathcal{I}[I]_{\sup}(x) \text{ or } \mathcal{I}[I]_{\sup}(x*y) \leq \mathcal{I}[I]_{\sup}(y),\\
&\mathcal{I}[F]_{\sup}(x*y) \leq \mathcal{I}[F]_{\sup}(x) \text{ or } \mathcal{I}[F]_{\sup}(x*y) \leq \mathcal{I}[F]_{\sup}(y).
\end{aligned}$$

Assume that $\mathcal{I}[T]_{\sup}(x*y) \leq \mathcal{I}[T]_{\sup}(x)$. Then,

$$\mathcal{I}[T]_\ell(x*y) = \mathcal{I}[T]_{\sup}(x*y) - \mathcal{I}[T]_{\inf}(x*y) \leq \mathcal{I}[T]_{\sup}(x) - \mathcal{I}[T]_{\inf}(x) = \mathcal{I}[T]_\ell(x).$$

If $\mathcal{I}[T]_{\sup}(x*y) \leq \mathcal{I}[T]_{\sup}(y)$, then

$$\mathcal{I}[T]_\ell(x*y) = \mathcal{I}[T]_{\sup}(x*y) - \mathcal{I}[T]_{\inf}(x*y) \leq \mathcal{I}[T]_{\sup}(y) - \mathcal{I}[T]_{\inf}(y) = \mathcal{I}[T]_\ell(y).$$

Hence, $\mathcal{I}[T]_\ell(x*y) \leq \max\{\mathcal{I}[T]_\ell(x), \mathcal{I}[T]_\ell(y)\}$ for all $x, y \in X$. By a similar way, we can prove that

$$\mathcal{I}[I]_\ell(x*y) \leq \max\{\mathcal{I}[I]_\ell(x), \mathcal{I}[I]_\ell(y)\}$$

and

$$\mathcal{I}[F]_\ell(x*y) \leq \max\{\mathcal{I}[F]_\ell(x), \mathcal{I}[F]_\ell(y)\}$$

for all $x, y \in X$. Therefore, $(X, \mathcal{I}[T]_\ell)$, $(X, \mathcal{I}[I]_\ell)$ and $(X, \mathcal{I}[F]_\ell)$ are 4-fuzzy subalgebra of $(X,*,0)$. □

**Theorem 11.** *If an interval neutrosophic set* $\mathcal{I} := (\mathcal{I}[T], \mathcal{I}[I], \mathcal{I}[F])$ *in* $X$ *is a* $(T(3,2), I(3,2), F(3,2))$*-interval neutrosophic subalgebra of* $(X,*,0)$*, then* $(X, \mathcal{I}[T]_\ell)$*,* $(X, \mathcal{I}[I]_\ell)$ *and* $(X, \mathcal{I}[F]_\ell)$ *are* 2*-fuzzy subalgebra of* $(X,*,0)$*.*

**Proof.** Assume that $\mathcal{I} := (\mathcal{I}[T], \mathcal{I}[I], \mathcal{I}[F])$ is a $(T(3,2), I(3,2), F(3,2))$-interval neutrosophic subalgebra of $(X,*,0)$. Then, $(X, \mathcal{I}[T]_{\inf})$, $(X, \mathcal{I}[I]_{\inf})$ and $(X, \mathcal{I}[F]_{\inf})$ are 3-fuzzy subalgebra of $X$, and $(X, \mathcal{I}[T]_{\sup})$, $(X, \mathcal{I}[I]_{\sup})$ and $(X, \mathcal{I}[F]_{\sup})$ are 2-fuzzy subalgebra of $X$. Hence,

$$\begin{aligned}
\mathcal{I}[T]_{\inf}(x*y) &\geq \max\{\mathcal{I}[T]_{\inf}(x), \mathcal{I}[T]_{\inf}(y)\},\\
\mathcal{I}[I]_{\inf}(x*y) &\geq \max\{\mathcal{I}[I]_{\inf}(x), \mathcal{I}[I]_{\inf}(y)\},\\
\mathcal{I}[F]_{\inf}(x*y) &\geq \max\{\mathcal{I}[F]_{\inf}(x), \mathcal{I}[F]_{\inf}(y)\},
\end{aligned}$$

and

$$\mathcal{I}[T]_{\sup}(x*y) \le \min\{\mathcal{I}[T]_{\sup}(x), \mathcal{I}[T]_{\sup}(y)\},$$
$$\mathcal{I}[I]_{\sup}(x*y) \le \min\{\mathcal{I}[I]_{\sup}(x), \mathcal{I}[I]_{\sup}(y)\},$$
$$\mathcal{I}[F]_{\sup}(x*y) \le \min\{\mathcal{I}[F]_{\sup}(x), \mathcal{I}[F]_{\sup}(y)\},$$

for all $x, y \in X$, which imply that

$$\mathcal{I}[T]_\ell(x*y) = \mathcal{I}[T]_{\sup}(x*y) - \mathcal{I}[T]_{\inf}(x*y) \le \mathcal{I}[T]_{\sup}(x) - \mathcal{I}[T]_{\inf}(x) = \mathcal{I}[T]_\ell(x),$$
$$\mathcal{I}[T]_\ell(x*y) = \mathcal{I}[T]_{\sup}(x*y) - \mathcal{I}[T]_{\inf}(x*y) \le \mathcal{I}[T]_{\sup}(y) - \mathcal{I}[T]_{\inf}(y) = \mathcal{I}[T]_\ell(y),$$
$$\mathcal{I}[I]_\ell(x*y) = \mathcal{I}[I]_{\sup}(x*y) - \mathcal{I}[I]_{\inf}(x*y) \le \mathcal{I}[I]_{\sup}(x) - \mathcal{I}[I]_{\inf}(x) = \mathcal{I}[I]_\ell(x),$$
$$\mathcal{I}[I]_\ell(x*y) = \mathcal{I}[I]_{\sup}(x*y) - \mathcal{I}[I]_{\inf}(x*y) \le \mathcal{I}[I]_{\sup}(y) - \mathcal{I}[I]_{\inf}(y) = \mathcal{I}[I]_\ell(y),$$

and

$$\mathcal{I}[F]_\ell(x*y) = \mathcal{I}[F]_{\sup}(x*y) - \mathcal{I}[F]_{\inf}(x*y) \le \mathcal{I}[F]_{\sup}(x) - \mathcal{I}[F]_{\inf}(x) = \mathcal{I}[F]_\ell(x),$$
$$\mathcal{I}[F]_\ell(x*y) = \mathcal{I}[F]_{\sup}(x*y) - \mathcal{I}[F]_{\inf}(x*y) \le \mathcal{I}[F]_{\sup}(y) - \mathcal{I}[F]_{\inf}(y) = \mathcal{I}[F]_\ell(y).$$

It follows that

$$\mathcal{I}[T]_\ell(x*y) \le \min\{\mathcal{I}[T]_\ell(x), \mathcal{I}[T]_\ell(y)\},$$
$$\mathcal{I}[I]_\ell(x*y) \le \min\{\mathcal{I}[I]_\ell(x), \mathcal{I}[I]_\ell(y)\},$$

and

$$\mathcal{I}[F]_\ell(x*y) \le \min\{\mathcal{I}[F]_\ell(x), \mathcal{I}[F]_\ell(y)\},$$

for all $x, y \in X$. Hence, $(X, \mathcal{I}[T]_\ell)$, $(X, \mathcal{I}[I]_\ell)$ and $(X, \mathcal{I}[F]_\ell)$ are 2-fuzzy subalgebra of $(X, *, 0)$. □

**Corollary 6.** *If an interval neutrosophic set* $\mathcal{I} := (\mathcal{I}[T], \mathcal{I}[I], \mathcal{I}[F])$ *in* $X$ *is a* $(T(3,2), I(3,2), F(3,2))$*-interval neutrosophic subalgebra of* $(X, *, 0)$*, then* $(X, \mathcal{I}[T]_\ell)$*,* $(X, \mathcal{I}[I]_\ell)$ *and* $(X, \mathcal{I}[F]_\ell)$ *are 4-fuzzy subalgebra of* $(X, *, 0)$*.*

**Theorem 12.** *If an interval neutrosophic set* $\mathcal{I} := (\mathcal{I}[T], \mathcal{I}[I], \mathcal{I}[F])$ *in* $X$ *is a* $(T(i,3), I(3,4), F(3,2))$*-interval neutrosophic subalgebra of* $(X, *, 0)$ *for* $i \in \{2, 4\}$*, then*

(1) $(X, \mathcal{I}[T]_\ell)$ *is a 3-fuzzy subalgebra of* $(X, *, 0)$.
(2) $(X, \mathcal{I}[I]_\ell)$ *is a 4-fuzzy subalgebra of* $(X, *, 0)$.
(3) $(X, \mathcal{I}[F]_\ell)$ *is a 2-fuzzy subalgebra of* $(X, *, 0)$.

**Proof.** Assume that $\mathcal{I} := (\mathcal{I}[T], \mathcal{I}[I], \mathcal{I}[F])$ is a $(T(4,3), I(3,4), F(3,2))$-interval neutrosophic subalgebra of $(X, *, 0)$. Then, $(X, \mathcal{I}[T]_{\inf})$ is a 4-fuzzy subalgebra of $X$, $(X, \mathcal{I}[T]_{\sup})$ is a 3-fuzzy subalgebra of $X$, $(X, \mathcal{I}[I]_{\inf})$ is a 3-fuzzy subalgebra of $X$, $(X, \mathcal{I}[I]_{\sup})$ is a 4-fuzzy subalgebra of $X$, $(X, \mathcal{I}[F]_{\inf})$ is a 3-fuzzy subalgebra of $X$, and $(X, \mathcal{I}[F]_{\sup})$ is a 2-fuzzy subalgebra of $X$. Hence,

$$\mathcal{I}[T]_{\inf}(x*y) \le \max\{\mathcal{I}[T]_{\inf}(x), \mathcal{I}[T]_{\inf}(y)\}, \tag{7}$$

$$\mathcal{I}[T]_{\sup}(x*y) \ge \max\{\mathcal{I}[T]_{\sup}(x), \mathcal{I}[T]_{\sup}(y)\}, \tag{8}$$

$$\mathcal{I}[I]_{\inf}(x*y) \ge \max\{\mathcal{I}[I]_{\inf}(x), \mathcal{I}[I]_{\inf}(y)\}, \tag{9}$$

$$\mathcal{I}[I]_{\sup}(x*y) \le \max\{\mathcal{I}[I]_{\sup}(x), \mathcal{I}[I]_{\sup}(y)\}, \tag{10}$$

$$\mathcal{I}[F]_{\inf}(x*y) \ge \max\{\mathcal{I}[F]_{\inf}(x), \mathcal{I}[F]_{\inf}(y)\}, \tag{11}$$

and

$$\mathcal{I}[F]_{\sup}(x*y) \leq \min\{\mathcal{I}[F]_{\sup}(x), \mathcal{I}[F]_{\sup}(y)\}, \tag{12}$$

for all $x, y \in X$. Then,

$$\mathcal{I}[T]_{\inf}(x*y) \leq \mathcal{I}[T]_{\inf}(x) \text{ or } \mathcal{I}[T]_{\inf}(x*y) \leq \mathcal{I}[T]_{\inf}(y)$$

by Label (7). It follows from Label (8) that

$$\mathcal{I}[T]_\ell(x*y) = \mathcal{I}[T]_{\sup}(x*y) - \mathcal{I}[T]_{\inf}(x*y) \geq \mathcal{I}[T]_{\sup}(x) - \mathcal{I}[T]_{\inf}(x) = \mathcal{I}[T]_\ell(x)$$

or

$$\mathcal{I}[T]_\ell(x*y) = \mathcal{I}[T]_{\sup}(x*y) - \mathcal{I}[T]_{\inf}(x*y) \geq \mathcal{I}[T]_{\sup}(y) - \mathcal{I}[T]_{\inf}(y) = \mathcal{I}[T]_\ell(y),$$

and so that $\mathcal{I}[T]_\ell(x*y) \geq \max\{\mathcal{I}[T]_\ell(x), \mathcal{I}[T]_\ell(y)\}$ for all $x, y \in X$. Thus, $(X, \mathcal{I}[T]_\ell)$ is a 3-fuzzy subalgebra of $(X, *, 0)$. The condition (10) implies that

$$\mathcal{I}[I]_{\sup}(x*y) \leq \mathcal{I}[I]_{\sup}(x) \text{ or } \mathcal{I}[I]_{\sup}(x*y) \leq \mathcal{I}[I]_{\sup}(y). \tag{13}$$

Combining Labels (9) and (13), we have

$$\mathcal{I}[I]_\ell(x*y) = \mathcal{I}[I]_{\sup}(x*y) - \mathcal{I}[I]_{\inf}(x*y) \leq \mathcal{I}[I]_{\sup}(x) - \mathcal{I}[I]_{\inf}(x) = \mathcal{I}[I]_\ell(x)$$

or

$$\mathcal{I}[I]_\ell(x*y) = \mathcal{I}[I]_{\sup}(x*y) - \mathcal{I}[I]_{\inf}(x*y) \leq \mathcal{I}[I]_{\sup}(y) - \mathcal{I}[I]_{\inf}(y) = \mathcal{I}[I]_\ell(y).$$

It follows that $\mathcal{I}[I]_\ell(x*y) \leq \max\{\mathcal{I}[I]_\ell(x), \mathcal{I}[I]_\ell(y)\}$ for all $x, y \in X$. Thus, $(X, \mathcal{I}[I]_\ell)$ is a 4-fuzzy subalgebra of $(X, *, 0)$. Using Labels (11) and (12), we have

$$\mathcal{I}[F]_\ell(x*y) = \mathcal{I}[F]_{\sup}(x*y) - \mathcal{I}[F]_{\inf}(x*y) \leq \mathcal{I}[F]_{\sup}(x) - \mathcal{I}[F]_{\inf}(x) = \mathcal{I}[F]_\ell(x)$$

and

$$\mathcal{I}[F]_\ell(x*y) = \mathcal{I}[F]_{\sup}(x*y) - \mathcal{I}[F]_{\inf}(x*y) \leq \mathcal{I}[F]_{\sup}(y) - \mathcal{I}[F]_{\inf}(y) = \mathcal{I}[F]_\ell(y),$$

and so $\mathcal{I}[F]_\ell(x*y) \leq \min\{\mathcal{I}[F]_\ell(x), \mathcal{I}[F]_\ell(y)\}$ for all $x, y \in X$. Therefore, $(X, \mathcal{I}[F]_\ell)$ is a 2-fuzzy subalgebra of $(X, *, 0)$. Similarly, we can prove the desired results for $i = 2$. □

**Corollary 7.** *If an interval neutrosophic set $\mathcal{I} := (\mathcal{I}[T], \mathcal{I}[I], \mathcal{I}[F])$ in $X$ is a $(T(i,3), I(3,4), F(3,2))$-interval neutrosophic subalgebra of $(X, *, 0)$ for $i \in \{2,4\}$, then*

(1) *$(X, \mathcal{I}[T]_\ell)$ is a 1-fuzzy subalgebra of $(X, *, 0)$.*
(2) *$(X, \mathcal{I}[I]_\ell)$ and $(X, \mathcal{I}[F]_\ell)$ are 4-fuzzy subalgebra of $(X, *, 0)$.*

By a similar way to the proof of Theorem 12, we have the following theorems.

**Theorem 13.** *If an interval neutrosophic set $\mathcal{I} := (\mathcal{I}[T], \mathcal{I}[I], \mathcal{I}[F])$ in $X$ is a $(T(i,3), I(3,2), F(3,2))$-interval neutrosophic subalgebra of $(X, *, 0)$ for $i \in \{2,4\}$, then*

(1) *$(X, \mathcal{I}[T]_\ell)$ is a 3-fuzzy subalgebra of $(X, *, 0)$.*
(2) *$(X, \mathcal{I}[I]_\ell)$ and $(X, \mathcal{I}[F]_\ell)$ are 2-fuzzy subalgebra of $(X, *, 0)$.*

**Corollary 8.** *If an interval neutrosophic set $\mathcal{I} := (\mathcal{I}[T], \mathcal{I}[I], \mathcal{I}[F])$ in $X$ is a $(T(i,3), I(3,2), F(3,2))$-interval neutrosophic subalgebra of $(X, *, 0)$ for $i \in \{2,4\}$, then*

(1) *$(X, \mathcal{I}[T]_\ell)$ is a 1-fuzzy subalgebra of $(X, *, 0)$.*
(2) *$(X, \mathcal{I}[I]_\ell)$ and $(X, \mathcal{I}[F]_\ell)$ are 4-fuzzy subalgebra of $(X, *, 0)$.*

**Theorem 14.** *If an interval neutrosophic set* $\mathcal{I} := (\mathcal{I}[T], \mathcal{I}[I], \mathcal{I}[F])$ *in* $X$ *is a* $(T(i,3), I(3,2), F(2,3))$*-interval neutrosophic subalgebra of* $(X,*,0)$ *for* $i \in \{2,4\}$, *then*

(1) $(X, \mathcal{I}[T]_\ell)$ *and* $(X, \mathcal{I}[F]_\ell)$ *are 3-fuzzy subalgebra of* $(X,*,0)$.
(2) $(X, \mathcal{I}[I]_\ell)$ *is a 2-fuzzy subalgebra of* $(X,*,0)$.

**Corollary 9.** *If an interval neutrosophic set* $\mathcal{I} := (\mathcal{I}[T], \mathcal{I}[I], \mathcal{I}[F])$ *in* $X$ *is a* $(T(i,3), I(3,2), F(2,3))$*-interval neutrosophic subalgebra of* $(X,*,0)$ *for* $i \in \{2,4\}$, *then*

(1) $(X, \mathcal{I}[T]_\ell)$ *and* $(X, \mathcal{I}[F]_\ell)$ *are 1-fuzzy subalgebra of* $(X,*,0)$.
(2) $(X, \mathcal{I}[I]_\ell)$ *is a 4-fuzzy subalgebra of* $(X,*,0)$.

**Acknowledgments:** The authors wish to thank the anonymous reviewers for their valuable suggestions.

**Author Contributions:** Young Bae Jun conceived and designed the main idea and wrote the paper, Seon Jeong Kim and Florentin Smarandache performed the idea, checking contents and finding examples. All authors have read and approved the final manuscript.

**Conflicts of Interest:** The authors declare no conflict of interest.

## References

1. Atanassov, K.T. Intuitionistic fuzzy sets. *Fuzzy Sets Syst.* **1986**, *20*, 87–96.
2. Zadeh, L.A. Fuzzy sets. *Inf. Control* **1965**, *8*, 338–353.
3. Wang, H.; Smarandache, F.; Zhang, Y.Q.; Sunderraman, R. *Interval Neutrosophic Sets and Logic: Theory and Applications in Computing*; Neutrosophic Book Series No. 5; Hexis: Phoenix, AZ, USA, 2005.
4. Smarandache, F. *A Unifying Field in Logics: Neutrosophic Logic. Neutrosophy, Neutrosophic Set, Neutrosophic Probability*; American Reserch Press: Rehoboth, NM, USA, 1999.
5. Smarandache, F. Neutrosophic set-a generalization of the intuitionistic fuzzy set. *Int. J. Pure Appl. Math.* **2005**, *24*, 287–297.
6. Wang, H.; Zhang, Y.; Sunderraman, R. Truth-value based interval neutrosophic sets. In Proceedings of the 2005 IEEE International Conference on Granular Computing, Beijing, China, 25–27 July 2005; Volume 1, pp. 274–277. doi:10.1109/GRC.2005.1547284.
7. Imai, Y.; Iséki, K. On axiom systems of propositional calculi. *Proc. Jpn. Acad.* **1966**, *42*, 19–21.
8. Iséki, K. An algebra related with a propositional calculus. *Proc. Jpn. Acad.* **1966**, *42*, 26–29.
9. Huang, Y.S. *BCI-Algebra*; Science Press: Beijing, China, 2006.
10. Meng, J.; Jun, Y.B. *BCK-Algebra*; Kyungmoon Sa Co.: Seoul, Korea, 1994.
11. Jun, Y.B.; Hur, K.; Lee, K.J. Hyperfuzzy subalgebra of *BCK*/*BCI*-algebra. *Ann. Fuzzy Math. Inf.* **2018**, *15*, 17–28.

axioms

MDPI

Article

# Cross Entropy Measures of Bipolar and Interval Bipolar Neutrosophic Sets and Their Application for Multi-Attribute Decision-Making

**Surapati Pramanik [1,]*, Partha Pratim Dey [2], Florentin Smarandache [3] and Jun Ye [4]**

1 Department of Mathematics, Nandalal Ghosh B.T. College, Panpur, P.O.-Narayanpur, District–North 24 Parganas, West Bengal 743126, India
2 Department of Mathematics, Patipukur Pallisree Vidyapith, Patipukur, Kolkata 700048, India; parsur.fuzz@gmail.com
3 Mathematics & Science Department, University of New Mexico, 705 Gurley Ave., Gallup, NM 87301, USA; smarand@unm.edu
4 Department of Electrical and Information Engineering, Shaoxing University, 508 Huancheng West Road, Shaoxing 312000, China; yehjun@aliyun.com
* Correspondence: sura_pati@yahoo.co.in; Tel.: +91-947-703-5544

Received: 7 January 2018; Accepted: 22 March 2018; Published: 24 March 2018

**Abstract:** The bipolar neutrosophic set is an important extension of the bipolar fuzzy set. The bipolar neutrosophic set is a hybridization of the bipolar fuzzy set and neutrosophic set. Every element of a bipolar neutrosophic set consists of three independent positive membership functions and three independent negative membership functions. In this paper, we develop cross entropy measures of bipolar neutrosophic sets and prove their basic properties. We also define cross entropy measures of interval bipolar neutrosophic sets and prove their basic properties. Thereafter, we develop two novel multi-attribute decision-making strategies based on the proposed cross entropy measures. In the decision-making framework, we calculate the weighted cross entropy measures between each alternative and the ideal alternative to rank the alternatives and choose the best one. We solve two illustrative examples of multi-attribute decision-making problems and compare the obtained result with the results of other existing strategies to show the applicability and effectiveness of the developed strategies. At the end, the main conclusion and future scope of research are summarized.

**Keywords:** neutrosophic set; bipolar neutrosophic set; interval bipolar neutrosophic set; multi-attribute decision-making; cross entropy measure

## 1. Introduction

Shannon and Weaver [1] and Shannon [2] proposed the entropy measure which dealt formally with communication systems at its inception. According to Shannon and Weaver [1] and Shannon [2], the entropy measure is an important decision-making apparatus for computing uncertain information. Shannon [2] introduced the concept of the cross entropy strategy in information theory.

The measure of a quantity of fuzzy information obtained from a fuzzy set or fuzzy system is termed fuzzy entropy. However, the meaning of fuzzy entropy is quite different from the classical Shannon entropy because it is defined based on a nonprobabilistic concept [3–5], while Shannon entropy is defined based on a randomness (probabilistic) concept. In 1968, Zadeh [6] extended the Shannon entropy to fuzzy entropy on a fuzzy subset with respect to the concerned probability distribution. In 1972, De Luca and Termini [7] proposed fuzzy entropy based on Shannon's function and introduced the axioms with which the fuzzy entropy should comply. Sander [8] presented Shannon fuzzy entropy and proved that the properties sharpness, valuation, and general additivity have to be imposed on fuzzy entropy. Xie and Bedrosian [9] proposed another form of total fuzzy entropy.

To overcome the drawbacks of total entropy [8,9], Pal and Pal [10] introduced hybrid entropy that can be used as an objective measure for a proper defuzzification of a certain fuzzy set. Hybrid entropy [10] considers both probabilistic entropies in the absence of fuzziness. In the same study, Pal and Pal [10] defined higher-order entropy. Kaufmann and Gupta [11] studied the degree of fuzziness of a fuzzy set by a metric distance between its membership function and the membership function (characteristic function) of its nearest crisp set. Yager [12,13] introduced a fuzzy entropy card as a fuzziness measure by observing that the intersection of a fuzzy set and its complement is not the void set. Kosko [14,15] studied the fuzzy entropy of a fuzzy set based on the fuzzy set geometry and distances between them. Parkash et al. [16] proposed two new measures of weighted fuzzy entropy.

Burillo and Bustince [17] presented an axiomatic definition of an intuitionistic fuzzy entropy measure. Szmidt and Kacprzyk [18] developed a new entropy measure based on a geometric interpretation of the intuitionistic fuzzy set (IFS). Wei et al. [19] proposed an entropy measure for interval-valued intuitionistic fuzzy sets (IVIFSs) and employed it in pattern recognition and multi criteria decision-making (MCDM). Li [20] presented a new multi-attribute decision-making (MADM) strategy combining entropy and technique for order of preference by similarity to ideal solution (TOPSIS) in the IVIFS environment.

Shang and Jiang [21] developed cross entropy in the fuzzy environment. Vlachos and Sergiadis [22] presented intuitionistic fuzzy cross entropy by extending fuzzy cross entropy [21]. Ye [23] proposed a new cross entropy in the IVIFS environment and developed an optimal decision-making strategy. Xia and Xu [24] defined a new entropy and a cross entropy and presented multi-attribute group decision-making (MAGDM) strategy in the IFS environment. Tong and Yu [25] defined cross entropy in the IVIFS environment and employed it to solve MADM problems.

Smarandache [26] introduced the neutrosophic set, which is a generalization of the fuzzy set [27] and intuitionistic fuzzy set [28]. The single-valued neutrosophic set (SVNS) [29], an instance of the neutrosophic set, has caught the attention of researchers due to its applicability in decision-making [30–61], conflict resolution [62], educational problems [63,64], image processing [65–67], cluster analysis [68,69], social problems [70,71], etc.

Majumdar and Samanta [72] proposed an entropy measure and presented an MCDM strategy in the SVNS environment. Ye [73] defined cross entropy for SVNS and proposed an MCDM strategy which bears undefined phenomena. To overcome the undefined phenomena, Ye [74] defined improved cross entropy measures for SVNSs and interval neutrosophic sets (INSs) [75], which are straightforward symmetric, and employed them to solve MADM problems. Since MADM strategies [73,74] are suitable for single-decision-maker-oriented problems, Pramanik et al. [76] defined NS-cross entropy and developed an MAGDM strategy which is straightforward symmetric and free from undefined phenomena and suitable for group decision making problem. Şahin [77] proposed two techniques to convert the interval neutrosophic information to single-valued neutrosophic information and fuzzy information. In the same study, Şahin [77] defined an interval neutrosophic cross entropy measure by utilizing two reduction methods and an MCDM strategy. Tian et al. [78] developed a transformation operator to convert interval neutrosophic numbers to single-valued neutrosophic numbers and defined cross entropy measures for two SVNSs. In the same study, Tian et al. [78] developed an MCDM strategy based on cross entropy and TOPSIS [79] where the weight of the criterion is incomplete. Tian et al. [78] defined a cross entropy for INSs and developed an MCDM strategy based on the cross entropy and TOPSIS. The MCDM strategies proposed by Sahin [77] and Tian et al. [78] are applicable for a single decision maker only. Therefore, multiple decision-makers cannot participate in the strategies in [77,78]. To tackle the problem, Dalapati et al. [80] proposed IN-cross entropy and weighted IN-cross entropy and developed an MAGDM strategy.

Deli et al. [81] proposed bipolar neutrosophic set (BNS) by hybridizing the concept of bipolar fuzzy sets [82,83] and neutrosophic sets [26]. A BNS has two fully independent parts, which are positive membership degree $T^+ \to [0, 1]$, $I^+ \to [0, 1]$, $F^+ \to [0, 1]$, and negative membership degree $T^- \to [-1, 0]$, $I^- \to [-1, 0]$, $F^- \to [-1, 0]$, where the positive membership degrees $T^+$, $I^+$, $F^+$ represent truth membership degree, indeterminacy membership degree, and false membership degree, respectively, of an element and the negative membership degrees $T^-, I^-, F^-$ represent truth membership degree, indeterminacy membership degree, and false membership degree, respectively, of an element to some implicit counter property corresponding to a BNS. Deli et al. [81] defined some operations, namely, score, accuracy, and certainty functions, to compare BNSs and provided some operators in order to aggregate BNSs. Deli and Subas [84] defined a correlation coefficient similarity measure for dealing with MCDM problems in a single-valued bipolar neutrosophic setting. Şahin et al. [85] proposed a Jaccard vector similarity measure for MCDM problems with single-valued neutrosophic information. Uluçay et al. [86] introduced a Dice similarity measure, weighted Dice similarity measure, hybrid vector similarity measure, and weighted hybrid vector similarity measure for BNSs and established an MCDM strategy. Dey et al. [87] investigated a TOPSIS strategy for solving multi-attribute decision-making (MADM) problems with bipolar neutrosophic information where the weights of the attributes are completely unknown to the decision-maker. Pramanik et al. [88] defined projection, bidirectional projection, and hybrid projection measures for BNSs and proved their basic properties. In the same study, Pramanik et al. [88] developed three new MADM strategies based on the proposed projection, bidirectional projection, and hybrid projection measures with bipolar neutrosophic information. Wang et al. [89] defined Frank operations of bipolar neutrosophic numbers (BNNs) and proposed Frank bipolar neutrosophic Choquet Bonferroni mean operators by combining Choquet integral operators and Bonferroni mean operators based on Frank operations of BNNs. In the same study, Wang et al. [89] established an MCDM strategy based on Frank Choquet Bonferroni operators of BNNs in a bipolar neutrosophic environment. Pramanik et al. [90] developed a Tomada de decisao interativa e multicritévio (TODIM) strategy for MAGDM in a bipolar neutrosophic environment. An MADM strategy based on cross entropy for BNSs is yet to appear in the literature.

Mahmood et al. [91] and Deli et al. [92] introduced the hybridized structure called interval bipolar neutrosophic sets (IBNSs) by combining BNSs and INSs and defined some operations and operators for IBNSs. An MADM strategy based on cross entropy for IBNSs is yet to appear in the literature.

**Research gap:**

An MADM strategy based on cross entropy for BNSs and an MADM strategy based on cross entropy for IBNSs.

**This paper answers the following research questions**:

i. Is it possible to define a new cross entropy measure for BNSs?
ii. Is it possible to define a new weighted cross entropy measure for BNSs?
iii. Is it possible to develop a new MADM strategy based on the proposed cross entropy measure in a BNS environment?
iv. Is it possible to develop a new MADM strategy based on the proposed weighted cross entropy measure in a BNS environment?
v. Is it possible to define a new cross entropy measure for IBNSs?
vi. Is it possible to define a new weighted cross entropy measure for IBNSs?
vii. Is it possible to develop a new MADM strategy based on the proposed cross entropy measure in an IBNS environment?
viii. Is it possible to develop a new MADM strategy based on the proposed weighted cross entropy measure in an IBNS environment?

**Motivation:**

The above-mentioned analysis presents the motivation behind proposing a cross-entropy-based strategy for tackling MADM in BNS and IBNS environments. This study develops two novel cross-entropy-based MADM strategies.

The objectives of the paper are:

1. To define a new cross entropy measure and prove its basic properties.
2. To define a new weighted cross measure and prove its basic properties.
3. To develop a new MADM strategy based on the weighted cross entropy measure in a BNS environment.
4. To develop a new MADM strategy based on the weighted cross entropy measure in an IBNS environment.

To fill the research gap, we propose a cross-entropy-based MADM strategy in the BNS environment and the IBNS environment.

**The main contributions of this paper are summarized below:**

1. We propose a new cross entropy measure in the BNS environment and prove its basic properties.
2. We propose a new weighted cross entropy measure in the IBNS environment and prove its basic properties.
3. We develop a new MADM strategy based on weighted cross entropy to solve MADM problems in a BNS environment.
4. We develop a new MADM strategy based on weighted cross entropy to solve MADM problems in an IBNS environment.
5. Two illustrative numerical examples are solved and a comparison analysis is provided.

The rest of the paper is organized as follows. In Section 2, we present some concepts regarding SVNSs, INSs, BNSs, and IBNSs. Section 3 proposes cross entropy and weighted cross entropy measures for BNSs and investigates their properties. In Section 4, we extend the cross entropy measures for BNSs to cross entropy measures for IBNSs and discuss their basic properties. Two novel MADM strategies based on the proposed cross entropy measures in bipolar and interval bipolar neutrosophic settings are presented in Section 5. In Section 6, two numerical examples are solved and a comparison with other existing methods is provided. In Section 7, conclusions and the scope of future work are provided.

## 2. Preliminary

In this section, we provide some basic definitions regarding SVNSs, INSs, BNSs, and IBNSs.

### 2.1. Single-Valued Neutrosophic Sets

An SVNS [29] $S$ in $U$ is characterized by a truth membership function $T_S(x)$, an indeterminate membership function $I_S(x)$, and a falsity membership function $F_S(x)$. An SVNS $S$ over $U$ is defined by

$$S = \{x, \langle T_S(x), I_S(x), F_S(x)\rangle | x \in U\}$$

where, $T_S(x), I_S(x), F_S(x)$: $U \to [0, 1]$ and $0 \leqslant T_S(x) + I_S(x) + F_S(x) \leqslant 3$ for each point $x \in U$.

### 2.2. Interval Neutrosophic Set

An interval neutrosophic set [75] $P$ in $U$ is expressed as given below:

$$\begin{aligned} P &= \{x, \langle T_P(x), I_P(x), F_P(x)\rangle | x \in U\} \\ &= \{x, [\inf T_p(x), \sup T_p(x)]; [\inf I_p(x), \sup I_p(x)]; [\inf F_p(x) \sup F_p(x)] | x \in U\} \end{aligned}$$

where $T_P(x), I_P(x), F_P(x)$ are the truth membership function, indeterminacy membership function, and falsity membership function, respectively. For each point $x$ in $U$, $T_P(x), I_P(x), F_P(x) \subseteq [0, 1]$ satisfying the condition $0 \leqslant \sup T_P(x) + \sup I_P(x) + \sup F_P(x) \leqslant 3$.

## 2.3. Bipolar Neutrosophic Set

A BNS [81] $E$ in $U$ is presented as given below:

$$E = \{x, \langle T_E^+(x), I_E^+(x), F_E^+(x), T_E^-(x), I_E^-(x), F_E^-(x)\rangle | x \in U\}$$

where $T_E^+(x)$, $I_E^+(x)$, $F_E^+(x)$: $U \to [0, 1]$ and $T_E^-(x)$, $I_E^-(x)$, $F_E^-(x)$: $U \to [-1, 0]$. Here, $T_E^+(x)$, $I_E^+(x)$, $F_E^+(x)$ denote the truth membership, indeterminate membership, and falsity membership functions corresponding to BNS $E$ on an element $x \in U$, and $T_E^-(x)$, $I_E^-(x)$, $F_E^-(x)$ denote the truth membership, indeterminate membership, and falsity membership of an element $x \in U$ to some implicit counter property corresponding to $E$.

**Definition 1.** Ref. [81]: *Let, $E_1 = \{x, \left\langle T_{E_1}^+(x), I_{E_1}^+(x), F_{E_1}^+(x), T_{E_1}^-(x), I_{E_1}^-(x), F_{E_1}^-(x)\right\rangle | x \in U\}$ and $E_2 = \{x, \left\langle T_{E_2}^+(x), I_{E_2}^+(x), F_{E_2}^+(x), T_{E_2}^-(x), I_{E_2}^-(x), F_{E_2}^-(x)\right\rangle | x \in X\}$ be any two BNSs. Then*

- *$E_1 \subseteq E_2$ if, and only if,*

  $T_{E_1}^+(x) \leqslant T_{E_2}^+(x)$, $I_{E_1}^+(x) \leqslant I_{E_2}^+(x)$, $F_{E_1}^+(x) \geqslant F_{E_2}^+(x)$; $T_{E_1}^-(x) \geqslant T_{E_2}^-(x)$, $I_{E_1}^-(x) \geqslant I_{E_2}^-(x)$, $F_{E_1}^-(x) \leqslant F_{E_2}^-(x)$ *for all $x \in U$.*

- *$E_1 = E_2$ if, and only if,*

  $T_{E_1}^+(x) = T_{E_2}^+(x)$, $I_{E_1}^+(x) = I_{E_2}^+(x)$, $F_{E_1}^+(x) = F_{E_2}^+(x)$; $T_{E_1}^-(x) = T_{E_2}^-(x)$, $I_{E_1}^-(x) = I_{E_2}^-(x)$, $F_{E_1}^-(x) = F_{E_2}^-(x)$ *for all $x \in U$.*

- *The complement of E is $E^c = \{x, \langle T_{E^c}^+(x), I_{E^c}^+(x), F_{E^c}^+(x), T_{E^c}^-(x), I_{E^c}^-(x), F_{E^c}^-(x)\rangle | x \in U\}$*

  *where*

  $$T_{E^c}^+(x) = F_E^+(x),\ I_{E^c}^+(x) = 1 - I_E^+(x),\ F_{E^c}^+(x) = T_E^+(x);$$
  $$T_{E^c}^-(x) = F_E^-(x),\ I_{E^c}^-(x) = -1 - I_E^-(x),\ F_{E^c}^-(x) = T_E^-(x).$$

- *The union $E_1 \cup E_2$ is defined as follows:*

  $E_1 \cup E_2 = \{Max\ (T_{E_1}^+(x), T_{E_2}^+(x)), Min\ (I_{E_1}^+(x), I_{E_2}^+(x)), Min\ (F_{E_1}^+(x), F_{E_2}^+(x)), Min\ (T_{E_1}^-(x), T_{E_2}^-(x)), Max\ (I_{E_1}^-(x), I_{E_2}^-(x)), Max\ (F_{E_1}^-(x), F_{E_2}^-(x))\}, \forall x \in U.$

- *The intersection $E_1 \cap E_2$ [88] is defined as follows:*

  $E_1 \cap E_2 = \{Min\ (T_{E_1}^+(x), T_{E_2}^+(x)), Max\ (I_{E_1}^+(x), I_{E_2}^+(x)), Max\ (F_{E_1}^+(x), F_{E_2}^+(x)), Max\ (T_{E_1}^-(x), T_{E_2}^-(x)), Min\ (I_{E_1}^-(x), I_{E_2}^-(x)), Min\ (F_{E_1}^-(x), F_{E_2}^-(x))\}, \forall x \in U.$

## 2.4. Interval Bipolar Neutrosophic Sets

An IBNS [91,92] $R = \{x, <[\inf T_R^+(x), \sup T_R^+(x)]; [\inf I_R^+(x), \sup I_R^+(x)]; [\inf F_R^+(x), \sup F_R^+(x)]; [\inf T_R^-(x), \sup T_R^-(x)]; [\inf I_R^-(x), \sup I_R^-(x)]; [\inf F_R^-(x), \sup F_R^-(x)]> | x \in U\}$ is characterized by positive and negative truth membership functions $T_R^+(x)$, $T_R^-(x)$, respectively; positive and negative indeterminacy membership functions $I_R^+(x)$, $I_R^-(x)$, respectively; and positive and negative falsity membership functions $F_R^+(x)$, $F_R^-(x)$, respectively. Here, for any $x \in U$, $T_R^+(x)$, $I_R^+(x)$, $F_R^+(x) \subseteq [0, 1]$ and $T_R^-(x)$, $I_R^-(x)$, $F_R^-(x) \subseteq [-1, 0]$ with the conditions $0 \leqslant \sup T_R^+(x) + \sup I_R^+(x) + \sup F_R^+(x) \leqslant 3$ and $-3 \leqslant \sup T_R^-(x) + \sup I_R^-(x) + \sup F_R^-(x) \leqslant 0$.

**Definition 2.** Ref. [91,92]: *Let $R = \{x, <[\inf T_R^+(x), \sup T_R^+(x)]; [\inf I_R^+(x), \sup I_R^+(x)]; [\inf F_R^+(x), \sup F_R^+(x)]; [\inf T_R^-(x), \sup T_R^-(x)]; [\inf I_R^-(x), \sup I_R^-(x)]; [\inf F_R^-(x), \sup F_R^-(x)]> | x \in U\}$ and $S = \{x, <[\inf T_S^+(x), \sup T_S^+(x)]; [\inf I_S^+(x), \sup I_S^+(x)]; [\inf F_S^+(x), \sup F_S^+(x)]; [\inf T_S^-(x), \sup T_S^-(x)]; [\inf I_S^-(x), \sup I_S^-(x)]; [\inf F_S^-(x), \sup F_S^-(x)]> | x \in U\}$ be two IBNSs in U. Then*

- *$R \subseteq S$ if, and only if,*

  $inf T_R^+(x) \leqslant inf T_S^+(x)$, $sup T_R^+(x) \leqslant sup T_S^+(x)$,

  $inf I_R^+(x) \geqslant inf I_S^+(x)$, $sup I_R^+(x) \geqslant sup I_S^+(x)$,

  $inf F_R^+(x) \geqslant inf F_S^+(x)$, $sup F_R^+(x) \geqslant sup F_S^+(x)$,

  $inf T_R^-(x) \geqslant inf T_S^-(x)$, $sup T_R^-(x) \geqslant sup T_S^-(x)$,

  $inf I_R^-(x) \leqslant inf\ I_S^-(x)$, $sup I_R^-(x) \leqslant sup I_S^-(x)$,

  $inf F_R^-(x) \leqslant inf F_S^-(x)$, $sup F_R^-(x) \leqslant sup F_S^-(x)$,

  *for all $x \in U$.*

- *$R = S$ if, and only if,*

  $inf T_R^+(x) = inf T_S^+(x)$, $sup T_R^+(x) = sup T_S^+(x)$, $inf I_R^+(x) = inf I_S^+(x)$, $sup I_R^+(x) = sup I_S^+(x)$,

  $inf F_R^+(x) = inf F_S^+(x)$, $sup F_R^+(x) = sup F_S^+(x)$, $inf T_R^-(x) = inf T_S^-(x)$, $sup T_R^-(x) = sup T_S^-(x)$,

  $inf I_R^-(x) = inf I_S^-(x)$, $sup I_R^-(x) = sup I_S^-(x)$, $inf F_R^-(x) = inf F_S^-(x)$, $sup F_R^-(x) = sup F_S^-(x)$,

  *for all $x \in U$.*

- *The complement of R is defined as* The complement of $R$ is defined as $R^C$ = {$x$, < [inf $T_{RC}^+(x)$, sup$T_{RC}^+(x)$]; [inf$I_{RC}^+(x)$, sup$I_{RC}^+(x)$]; [inf$F_{RC}^+(x)$, sup$F_{RC}^+(x)$]; [inf$T_{RC}^-(x)$, sup$T_{RC}^-(x)$]; [inf$I_{RC}^-(x)$, sup$I_{RC}^-(x)$]; [inf $F_{RC}^-(x)$, sup$F_{RC}^-(x)$] > | $x \in U$} where

  inf $T_{RC}^+(x)$ = inf$F_R^+(x)$, sup$T_{RC}^+(x)$ = sup$F_R^+(x)$

  inf $I_{RC}^+(x) = 1 -$ sup$I_R^+(x)$, sup$I_{RC}^+(x) = 1 -$ inf$I_R^+(x)$

  inf$F_{RC}^+(x)$ = inf$T_R^+$, sup$F_{RC}^+(x)$ = sup$T_R^+$

  inf$T_{RC}^-(x)$ = inf$F_R^-$, sup$T_{RC}^-(x)$ = sup$F_R^-$

  inf$I_{RC}^-(x) = -1 -$ sup$I_R^-(x)$, sup$I_{RC}^-(x) = -1 -$ inf$I_R^-(x)$

  inf$F_{RC}^-(x)$ = inf$T_R^-(x)$, sup$F_{RC}^-(x)$ = sup$T_R^-(x)$

  for all $x \in U$.

## 3. Cross Entropy Measures of Bipolar Neutrosophic Sets

In this section we define a cross entropy measure between two BNSs and establish some of its basic properties.

**Definition 3.** *For any two BNSs M and N in U, the cross entropy measure can be defined as follows.*

$$C_B(M,N) = \sum_{i=1}^{n}\left[\begin{array}{l} \sqrt{\frac{T_M^+(x_i)+T_N^+(x_i)}{2}} - \left(\frac{\sqrt{T_M^+(x_i)}+\sqrt{T_N^+(x_i)}}{2}\right) + \sqrt{\frac{I_M^+(x_i)+I_N^+(x_i)}{2}} - \left(\frac{\sqrt{I_M^+(x_i)}+\sqrt{I_N^+(x_i)}}{2}\right) + \\ \sqrt{\frac{\left(1-I_M^+(x_i)\right)+\left(1-I_N^+(x_i)\right)}{2}} - \left(\frac{\sqrt{\left(1-I_M^+(x_i)\right)}+\sqrt{\left(1-I_N^+(x_i)\right)}}{2}\right) + \sqrt{\frac{F_M^+(x_i)+F_N^+(x_i)}{2}} - \left(\frac{\sqrt{F_M^+(x_i)}+\sqrt{F_N^+(x_i)}}{2}\right) + \\ \sqrt{\frac{-\left(T_M^-(x_i)+T_N^-(x_i)\right)}{2}} - \left(\frac{\sqrt{\left(-T_M^-(x_i)\right)}+\sqrt{\left(-T_N^-(x_i)\right)}}{2}\right) + \sqrt{\frac{-\left(I_M^-(x_i)+I_N^-(x_i)\right)}{2}} - \left(\frac{\sqrt{\left(-I_M^-(x_i)\right)}+\sqrt{\left(-I_N^-(x_i)\right)}}{2}\right) + \\ \sqrt{\frac{\left(1+I_M^-(x_i)\right)+\left(1+I_N^-(x_i)\right)}{2}} - \left(\frac{\sqrt{1+I_M^-(x_i)}+\sqrt{1+I_N^-(x_i)}}{2}\right) + \sqrt{\frac{-\left(F_M^-(x_i)+F_N^-(x_i)\right)}{2}} - \left(\frac{\sqrt{\left(-F_M^-(x_i)\right)}+\sqrt{\left(-F_N^-(x_i)\right)}}{2}\right) \end{array}\right] \quad (1)$$

**Theorem 1.** *If* $M = \langle T_M^+(x_i), I_M^+(x_i), F_M^+(x_i), T_M^-(x_i), I_M^-(x_i), F_M^-(x_i)\rangle$ *and* $N \langle T_N^+(x_i), I_N^+(x_i), F_N^+(x_i), T_N^-(x_i), I_N^-(x_i), F_N^-(x_i)\rangle$ *are two BNSs in U, then the cross entropy measure* $C_B(M, N)$ *satisfies the following properties:*

(1) $C_B(M, N) \geqslant 0$;

(2) $C_B(M, N) = 0$ *if, and only if,* $T_M^+(x_i) = T_N^+(x_i)$, $I_M^+(x_i) = I_N^+(x_i)$, $F_M^+(x_i) = F_N^+(x_i)$, $T_M^-(x_i) = T_N^-(x_i)$, $I_M^-(x_i) = I_N^-(x_i)$, $F_M^-(x_i) = F_N^-(x_i)$, $\forall\, x \in U$;

(3) $C_B(M, N) = C_B(N, M)$;

(4) $C_B(M, N) = C_B(M^C, N^C)$.

**Proof**

(1) We have the inequality $\left(\frac{a+b}{2}\right)^{\frac{1}{2}} \geqslant \frac{a^{\frac{1}{2}}+b^{\frac{1}{2}}}{2}$ for all positive numbers $a$ and $b$. From the inequality we can easily obtain $C_B(M, N) \geqslant 0$.

(2) The inequality $\left(\frac{a+b}{2}\right)^{\frac{1}{2}} \geqslant \frac{a^{\frac{1}{2}}+b^{\frac{1}{2}}}{2}$ becomes the equality $\left(\frac{a+b}{2}\right)^{\frac{1}{2}} = \frac{a^{\frac{1}{2}}+b^{\frac{1}{2}}}{2}$ if, and only if, $a = b$ and therefore $C_B(M, N) = 0$ if, and only if, $M = N$, i.e., $T_M^+(x_i) = T_N^+(x_i)$, $I_M^+(x_i) = I_N^+(x_i)$, $F_M^+(x_i) = F_N^+(x_i)$, $T_M^-(x_i) = T_N^-(x_i)$, $I_M^-(x_i) = I_N^-(x_i)$, $F_M^-(x_i) = F_N^-(x_i)$ $\forall\, x \in U$.

(3)
$$C_B(M,N) = \sum_{i=1}^{n} \left[\begin{array}{l}
\sqrt{\frac{T_M^+(x_i)+T_N^+(x_i)}{2}} - \left(\frac{\sqrt{T_M^+(x_i)}+\sqrt{T_N^+(x_i)}}{2}\right) + \sqrt{\frac{I_M^+(x_i)+I_N^+(x_i)}{2}} - \left(\frac{\sqrt{I_M^+(x_i)}+\sqrt{I_N^+(x_i)}}{2}\right) + \\
\sqrt{\frac{(1-I_M^+(x_i))+(1-I_N^+(x_i))}{2}} - \left(\frac{\sqrt{(1-I_M^+(x_i))}+\sqrt{(1-I_N^+(x_i))}}{2}\right) + \sqrt{\frac{F_M^+(x_i)+F_N^+(x_i)}{2}} - \left(\frac{\sqrt{F_M^+(x_i)}+\sqrt{F_N^+(x_i)}}{2}\right) + \\
\sqrt{\frac{-(T_M^-(x_i)+T_N^-(x_i))}{2}} - \left(\frac{\sqrt{(-T_M^-(x_i))}+\sqrt{(-T_N^-(x_i))}}{2}\right) + \sqrt{\frac{-(I_M^-(x_i)+I_N^-(x_i))}{2}} - \left(\frac{\sqrt{(-I_M^-(x_i))}+\sqrt{(-I_N^-(x_i))}}{2}\right) + \\
\sqrt{\frac{(1+I_M^-(x_i))+(1+I_N^-(x_i))}{2}} - \left(\frac{\sqrt{1+I_M^-(x_i)}+\sqrt{1+I_N^-(x_i)}}{2}\right) + \sqrt{\frac{-(F_M^-(x_i)+F_N^-(x_i))}{2}} - \left(\frac{\sqrt{(-F_M^-(x_i))}+\sqrt{(-F_N^-(x_i))}}{2}\right)
\end{array}\right]$$

$$= \sum_{i=1}^{n} \left[\begin{array}{l}
\sqrt{\frac{T_N^+(x_i)+T_M^+(x_i)}{2}} - \left(\frac{\sqrt{T_N^+(x_i)}+\sqrt{T_M^+(x_i)}}{2}\right) + \sqrt{\frac{I_N^+(x_i)+I_M^+(x_i)}{2}} - \left(\frac{\sqrt{I_N^+(x_i)}+\sqrt{I_M^+(x_i)}}{2}\right) + \\
\sqrt{\frac{(1-I_N^+(x_i))+(1-I_M^+(x_i))}{2}} - \left(\frac{\sqrt{(1-I_N^+(x_i))}+\sqrt{(1-I_M^+(x_i))}}{2}\right) + \sqrt{\frac{F_N^+(x_i)+F_M^+(x_i)}{2}} - \left(\frac{\sqrt{F_N^+(x_i)}+\sqrt{F_M^+(x_i)}}{2}\right) + \\
\sqrt{\frac{-(T_N^-(x_i)+T_M^-(x_i))}{2}} - \left(\frac{\sqrt{(-T_N^-(x_i))}+\sqrt{(-T_M^-(x_i))}}{2}\right) + \sqrt{\frac{-(I_N^-(x_i)+I_M^-(x_i))}{2}} - \left(\frac{\sqrt{(-I_N^-(x_i))}+\sqrt{(-I_M^-(x_i))}}{2}\right) + \\
\sqrt{\frac{(1+I_N^-(x_i))+(1+I_M^-(x_i))}{2}} - \left(\frac{\sqrt{1+I_N^-(x_i)}+\sqrt{1+I_M^-(x_i)}}{2}\right) + \sqrt{\frac{-(F_N^-(x_i)+F_M^-(x_i))}{2}} - \left(\frac{\sqrt{(-F_N^-(x_i))}+\sqrt{(-F_M^-(x_i))}}{2}\right)
\end{array}\right] = C_B(N, M).$$

(4) $C_B(M^C, N^C)$
$$= \sum_{i=1}^{n} \left[\begin{array}{l}
\sqrt{\frac{F_M^+(x_i)+F_N^+(x_i)}{2}} - \left(\frac{\sqrt{F_M^+(x_i)}+\sqrt{F_N^+(x_i)}}{2}\right) + \sqrt{\frac{(1-I_M^+(x_i))+(1-I_N^+(x_i))}{2}} - \left(\frac{\sqrt{(1-I_M^+(x_i))}+\sqrt{1-I_N^+(x_i))}}{2}\right) + \\
\sqrt{\frac{1-(1-I_M^+(x_i))+1-(1-I_N^+(x_i))}{2}} - \left(\frac{\sqrt{1-(1-I_M^+(x_i))}+\sqrt{1-(1-I_N^+(x_i))}}{2}\right) + \sqrt{\frac{T_M^+(x_i)+T_N^+(x_i)}{2}} - \left(\frac{\sqrt{T_M^+(x_i)}+\sqrt{T_N^+(x_i)}}{2}\right) + \\
\sqrt{\frac{-(F_M^-(x_i)+F_N^-(x_i))}{2}} - \left(\frac{\sqrt{(-F_M^-(x_i))}+\sqrt{(-F_N^-(x_i))}}{2}\right) + \sqrt{\frac{-(-1-I_M^-(x_i))-(-1-I_N^-(x_i))}{2}} - \left(\frac{\sqrt{-(-1-I_M^-(x_i))}+\sqrt{-(-1-I_N^-(x_i))}}{2}\right) + \\
\sqrt{\frac{1+(-1-I_M^-(x_i))+1+(-1-I_N^-(x_i))}{2}} - \left(\frac{\sqrt{1+(-1-I_M^-(x_i))}+\sqrt{1+(-1-I_N^-(x_i))}}{2}\right) + \sqrt{\frac{-(T_M^-(x_i)+T_N^-(x_i))}{2}} - \left(\frac{\sqrt{(-T_M^-(x_i))}+\sqrt{(-T_N^-(x_i))}}{2}\right)
\end{array}\right]$$

$$= \sum_{i=1}^{n} \left[\begin{array}{l}
\sqrt{\frac{T_M^+(x_i)+T_N^+(x_i)}{2}} - \left(\frac{\sqrt{T_M^+(x_i)}+\sqrt{T_N^+(x_i)}}{2}\right) + \sqrt{\frac{I_M^+(x_i)+I_N^+(x_i)}{2}} - \left(\frac{\sqrt{I_M^+(x_i)}+\sqrt{I_N^+(x_i)}}{2}\right) + \\
\sqrt{\frac{(1-I_M^+(x_i))+(1-I_N^+(x_i))}{2}} - \left(\frac{\sqrt{(1-I_M^+(x_i))}+\sqrt{(1-I_N^+(x_i))}}{2}\right) + \sqrt{\frac{F_M^+(x_i)+F_N^+(x_i)}{2}} - \left(\frac{\sqrt{F_M^+(x_i)}+\sqrt{F_N^+(x_i)}}{2}\right) + \\
\sqrt{\frac{-(T_M^-(x_i)+T_N^-(x_i))}{2}} - \left(\frac{\sqrt{(-T_M^-(x_i))}+\sqrt{(-T_N^-(x_i))}}{2}\right) + \sqrt{\frac{-(I_M^-(x_i)+I_N^-(x_i))}{2}} - \left(\frac{\sqrt{(-I_M^-(x_i))}+\sqrt{(-I_N^-(x_i))}}{2}\right) + \\
\sqrt{\frac{(1+I_M^-(x_i))+(1+I_N^-(x_i))}{2}} - \left(\frac{\sqrt{1+I_M^-(x_i)}+\sqrt{1+I_N^-(x_i)}}{2}\right) + \sqrt{\frac{-(F_M^-(x_i)+F_N^-(x_i))}{2}} - \left(\frac{\sqrt{(-F_M^-(x_i))}+\sqrt{(-F_N^-(x_i))}}{2}\right)
\end{array}\right] = C_B(M, N).$$

The proof is completed. □

**Example 1.** *Suppose that M = <0.7, 0.3, 0.4, −0.3, −0.5, −0.1> and N = <0.5, 0.2, 0.5, −0.3, −0.3, −0.2> are two BNSs; then the cross entropy between M and N is calculated as follows:*

$$C_B(M,N) = \left[\begin{array}{l} \sqrt{\frac{0.7+0.5}{2}} - \left(\frac{\sqrt{0.7}+\sqrt{0.5}}{2}\right) + \sqrt{\frac{0.3+0.2}{2}} - \left(\frac{\sqrt{0.3}+\sqrt{0.2}}{2}\right) + \sqrt{\frac{(1-0.3)+(1-0.2)}{2}} - \left(\frac{\sqrt{1-0.3}+\sqrt{1-0.2}}{2}\right) + \sqrt{\frac{0.4+0.5}{2}} - \\ \left(\frac{\sqrt{0.4}+\sqrt{0.5}}{2}\right) + \sqrt{\frac{-(-0.3-0.3)}{2}} - \left(\frac{\sqrt{-(-0.3)}+\sqrt{-(-0.3)}}{2}\right) + \sqrt{\frac{-(-0.5-0.3)}{2}} - \left(\frac{\sqrt{-(-0.5)}+\sqrt{-(-0.3)}}{2}\right) \\ +\sqrt{\frac{(1-0.5)+(1-0.3)}{2}} - \left(\frac{\sqrt{1-0.5}+\sqrt{[1-0.3]}}{2}\right) + \sqrt{\frac{-(-0.1-0.2)}{2}} - \left(\frac{\sqrt{-(-0.1)}+\sqrt{-(-0.2)}}{2}\right) \end{array}\right] = 0.01738474.$$

**Definition 4.** *Suppose that $w_i$ is the weight of each element $x_i$, $i = 1, 2, \ldots, n$, where $w_i \in [0, 1]$ and $\sum_{i=1}^{n} w_i = 1$; then the weighted cross entropy measure between any two BNSs M and N in U can be defined as follows.*

$$C_B(M,N)_w = \sum_{i=1}^{n} w_i \left[\begin{array}{l} \sqrt{\frac{T_M^+(x_i)+T_N^+(x_i)}{2}} - \left(\frac{\sqrt{T_M^+(x_i)}+\sqrt{T_N^+(x_i)}}{2}\right) + \sqrt{\frac{I_M^+(x_i)+I_N^+(x_i)}{2}} - \left(\frac{\sqrt{I_M^+(x_i)}+\sqrt{I_N^+(x_i)}}{2}\right) + \\ \sqrt{\frac{(1-I_M^+(x_i))+(1-I_N^+(x_i))}{2}} - \left(\frac{\sqrt{1-I_M^+(x_i)}+\sqrt{1-I_N^+(x_i)}}{2}\right) + \sqrt{\frac{F_M^+(x_i)+F_N^+(x_i)}{2}} - \left(\frac{\sqrt{F_M^+(x_i)}+\sqrt{F_N^+(x_i)}}{2}\right) + \\ \sqrt{\frac{-(T_M^-(x_i)+T_N^-(x_i))}{2}} - \left(\frac{\sqrt{(-T_M^-(x_i))}+\sqrt{(-T_N^-(x_i))}}{2}\right) + \sqrt{\frac{-(I_M^-(x_i)+I_N^-(x_i))}{2}} - \left(\frac{\sqrt{(-I_M^-(x_i))}+\sqrt{(-I_N^-(x_i))}}{2}\right) + \\ \sqrt{\frac{(1+I_M^-(x_i))+(1+I_N^-(x_i))}{2}} - \left(\frac{\sqrt{1+I_M^-(x_i)}+\sqrt{1+I_N^-(x_i)}}{2}\right) + \sqrt{\frac{-(F_M^-(x_i)+F_N^-(x_i))}{2}} - \left(\frac{\sqrt{(-F_M^-(x_i))}+\sqrt{(-F_N^-(x_i))}}{2}\right) \end{array}\right] \quad (2)$$

**Theorem 2.** *If M = <$T_M^+(x_i)$, $I_M^+(x_i)$, $F_M^+(x_i)$, $T_M^-(x_i)$, $I_M^-(x_i)$, $F_M^-(x_i)$> and N <$T_N^+(x_i)$, $I_N^+(x_i)$, $F_N^+(x_i)$, $T_N^-(x_i)$, $I_N^-(x_i)$, $F_N^-(x_i)$> are two BNSs in U, then the weighted cross entropy measure $C_B(M, N)_w$ satisfies the following properties:*

(1) $C_B(M, N)_w \geqslant 0$;
(2) $C_B(M, N)_w = 0$ *if, and only if,* $T_M^+(x_i) = T_N^+(x_i)$, $I_M^+(x_i) = I_N^+(x_i)$, $F_M^+(x_i) = F_N^+(x_i)$, $T_M^-(x_i) = T_N^-(x_i)$, $I_M^-(x_i) = I_N^-(x_i)$, $F_M^-(x_i) = F_N^-(x_i)$, $\forall\, x \in U$;
(3) $C_B(M, N)_w = C_B(N, M)_w$;
(4) $C_B(M^C, N^C)_w = C_B(M, N)_w$.

Proof is given in Appendix A.

**Example 2.** *Suppose that M = <0.7, 0.3, 0.4, −0.3, −0.5, −0.1> and N = <0.5, 0.2, 0.5, −0.3, −0.3, −0.2> are two BNSs and w = 0.4; then the weighted cross entropy between M and N is calculated as given below.*

$$C_B(M,N)_w = 0.4 \times \left[\begin{array}{l} \sqrt{\frac{0.7+0.5}{2}} - \left(\frac{\sqrt{0.7}+\sqrt{0.5}}{2}\right) + \sqrt{\frac{0.3+0.2}{2}} - \left(\frac{\sqrt{0.3}+\sqrt{0.2}}{2}\right) + \sqrt{\frac{(1-0.3)+(1-0.2)}{2}} - \left(\frac{\sqrt{1-0.3}+\sqrt{1-0.2}}{2}\right) \\ +\sqrt{\frac{0.4+0.5}{2}} - \left(\frac{\sqrt{0.4}+\sqrt{0.5}}{2}\right) + \sqrt{\frac{-(-0.3-0.3)}{2}} - \left(\frac{\sqrt{-(-0.3)}+\sqrt{-(-0.3)}}{2}\right) + \sqrt{\frac{-(-0.5-0.3)}{2}} - \left(\frac{\sqrt{-(-0.5)}+\sqrt{-(-0.3)}}{2}\right) \\ +\sqrt{\frac{(1-0.5+(1-0.3)}{2}} - \left(\frac{\sqrt{1-0.5}+\sqrt{1-0.3}}{2}\right) + \sqrt{\frac{-(-0.1-0.2)}{2}} - \left(\frac{\sqrt{-(-0.1)}+\sqrt{-(-0.2)}}{2}\right) \end{array}\right] = 0.006953896.$$

## 4. Cross Entropy Measure of IBNSs

This section extends the concepts of cross entropy and weighted cross entropy measures of BNSs to IBNSs.

**Definition 5.** *The cross entropy measure between any two IBNSs R = <[inf$T_R^+(x_i)$, sup$T_R^+(x_i)$], [inf$I_R^+(x_i)$, sup$I_R^+(x_i)$], [inf$F_R^+(x_i)$, sup$F_R^+(x_i)$], [inf$T_R^-(x_i)$, sup$T_R^-(x_i)$], [inf$I_R^-(x_i)$, sup$I_R^-(x_i)$], [inf$F_R^-(x_i)$,*

*$supF_R^-(x_i)$]> and S = <[$infT_S^+(x_i)$, $supT_S^+(x_i)$], [$infI_S^+(x_i)$, $supI_S^+(x_i)$], [$infF_S^+(x_i)$, $supF_S^+(x_i)$], [$infT_S^-(x_i)$, $supT_S^-(x_i)$], [$infI_S^-(x_i)$, sup $I_S^-(x_i)$], [$infF_S^-(x_i)$, $supF_S^-(x_i)$]> in U can be defined as follows.*

$$
C_{IB}(R,S)=\frac{1}{2}\sum_{i=1}^{n}\left\|\left[,\begin{array}{l}
\sqrt{\frac{\mathrm{inf}T_R^+(x_i)+\mathrm{inf}T_S^+(x_i)}{2}}-\left(\frac{\sqrt{\mathrm{inf}T_R^+(x_i)}+\sqrt{\mathrm{inf}T_S^+(x_i)}}{2}\right)+\sqrt{\frac{\mathrm{sup}T_R^+(x_i)+\mathrm{sup}T_S^+(x_i)}{2}}-\left(\frac{\sqrt{\mathrm{sup}T_R^+(x_i)}+\sqrt{\mathrm{sup}T_S^+(x_i)}}{2}\right)+\\
\sqrt{\frac{\mathrm{inf}I_R^+(x_i)+\mathrm{inf}I_S^+(x_i)}{2}}-\left(\frac{\sqrt{\mathrm{inf}I_R^+(x_i)}+\sqrt{\mathrm{inf}I_S^+(x_i)}}{2}\right)+\sqrt{\frac{\mathrm{sup}I_R^+(x_i)+\mathrm{sup}I_S^+(x_i)}{2}}-\left(\frac{\sqrt{\mathrm{sup}I_R^+(x_i)}+\sqrt{\mathrm{sup}I_S^+(x_i)}}{2}\right)+\\
\sqrt{\frac{\left(1-\mathrm{inf}I_R^+(x_i)\right)+\left(1-\mathrm{inf}I_S^+(x_i)\right)}{2}}-\left(\frac{\sqrt{1-\mathrm{inf}I_R^+(x_i)}+\sqrt{[1-\mathrm{inf}I_S^+(x_i)]}}{2}\right)+\sqrt{\frac{\left(1-\mathrm{sup}I_R^+(x_i)\right)+\left(1-\mathrm{sup}I_S^+(x_i)\right)}{2}}-\\
\left(\frac{\sqrt{1-\mathrm{sup}I_R^+(x_i)}+\sqrt{1-\mathrm{sup}I_S^+(x_i)}}{2}\right)+\sqrt{\frac{\mathrm{inf}F_R^+(x_i)+\mathrm{inf}F_S^+(x_i)}{2}}-\left(\frac{\sqrt{\mathrm{inf}F_R^+(x_i)}+\sqrt{\mathrm{inf}F_S^+(x_i)}}{2}\right)+\\
\sqrt{\frac{\mathrm{sup}F_R^+(x_i)+\mathrm{sup}F_S^+(x_i)}{2}}-\left(\frac{\sqrt{\mathrm{sup}F_R^+(x_i)}+\sqrt{\mathrm{sup}F_S^+(x_i)}}{2}\right)+\sqrt{\frac{-\left(\mathrm{inf}T_R^-(x_i)+\mathrm{inf}T_S^-(x_i)\right)}{2}}-\\
\left(\frac{\sqrt{\left(-\mathrm{inf}T_R^-(x_i)\right)}+\sqrt{\left(-\mathrm{inf}T_S^-(x_i)\right)}}{2}\right)+\sqrt{\frac{-\left(\mathrm{sup}T_R^-(x_i)+\mathrm{sup}T_S^+(x_i)\right)}{2}}-\left(\frac{\sqrt{\left(-\mathrm{sup}T_R^-(x_i)\right)}+\sqrt{\left(-\mathrm{sup}T_S^-(x_i)\right)}}{2}\right)+\\
\sqrt{\frac{-\left(\mathrm{inf}I_R^-(x_i)+\mathrm{inf}I_S^-(x_i)\right)}{2}}-\left(\frac{\sqrt{\left(-\mathrm{inf}I_R^-(x_i)\right)}+\sqrt{\left(-\mathrm{inf}I_S^-(x_i)\right)}}{2}\right)+\sqrt{\frac{-\left(\mathrm{sup}I_R^-(x_i)+\mathrm{sup}I_S^-(x_i)\right)}{2}}-\\
\left(\frac{\sqrt{\left(-\mathrm{sup}I_R^-(x_i)\right)}+\sqrt{\left(-\mathrm{sup}I_S^-(x_i)\right)}}{2}\right)+\sqrt{\frac{\left(1+\mathrm{inf}I_R^-(x_i)\right)+\left(1+\mathrm{inf}I_S^-(x_i)\right)}{2}}-\left(\frac{\sqrt{1+\mathrm{inf}I_R^-(x_i)}+\sqrt{1+\mathrm{inf}I_S^-(x_i)}}{2}\right)+\\
\sqrt{\frac{\left(1+\mathrm{sup}I_R^-(x_i)\right)+\left(1+\mathrm{sup}I_S^-(x_i)\right)}{2}}-\left(\frac{\sqrt{1+\mathrm{sup}I_R^-(x_i)}+\sqrt{1+\mathrm{sup}I_S^-(x_i)}}{2}\right)+\sqrt{\frac{-\left(\mathrm{inf}F_R^-(x_i)+\mathrm{inf}F_S^-(x_i)\right)}{2}}-\\
\left(\frac{\sqrt{\left(-\mathrm{inf}F_R^-(x_i)\right)}+\sqrt{\left(-\mathrm{inf}F_S^-(x_i)\right)}}{2}\right)+\sqrt{\frac{-\left(\mathrm{sup}F_R^-(x_i)+\mathrm{sup}F_S^-(x_i)\right)}{2}}-\left(\frac{\sqrt{\left(-\mathrm{sup}F_R^-(x_i)\right)}+\sqrt{\left(-\mathrm{sup}F_S^-(x_i)\right)}}{2}\right)
\end{array},\right]\right\| \tag{3}
$$

**Theorem 3.** *If R = <[$infT_R^+(x_i)$, $supT_R^+(x_i)$], [inf, $supI_R^+(x_i)$], [$infF_R^+(x_i)$, $supF_R^+(x_i)$], [$infT_R^-(x_i)$, $supT_R^-(x_i)$], [$infI_R^-(x_i)$, $supI_R^-(x_i)$], [$infF_R^-(x_i)$, $supF_R^-(x_i)$]> and S = <[$infT_S^+(x_i)$, $supT_S^+(x_i)$], [$infI_S^+(x_i)$, $supI_S^+(x_i)$], [$infF_S^+(x_i)$, $supF_S^+(x_i)$], [$infT_S^-(x_i)$, $supT_S^-(x_i)$], [$infI_S^-(x_i)$, $supI_S^-(x_i)$], [$infF_S^-(x_i)$, $supF_S^-(x_i)$]> are two IBNSs in U, then the cross entropy measure $C_{IB}(R, S)$ satisfies the following properties:*

(1) *$C_{IB}(R, S) \geqslant 0$;*
(2) *$C_{IB}(R, S) = 0$ for R = S i.e., inf $T_R^+(x_i) = infT_S^+(x_i)$, $supT_R^+(x_i) = supT_S^+(x_i)$, $infI_R^+(x_i) = infI_S^+(x_i)$, $supI_R^+(x_i) = supI_S^+(x_i)$, $infF_R^+(x_i) = infF_S^+(x_i)$, $supF_R^+(x_i) = supF_S^+(x_i)$, $infT_R^-(x_i) = infT_S^-(x_i)$, $supT_R^-(x_i) = supT_S^-(x_i)$, $infI_R^-(x_i) = infI_S^-(x_i)$, $supI_R^-(x_i) = supI_S^-(x_i)$, $infF_R^-(x_i) = infF_S^-(x_i)$, $supF_R^-(x_i) = supF_S^-(x_i)$ $\forall$ $x \in U$;*
(3) *$C_{IB}(R, S) = C_{IB}(S, R)$;*
(4) *$C_{IB}(R^C, S^C) = C_{IB}(R, S)$.*

**Proof**

(1) From the inequality stated in Theorem 1, we can easily get $C_{IB}(R, S) \geqslant 0$.
(2) Since $\mathrm{inf}T_R^+(x_i) = \mathrm{inf}T_S^+(x_i)$, $\mathrm{sup}T_R^+(x_i) = \mathrm{sup}T_S^+(x_i)$, $\mathrm{inf}I_R^+(x_i) = \mathrm{inf}I_S^+(x_i)$, $\mathrm{sup}I_R^+(x_i) = \mathrm{sup}I_S^+(x_i)$, $\mathrm{inf}F_R^+(x_i) = \mathrm{inf}F_S^+(x_i)$, $\mathrm{sup}F_R^+(x_i) = \mathrm{sup}F_S^+(x_i)$, $\mathrm{inf}T_R^-(x_i) = \mathrm{inf}T_S^-(x_i)$, $\mathrm{sup}T_R^-(x_i) = \mathrm{sup}T_S^-(x_i)$, $\mathrm{inf}I_R^-(x_i) = \mathrm{inf}I_S^-(x_i)$, $\mathrm{sup}I_R^-(x_i) = \mathrm{sup}I_S^-(x_i)$, $\mathrm{inf}F_R^-(x_i) = \mathrm{inf}F_S^-(x_i)$, $\mathrm{sup}F_R^-(x_i) = \mathrm{sup}F_S^-(x_i)$ $\forall$ $x \in U$, we have $C_{IB}(R, S) = 0$.

$$
\begin{aligned}
(3)\quad C_{IB}(R,\ S) = \frac{1}{2}\sum_{i=1}^{n}
&\left[\!\!\left[\begin{array}{l}
\sqrt{\frac{\inf T_R^+(x_i)+\inf T_S^+(x_i)}{2}}-\left(\frac{\sqrt{\inf T_R^+(x_i)}+\sqrt{\inf T_S^+(x_i)}}{2}\right)+\sqrt{\frac{\sup T_R^+(x_i)+\sup T_S^+(x_i)}{2}}-\left(\frac{\sqrt{\sup T_R^+(x_i)}+\sqrt{\sup T_S^+(x_i)}}{2}\right)+\\
\sqrt{\frac{\inf I_R^+(x_i)+\inf I_S^+(x_i)}{2}}-\left(\frac{\sqrt{\inf I_R^+(x_i)}+\sqrt{\inf I_S^+(x_i)}}{2}\right)+\sqrt{\frac{\sup I_R^+(x_i)+\sup I_S^+(x_i)}{2}}-\left(\frac{\sqrt{\sup I_R^+(x_i)}+\sqrt{\sup I_S^+(x_i)}}{2}\right)+\\
\sqrt{\frac{\left([1-\inf I_R^+(x_i)\right)+\left(1-\inf I_S^+(x_i)\right)}{2}}-\left(\frac{\sqrt{1-\inf I_R^+(x_i)}+\sqrt{1-\inf I_S^+(x_i)}}{2}\right)+\sqrt{\frac{\left(1-\sup I_R^+(x_i)\right)+\left(1-\sup I_S^+(x_i)\right)}{2}}-\\
\left(\frac{\sqrt{1-\sup I_R^+(x_i)}+\sqrt{1-\sup I_S^+(x_i)}}{2}\right)+\sqrt{\frac{\inf F_R^+(x_i)+\inf F_S^+(x_i)}{2}}-\left(\frac{\sqrt{\inf F_R^+(x_i)}+\sqrt{\inf F_S^+(x_i)}}{2}\right)+\\
\sqrt{\frac{\sup F_R^+(x_i)+\sup F_S^+(x_i)}{2}}-\left(\frac{\sqrt{\sup F_R^+(x_i)}+\sqrt{\sup F_S^+(x_i)}}{2}\right)+\sqrt{\frac{-\left(\inf T_R^-(x_i)+\inf T_S^-(x_i)\right)}{2}}-\\
\left(\frac{\sqrt{\left(-\inf T_R^-(x_i)\right)}+\sqrt{\left(-\inf T_S^-(x_i)\right)}}{2}\right)+\sqrt{\frac{-\left(\sup T_R^-(x_i)+\sup T_S^-(x_i)\right)}{2}}-\left(\frac{\sqrt{\left(-\sup T_R^-(x_i)\right)}+\sqrt{\left(-\sup T_S^-(x_i)\right)}}{2}\right)+\\
\sqrt{\frac{-\left(\inf I_R^-(x_i)+\inf I_S^-(x_i)\right)}{2}}-\left(\frac{\sqrt{\left(-\inf I_R^-(x_i)\right)}+\sqrt{\left(-\inf I_S^-(x_i)\right)}}{2}\right)+\sqrt{\frac{-\left(\sup I_R^-(x_i)+\sup I_S^-(x_i)\right)}{2}}-\\
\left(\frac{\sqrt{\left(-\sup I_R^-(x_i)\right)}+\sqrt{\left(-\sup I_S^-(x_i)\right)}}{2}\right)+\sqrt{\frac{\left(1+\inf I_R^-(x_i)\right)+\left(1+\inf I_S^-(x_i)\right)}{2}}-\left(\frac{\sqrt{1+\inf I_R^-(x_i)}+\sqrt{1+\inf I_S^-(x_i)}}{2}\right)+\\
\sqrt{\frac{\left(1+\sup I_R^-(x_i)\right)+\left(1+\sup I_S^-(x_i)\right)}{2}}-\left(\frac{\sqrt{1+\sup I_R^-(x_i)}+\sqrt{1+\sup I_S^-(x_i)}}{2}\right)+\sqrt{\frac{-\left(\inf F_R^-(x_i)+\inf F_S^-(x_i)\right)}{2}}-\\
\left(\frac{\sqrt{\left(-\inf F_R^-(x_i)\right)}+\sqrt{\left(-\inf F_S^-(x_i)\right)}}{2}\right)+\sqrt{\frac{-\left(\sup F_R^-(x_i)+\sup F_S^-(x_i)\right)}{2}}-\left(\frac{\sqrt{\left(-\sup F_R^-(x_i)\right)}+\sqrt{\left(-\sup F_S^-(x_i)\right)}}{2}\right)
\end{array}\right]\!\!\right]\\
= \frac{1}{2}\sum_{i=1}^{n}
&\left[\!\!\left[\begin{array}{l}
\sqrt{\frac{\inf T_S^+(x_i)+\inf T_R^+(x_i)}{2}}-\left(\frac{\sqrt{\inf T_S^+(x_i)}+\sqrt{\inf T_R^+(x_i)}}{2}\right)+\sqrt{\frac{\sup T_S^+(x_i)+\sup T_R^+(x_i)}{2}}-\left(\frac{\sqrt{\sup T_S^+(x_i)}+\sqrt{\sup T_R^+(x_i)}}{2}\right)+\\
\sqrt{\frac{\inf I_S^+(x_i)+\inf I_R^+(x_i)}{2}}-\left(\frac{\sqrt{\inf I_S^+(x_i)}+\sqrt{\inf I_R^+(x_i)}}{2}\right)+\sqrt{\frac{\sup I_S^+(x_i)+\sup I_R^+(x_i)}{2}}-\left(\frac{\sqrt{\sup I_S^+(x_i)}+\sqrt{\sup I_R^+(x_i)}}{2}\right)+\\
\sqrt{\frac{\left(1-\inf I_S^+(x_i)\right)+\left(1-\inf I_R^+(x_i)\right)}{2}}-\left(\frac{\sqrt{1-\inf I_S^+(x_i)}+\sqrt{1-\inf I_R^+(x_i)}}{2}\right)+\sqrt{\frac{\left(1-\sup I_S^+(x_i)\right)+\left(1-\sup I_R^+(x_i)\right)}{2}}-\\
\left(\frac{\sqrt{1-\sup I_S^+(x_i)}+\sqrt{1-\sup I_R^+(x_i)}}{2}\right)+\sqrt{\frac{\inf F_S^+(x_i)+\inf F_R^+(x_i)}{2}}-\left(\frac{\sqrt{\inf F_S^+(x_i)}+\sqrt{\inf F_R^+(x_i)}}{2}\right)+\\
\sqrt{\frac{\sup F_S^+(x_i)+\sup F_R^+(x_i)}{2}}-\left(\frac{\sqrt{\sup F_S^+(x_i)}+\sqrt{\sup F_R^+(x_i)}}{2}\right)+\sqrt{\frac{-\left(\inf T_S^-(x_i)+\inf T_R^-(x_i)\right)}{2}}-\\
\left(\frac{\sqrt{\left(-\inf T_S^-(x_i)\right)}+\sqrt{\left(-\inf T_R^-(x_i)\right)}}{2}\right)+\sqrt{\frac{-\left(\sup T_S^-(x_i)+\sup T_R^-(x_i)\right)}{2}}-\left(\frac{\sqrt{\left(-\sup T_S^-(x_i)\right)}+\sqrt{\left(-\sup T_R^-(x_i)\right)}}{2}\right)+\\
\sqrt{\frac{-\left(\inf I_S^-(x_i)+\inf I_R^-(x_i)\right)}{2}}-\left(\frac{\sqrt{\left(-\inf I_S^-(x_i)\right)}+\sqrt{\left(-\inf I_R^-(x_i)\right)}}{2}\right)+\sqrt{\frac{-\left(\sup I_S^-(x_i)+\sup I_R^-(x_i)\right)}{2}}-\\
\left(\frac{\sqrt{\left(-\sup I_S^-(x_i)\right)}+\sqrt{\left(-\sup I_R^-(x_i)\right)}}{2}\right)+\sqrt{\frac{\left(1+\inf I_S^-(x_i)\right)+\left(1+\inf I_R^-(x_i)\right)}{2}}-\left(\frac{\sqrt{1+\inf I_S^-(x_i)}+\sqrt{1+\inf I_R^-(x_i)}}{2}\right)+\\
\sqrt{\frac{\left(1+\sup I_S^-(x_i)\right)+\left(1+\sup I_R^-(x_i)\right)}{2}}-\left(\frac{\sqrt{1+\sup I_S^-(x_i)}+\sqrt{[1+\sup I_R^-(x_i)]}}{2}\right)+\sqrt{\frac{-\left(\inf F_S^-(x_i)+\inf F_R^-(x_i)\right)}{2}}-\\
\left(\frac{\sqrt{\left(-\inf F_S^-(x_i)\right)}+\sqrt{\left(-\inf F_R^-(x_i)\right)}}{2}\right)+\sqrt{\frac{-\left(\sup F_S^-(x_i)+\sup F_R^-(x_i)\right)}{2}}-\left(\frac{\sqrt{\left(-\sup F_S^-(x_i)\right)}+\sqrt{\left(-\sup F_R^-(x_i)\right)}}{2}\right)
\end{array}\right]\!\!\right]\\
= C_{IB}(S,\ R)&.
\end{aligned}
$$

(4) $C_{IB}(R^C, S^C) =$

$$
\frac{1}{2}\sum_{i=1}^{n}\left(\begin{aligned}
&\Bigg[\sqrt{\frac{\inf F_R^{+}(x_i)+\inf F_S^{+}(x_i)}{2}}-\left(\frac{\sqrt{\inf F_R^{+}(x_i)}+\sqrt{\inf F_S^{+}(x_i)}}{2}\right)+\sqrt{\frac{\sup F_R^{+}(x_i)+\sup F_S^{+}(x_i)}{2}}-\left(\frac{\sqrt{\sup F_R^{+}(x_i)}+\sqrt{\sup F_S^{+}(x_i)}}{2}\right)+\\
&\sqrt{\frac{\left(1-\inf I_R^{+}(x_i)\right)+\left(1-\inf I_S^{+}(x_i)\right)}{2}}-\left(\frac{\sqrt{\left(1-\inf I_R^{+}(x_i)\right)}+\sqrt{\left(1-\inf I_S^{+}(x_i)\right)}}{2}\right)+\sqrt{\frac{\left(1-\sup I_R^{+}(x_i)\right)+\left(1-\sup I_S^{+}(x_i)\right)}{2}}-\left(\frac{\sqrt{\left(1-\sup I_R^{+}(x_i)\right)}+\sqrt{\left(1-\sup I_S^{+}(x_i)\right)}}{2}\right)+\\
&\sqrt{\frac{1-\left(1-\inf I_R^{+}(x_i)\right)+1-\left(1-\inf I_S^{+}(x_i)\right)}{2}}-\left(\frac{\sqrt{1-\left(1-\inf I_R^{+}(x_i)\right)}+\sqrt{1-\left(1-\inf I_S^{+}(x_i)\right)}}{2}\right)+\sqrt{\frac{1-\left(1-\sup I_R^{+}(x_i)\right)+1-\left(1-\sup I_S^{+}(x_i)\right)}{2}}-\\
&\left(\frac{\sqrt{1-\left(1-\sup I_R^{+}(x_i)\right)}+\sqrt{1-\left(1-\sup I_S^{+}(x_i)\right)}}{2}\right)+\sqrt{\frac{\inf T_R^{+}(x_i)+\inf T_S^{+}(x_i)}{2}}-\left(\frac{\sqrt{\inf T_R^{+}(x_i)}+\sqrt{\inf T_S^{+}(x_i)}}{2}\right)+\sqrt{\frac{\sup T_R^{+}(x_i)+\sup T_S^{+}(x_i)}{2}}-\left(\frac{\sqrt{\sup T_R^{+}(x_i)}+\sqrt{\sup T_S^{+}(x_i)}}{2}\right)\\
&+\sqrt{\frac{-\left(\inf F_R^{-}(x_i)+\inf F_S^{-}(x_i)\right)}{2}}-\left(\frac{\sqrt{\left(-\inf F_R^{-}(x_i)\right)}+\sqrt{\left(-\inf F_S^{-}(x_i)\right)}}{2}\right)+\sqrt{\frac{-\left(\sup F_R^{-}(x_i)+\sup F_S^{-}(x_i)\right)}{2}}-\left(\frac{\sqrt{\left(-\sup F_R^{-}(x_i)\right)}+\sqrt{\left(-\sup F_S^{-}(x_i)\right)}}{2}\right)+\\
&\sqrt{\frac{-\left(\left(-1-\inf I_R^{-}(x_i)\right)+\left(-1-\inf I_S^{-}(x_i)\right)\right)}{2}}-\left(\frac{\sqrt{-\left(-1-\inf I_R^{-}(x_i)\right)}+\sqrt{-\left(-1-\inf I_S^{-}(x_i)\right)}}{2}\right)+\sqrt{\frac{-\left(\left(-1-\sup I_R^{-}(x_i)\right)+\left(-1-\sup I_S^{-}(x_i)\right)\right)}{2}}-\\
&\left(\frac{\sqrt{-\left(-1-\sup I_R^{-}(x_i)\right)}+\sqrt{-\left(-1-\sup I_S^{-}(x_i)\right)}}{2}\right)+\sqrt{\frac{1+\left(-1-\inf I_R^{-}(x_i)\right)+1+\left(-1-\inf I_S^{-}(x_i)\right)}{2}}-\left(\frac{\sqrt{1+\left(-1-\inf I_R^{-}(x_i)\right)}+\sqrt{1+\left(-1-\inf I_S^{-}(x_i)\right)}}{2}\right)+\\
&\sqrt{\frac{1+\left(-1-\sup I_R^{-}(x_i)\right)+1+\left(-1-\sup I_S^{-}(x_i)\right)}{2}}-\left(\frac{\sqrt{1+\left(-1-\sup I_R^{-}(x_i)\right)}+\sqrt{1+\left(-1-\sup I_S^{-}(x_i)\right)}}{2}\right)+\sqrt{\frac{-\left(\inf T_R^{-}(x_i)+\inf T_S(x_i)\right)}{2}}-\\
&\left(\frac{\sqrt{\left(-\inf T_R^{-}(x_i)\right)}+\sqrt{\left(-\inf T_S^{-}(x_i)\right)}}{2}\right)+\sqrt{\frac{-\left(\sup T_R^{-}(x_i)+\sup T_S^{-}(x_i)\right)}{2}}-\left(\frac{\sqrt{\left(-\sup T_R^{-}(x_i)\right)}+\sqrt{\left(-\sup T_S^{-}(x_i)\right)}}{2}\right)\Bigg]
\end{aligned}\right)
$$

$$
=\frac{1}{2}\sum_{i=1}^{n}\left(\begin{aligned}
&\Bigg[\sqrt{\frac{\inf T_R^{+}(x_i)+\inf T_S^{+}(x_i)}{2}}-\left(\frac{\sqrt{\inf T_R^{+}(x_i)}+\sqrt{\inf T_S^{+}(x_i)}}{2}\right)+\sqrt{\frac{\sup T_R^{+}(x_i)+\sup T_S^{+}(x_i)}{2}}-\left(\frac{\sqrt{\sup T_R^{+}(x_i)}+\sqrt{\sup T_S^{+}(x_i)}}{2}\right)+\\
&\sqrt{\frac{\inf I_R^{+}(x_i)+\inf I_S^{+}(x_i)}{2}}-\left(\frac{\sqrt{\inf I_R^{+}(x_i)}+\sqrt{\inf I_S^{+}(x_i)}}{2}\right)+\sqrt{\frac{\sup I_R^{+}(x_i)+\sup I_S^{+}(x_i)}{2}}-\left(\frac{\sqrt{\sup I_R^{+}(x_i)}+\sqrt{\sup I_S^{+}(x_i)}}{2}\right)+\\
&\sqrt{\frac{\left([1-\inf I_R^{+}(x_i)\right)+\left(1-\inf I_S^{+}(x_i)\right)}{2}}-\left(\frac{\sqrt{1-\inf I_R^{+}(x_i)}+\sqrt{1-\inf I_S^{+}(x_i)}}{2}\right)+\sqrt{\frac{\left(1-\sup I_R^{+}(x_i)\right)+\left(1-\sup I_S^{+}(x_i)\right)}{2}}-\\
&\left(\frac{\sqrt{1-\sup I_R^{+}(x_i)}+\sqrt{1-\sup I_S^{+}(x_i)}}{2}\right)+\sqrt{\frac{\inf F_R^{+}(x_i)+\inf F_S^{+}(x_i)}{2}}-\left(\frac{\sqrt{\inf F_R^{+}(x_i)}+\sqrt{\inf F_S^{+}(x_i)}}{2}\right)+\\
&\sqrt{\frac{\sup F_R^{+}(x_i)+\sup F_S^{+}(x_i)}{2}}-\left(\frac{\sqrt{\sup F_R^{+}(x_i)}+\sqrt{\sup F_S^{+}(x_i)}}{2}\right)+\sqrt{\frac{-\left(\inf T_R^{-}(x_i)+\inf T_S^{-}(x_i)\right)}{2}}-\\
&\left(\frac{\sqrt{\left(-\inf T_R^{-}(x_i)\right)}+\sqrt{\left(-\inf T_S^{-}(x_i)\right)}}{2}\right)+\sqrt{\frac{-\left(\sup T_R^{-}(x_i)+\sup T_S^{-}(x_i)\right)}{2}}-\left(\frac{\sqrt{\left(-\sup T_R^{-}(x_i)\right)}+\sqrt{\left(-\sup T_S^{-}(x_i)\right)}}{2}\right)+\\
&\sqrt{\frac{-\left(\inf I_R^{-}(x_i)+\inf I_S^{-}(x_i)\right)}{2}}-\left(\frac{\sqrt{\left(-\inf I_R^{-}(x_i)\right)}+\sqrt{\left(-\inf I_S^{-}(x_i)\right)}}{2}\right)+\sqrt{\frac{-\left(\sup I_R^{-}(x_i)+\sup I_S^{-}(x_i)\right)}{2}}-\\
&\left(\frac{\sqrt{\left(-\sup I_R^{-}(x_i)\right)}+\sqrt{\left(-\sup I_S^{-}(x_i)\right)}}{2}\right)+\sqrt{\frac{\left(1+\inf I_R^{-}(x_i)\right)+\left(1+\inf I_S^{-}(x_i)\right)}{2}}-\left(\frac{\sqrt{1+\inf I_R^{-}(x_i)}+\sqrt{1+\inf I_S^{-}(x_i)}}{2}\right)+\\
&\sqrt{\frac{\left(1+\sup I_R^{-}(x_i)\right)+\left(1+\sup I_S^{-}(x_i)\right)}{2}}-\left(\frac{\sqrt{1+\sup I_R^{-}(x_i)}+\sqrt{1+\sup I_S^{-}(x_i)}}{2}\right)+\sqrt{\frac{-\left(\inf F_R^{-}(x_i)+\inf F_S^{-}(x_i)\right)}{2}}-\\
&\left(\frac{\sqrt{\left(-\inf F_R^{-}(x_i)\right)}+\sqrt{\left(-\inf F_S^{-}(x_i)\right)}}{2}\right)+\sqrt{\frac{-\left(\sup F_R^{-}(x_i)+\sup F_S^{-}(x_i)\right)}{2}}-\left(\frac{\sqrt{\left(-\sup F_R^{-}(x_i)\right)}+\sqrt{\left(-\sup F_S^{-}(x_i)\right)}}{2}\right)\Bigg]
\end{aligned}\right)
$$

$= C_{IB}(R, S)$. □

**Example 3.** *Suppose that R = <[0.5, 0.8], [0.4, 0.6], [0.2, 0.6], [−0.3, −0.1], [−0.5, −0.1], [−0.5, −0.2]> and S = <[0.5, 0.9], [0.4, 0.5], [0.1, 0.4], [−0.5, −0.3], [−0.7, −0.3], [−0.6, −0.3]> are two IBNSs; the cross entropy between R and S is computed as follows:*

$$
C_{IB}(R,S)=\frac{1}{2}\left[\begin{aligned}
&\sqrt{\tfrac{0.5+0.5}{2}}-\left(\tfrac{\sqrt{0.5}+\sqrt{0.5}}{2}\right)+\sqrt{\tfrac{0.8+0.9}{2}}-\left(\tfrac{\sqrt{0.8}+\sqrt{0.9}}{2}\right)+\sqrt{\tfrac{0.4+0.4}{2}}-\left(\tfrac{\sqrt{0.4}+\sqrt{0.4}}{2}\right)+\sqrt{\tfrac{0.6+0.5}{2}}-\left(\tfrac{\sqrt{0.6}+\sqrt{0.5}}{2}\right)+\\
&\sqrt{\tfrac{[1-0.4]+[1-0.4]}{2}}-\left(\tfrac{\sqrt{1-0.4}+\sqrt{[1-0.4]}}{2}\right)+\sqrt{\tfrac{[1-0.6]+[1-0.5]}{2}}-\left(\tfrac{\sqrt{1-0.6}+\sqrt{[1-0.5]}}{2}\right)+\sqrt{\tfrac{0.2+0.1}{2}}-\left(\tfrac{\sqrt{0.2}+\sqrt{0.1}}{2}\right)+\\
&\sqrt{\tfrac{0.6+0.4}{2}}-\left(\tfrac{\sqrt{0.6}+\sqrt{0.4}}{2}\right)+\sqrt{\tfrac{-(-0.3-0.5)}{2}}-\left(\tfrac{\sqrt{-(-0.3)}+\sqrt{-(-0.5)}}{2}\right)+\sqrt{\tfrac{-(-0.1-0.3)}{2}}-\left(\tfrac{\sqrt{-(-0.1)}+\sqrt{-(-0.3)}}{2}\right)+\\
&\sqrt{\tfrac{-(-0.5-0.7)}{2}}-\left(\tfrac{\sqrt{-(-0.5)}+\sqrt{-(-0.7)}}{2}\right)+\sqrt{\tfrac{-(-0.1-0.3)}{2}}-\left(\tfrac{\sqrt{-(-0.1)}+\sqrt{-(-0.3)}}{2}\right)+\sqrt{\tfrac{[1-0.5]+[1-0.7]}{2}}-\\
&\left(\tfrac{\sqrt{1-0.5}+\sqrt{[1-0.7]}}{2}\right)+\sqrt{\tfrac{[1-0.1]+[1-0.3]}{2}}-\left(\tfrac{\sqrt{1-0.1}+\sqrt{[1-0.3]}}{2}\right)+\sqrt{\tfrac{-(-0.5-0.6)}{2}}-\left(\tfrac{\sqrt{-(-0.5)}+\sqrt{-(-0.6)}}{2}\right)+\\
&\sqrt{\tfrac{-(-0.2-0.3)}{2}}-\left(\tfrac{\sqrt{-(-0.2)}+\sqrt{-(-0.3)}}{2}\right)
\end{aligned}\right]=0.02984616.
$$

**Definition 6.** *Let $w_i$ be the weight of each element $x_i$, $i = 1, 2, \ldots, n$, and $w_i \in [0, 1]$ with $\sum\limits_{i=1}^{n} w_i = 1$; then the weighted cross entropy measure between any two IBNSs R = <$[infT_R^+(x_i), supT_R^+(x_i)]$, $[infI_R^+(x_i), supI_R^+(x_i)]$, $[infF_R^+(x_i), supF_R^+(x_i)]$, $[infT_R^-(x_i), supT_R^-(x_i)]$, $[infI_R^-(x_i), supI_R^-(x_i)]$, $[infF_R^-(x_i), supF_R^-(x_i)]$> and S = <$[infT_S^+(x_i), supT_S^+(x_i)]$, $[infI_S^+(x_i), supI_S^+(x_i)]$, $[infF_S^+(x_i), supF_S^+(x_i)]$, $[infT_S^-(x_i), supT_S^-(x_i)]$, $[infI_S^-(x_i), supI_S^-(x_i)]$, $[infF_S^-(x_i), supF_S^-(x_i)]$> in U can be defined as follows.*

$$
\begin{aligned}
C_{IB}(R,S)_w = \frac{1}{2}\sum_{i=1}^{n} w_i \Bigg[\Bigg[ & \sqrt{\frac{\inf T_R^+(x_i)+\inf T_S^+(x_i)}{2}} - \left(\frac{\sqrt{\inf T_R^+(x_i)}+\sqrt{\inf T_S^+(x_i)}}{2}\right) + \sqrt{\frac{\sup T_R^+(x_i)+\sup T_S^+(x_i)}{2}} - \left(\frac{\sqrt{\sup T_R^+(x_i)}+\sqrt{\sup T_S^+(x_i)}}{2}\right) + \\
& \sqrt{\frac{\inf I_R^+(x_i)+\inf I_S^+(x_i)}{2}} - \left(\frac{\sqrt{\inf I_R^+(x_i)}+\sqrt{\inf I_S^+(x_i)}}{2}\right) + \sqrt{\frac{\sup I_R^+(x_i)+\sup I_S^+(x_i)}{2}} - \left(\frac{\sqrt{\sup I_R^+(x_i)}+\sqrt{\sup I_S^+(x_i)}}{2}\right) + \\
& \sqrt{\frac{\left(1-\inf I_R^+(x_i)\right)+\left(1-\inf I_S^+(x_i)\right)}{2}} - \left(\frac{\sqrt{1-\inf I_R^+(x_i)}+\sqrt{1-\inf I_S^+(x_i)}}{2}\right) + \sqrt{\frac{\left(1-\sup I_R^+(x_i)\right)+\left(1-\sup I_S^+(x_i)\right)}{2}} - \\
& \left(\frac{\sqrt{1-\sup I_R^+(x_i)}+\sqrt{\left(1-\sup I_S^+(x_i)\right)}}{2}\right) + \sqrt{\frac{\inf F_R^+(x_i)+\inf F_S^+(x_i)}{2}} - \left(\frac{\sqrt{\inf F_R^+(x_i)}+\sqrt{\inf F_S^+(x_i)}}{2}\right) + \\
& \sqrt{\frac{\sup F_R^+(x_i)+\sup F_S^+(x_i)}{2}} - \left(\frac{\sqrt{\sup F_R^+(x_i)}+\sqrt{\sup F_S^+(x_i)}}{2}\right) + \sqrt{\frac{-\left(\inf T_R^-(x_i)+\inf T_S^-(x_i)\right)}{2}} - \\
& \left(\frac{\sqrt{\left(-\inf T_R^-(x_i)\right)}+\sqrt{\left(-\inf T_S^-(x_i)\right)}}{2}\right) + \sqrt{\frac{-\left(\sup T_R^-(x_i)+\sup T_S^-(x_i)\right)}{2}} - \left(\frac{\sqrt{\left(-\sup T_R^-(x_i)\right)}+\sqrt{\left(-\sup T_S^-(x_i)\right)}}{2}\right) + \\
& \sqrt{\frac{-\left(\inf I_R^-(x_i)+\inf I_S^-(x_i)\right)}{2}} - \left(\frac{\sqrt{\left(-\inf I_R^-(x_i)\right)}+\sqrt{\left(-\inf I_S^-(x_i)\right)}}{2}\right) + \sqrt{\frac{-\left(\sup I_R^-(x_i)+\sup I_S^-(x_i)\right)}{2}} - \\
& \left(\frac{\sqrt{\left(-\sup I_R^-(x_i)\right)}+\sqrt{\left(-\sup I_S^-(x_i)\right)}}{2}\right) + \sqrt{\frac{\left(1+\inf I_R^-(x_i)\right)+\left(1+\inf I_S^-(x_i)\right)}{2}} - \left(\frac{\sqrt{1+\inf I_R^-(x_i)}+\sqrt{1+\inf I_S^-(x_i)}}{2}\right) + \\
& \sqrt{\frac{\left(1+\sup I_R^-(x_i)\right)+\left(1+\sup I_S^-(x_i)\right)}{2}} - \left(\frac{\sqrt{1+\sup I_R^-(x_i)}+\sqrt{[1+\sup I_S^-(x_i)]}}{2}\right) + \sqrt{\frac{-\left(\inf F_R^-(x_i)+\inf F_S^-(x_i)\right)}{2}} - \\
& \left(\frac{\sqrt{\left(-\inf F_R^-(x_i)\right)}+\sqrt{\left(-\inf F_S^-(x_i)\right)}}{2}\right) + \sqrt{\frac{-\left(\sup F_R^-(x_i)+\sup F_S^-(x_i)\right)}{2}} - \left(\frac{\sqrt{\left(-\sup F_R^-(x_i)\right)}+\sqrt{\left(-\sup F_S^-(x_i)\right)}}{2}\right) \Bigg]\Bigg]
\end{aligned} \tag{4}
$$

**Theorem 4.** *For any two IBNSs R = <$[infT_R^+(x_i), sup\ T_R^+(x_i)]$, $[infI_R^+(x_i), supI_R^+(x_i)]$, $[infF_R^+(x_i), supF_R^+(x_i)]$, $[infT_R^-(x_i), supT_R^-(x_i)]$, $[infI_R^-(x_i), supI_R^-(x_i)]$, $[infF_R^-(x_i), supF_R^-(x_i)]$> and S = <$[infT_S^+(x_i), supT_S^+(x_i)]$, $[infI_S^+(x_i), supI_S^+(x_i)]$, $[infF_S^+(x_i), supF_S^+(x_i)]$, $[infT_S^-(x_i), supT_S^-(x_i)]$, $[infI_S^-(x_i), supI_S^-(x_i)]$, $[infF_S^-(x_i), supF_S^-(x_i)]$> in U, the weighted cross entropy measure $C_{IB}(R, S)_w$ also satisfies the following properties:*

(1) *$C_{IB}(R, S)_w \geqslant 0$;*

(2) *$C_{IB}(R, S)_w = 0$ if, and only if, R = S i.e., $inf\ T_R^+(x_i) = infT_S^+(x_i)$, $supT_R^+(x_i) = supT_S^+(x_i)$, $infI_R^+(x_i) = infI_S^+(x_i)$, $supI_R^+(x_i) = supI_S^+(x_i)$, $infF_R^+(x_i) = infF_S^+(x_i)$, $supF_R^+(x_i) = supF_S^+(x_i)$, $infT_R^-(x_i) = infT_S^-(x_i)$, $supT_R^-(x_i) = supT_S^-(x_i)$, $infI_R^-(x_i) = infI_S^-(x_i)$, $supI_R^-(x_i) = supI_S^-(x_i)$, $infF_R^-(x_i) = infF_S^-(x_i)$, $supF_R^-(x_i) = supF_S^-(x_i)$ $\forall\, x \in U$;*

(3) *$C_{IB}(R, S)_w = C_{IB}(S, R)_w$;*

(4) *$C_{IB}(R^C, S^C)_w = C_{IB}(R, S)_w$.*

The proofs are presented in Appendix B.

**Example 4.** *Consider the two IBNSs R = <[0.5, 0.8], [0.4, 0.6], [0.2, 0.6], [−0.3, −0.1], [−0.5, −0.1], [−05, −0.2]> and S = <[0.5, 0.9], [0.4, 0.5], [0.1, 0.4], [−0.5, −0.3], [−0.7, −0.3], [−0.6, −0.3]>, and let w = 0.3; then the weighted cross entropy between R and S is calculated as follows:*

$$C_{IB}(R,S) = \frac{1}{2} \times 0.3 \times \left[ \begin{array}{l} \sqrt{\frac{0.5+0.5}{2}} - \left(\frac{\sqrt{0.5}+\sqrt{0.5}}{2}\right) + \sqrt{\frac{0.8+0.9}{2}} - \left(\frac{\sqrt{0.8}+\sqrt{0.9}}{2}\right) + \sqrt{\frac{0.4+0.4}{2}} - \left(\frac{\sqrt{0.4}+\sqrt{0.4}}{2}\right) + \sqrt{\frac{0.6+0.5}{2}} - \left(\frac{\sqrt{0.6}+\sqrt{0.5}}{2}\right) + \\ \sqrt{\frac{[1-0.4]+[1-0.4]}{2}} - \left(\frac{\sqrt{1-0.4}+\sqrt{[1-0.4]}}{2}\right) + \sqrt{\frac{[1-0.6]+[1-0.5]}{2}} - \left(\frac{\sqrt{1-0.6}+\sqrt{[1-0.5]}}{2}\right) + \sqrt{\frac{0.2+0.1}{2}} - \left(\frac{\sqrt{0.2}+\sqrt{0.1}}{2}\right) + \\ \sqrt{\frac{0.6+0.4}{2}} - \left(\frac{\sqrt{0.6}+\sqrt{0.4}}{2}\right) + \sqrt{\frac{-(-0.3-0.5)}{2}} - \left(\frac{\sqrt{-(-0.3)}+\sqrt{-(-0.5)}}{2}\right) + \sqrt{\frac{-(-0.1-0.3)}{2}} - \left(\frac{\sqrt{-(-0.1)}+\sqrt{-(-0.3)}}{2}\right) + \\ \sqrt{\frac{-(-0.5-0.7)}{2}} - \left(\frac{\sqrt{-(-0.5)}+\sqrt{-(-0.7)}}{2}\right) + \sqrt{\frac{-(-0.1-0.3)}{2}} - \left(\frac{\sqrt{-(-0.1)}+\sqrt{-(-0.3)}}{2}\right) + \sqrt{\frac{[1-0.5]+[1-0.7]}{2}} - \\ \left(\frac{\sqrt{1-0.5}+\sqrt{[1-0.7]}}{2}\right) + \sqrt{\frac{[1-0.1]+[1-0.3]}{2}} - \left(\frac{\sqrt{1-0.1}+\sqrt{[1-0.3]}}{2}\right) + \sqrt{\frac{-(-0.5-0.6)}{2}} - \left(\frac{\sqrt{-(-0.5)}+\sqrt{-(-0.6)}}{2}\right) + \\ \sqrt{\frac{-(-0.2-0.3)}{2}} - \left(\frac{\sqrt{-(-0.2)}+\sqrt{-(-0.3)}}{2}\right) \end{array} \right] = 0.00895385.$$

## 5. MADM Strategies Based on Cross Entropy Measures

In this section, we propose two new MADM strategies based on weighted cross entropy measures in bipolar neutrosophic and interval bipolar neutrosophic environments. Let $B = \{B_1, B_2, \ldots, B_m\}$ $(m \geqslant 2)$ be a discrete set of $m$ feasible alternatives which are to be evaluated based on $n$ attributes $C = \{C_1, C_2, \ldots, C_n\}$ $(n \geqslant 2)$ and let $w_j$ be the weight vector of the attributes such that $0 \leqslant w_j \leqslant 1$ and $\sum_{j=1}^{n} w_j = 1$.

### 5.1. MADM Strategy Based on Weighted Cross Entropy Measures of BNS

The procedure for solving MADM problems in a bipolar neutrosophic environment is presented in the following steps:

**Step 1.** The rating of the performance value of alternative $B_i$ $(i = 1, 2, \ldots, m)$ with respect to the predefined attribute $C_j$ $(j = 1, 2, \ldots, n)$ can be expressed in terms of bipolar neutrosophic information as follows:

$$B_i = \{C_j, < T^+_{B_i}(C_j), I^+_{B_i}(C_j), F^+_{B_i}(C_j), T^-_{B_i}(C_j), I^-_{B_i}(C_j), F^-_{B_i}(C_j) > | C_j \in C_j, j = 1, 2, \ldots, n\},$$

where $0 \leqslant T^+_{B_i}(C_j) + I^+_{B_i}(C_j) + F^+_{B_i}(C_j) \leqslant 3$ and $-3 \leqslant T^-_{B_i}(C_j) + I^-_{B_i}(C_j) + F^-_{B_i}(C_j) \leqslant 0$, $i = 1, 2, \ldots, m$; $j = 1, 2, \ldots, n$.

Assume that $\tilde{d}_{ij} = <T^+_{ij}, I^+_{ij}, F^+_{ij}, T^-_{ij}, I^-_{ij}, F^-_{ij}>$ is the bipolar neutrosophic decision matrix whose entries are the rating values of the alternatives with respect to the attributes provided by the expert or decision-maker. The bipolar neutrosophic decision matrix $[\tilde{d}_{ij}]_{m \times n}$ can be expressed as follows:

$$[\tilde{d}_{ij}]_{m \times n} = \begin{array}{c} \\ B_1 \\ B_2 \\ . \\ . \\ B_m \end{array} \begin{array}{c} \begin{array}{cccc} C_1 & C_2 & \ldots & C_n \end{array} \\ \left( \begin{array}{cccc} d_{11} & d_{12} & \ldots & d_{1n} \\ d_{21} & d_{22} & \ldots & d_{2n} \\ . & . & \ldots & . \\ . & . & \ldots & . \\ d_{m1} & d_{m2} & \ldots & d_{mn} \end{array} \right) \end{array}.$$

**Step 2.** The positive ideal solution (PIS) $<p^* = (d_1^*, d_2^*, \ldots, d_n^*)>$ of the bipolar neutrosophic information is obtained as follows:

$$p_j^* = \left\langle T_j^{*+}, I_j^{*+}, F_j^{*+}, T_j^{*-}, I_j^{*-}, F_j^{*-} \right\rangle = < [\{\underset{i}{Max}(T_{ij}^{+})|j \in H_1\}; \{\underset{i}{Min}(T_{ij}^{+})|j \in H_2\}],$$
$$[\{\underset{i}{Min}(I_{ij}^{+})|j \in H_1\}; \{\underset{i}{Max}(I_{ij}^{+})|j \in H_2\}], [\{\underset{i}{Min}(F_{ij}^{+})|j \in H_1\}; \{\underset{i}{Max}(F_{ij}^{+})|j \in H_2\}],$$
$$[\{\underset{i}{Min}(T_{ij}^{-})|j \in H_1\}; \{\underset{i}{Max}(T_{ij}^{-})|j \in H_2\}], [\{\underset{i}{Max}(I_{ij}^{-})|j \in H_1\}; \{\underset{i}{Min}(I_{ij}^{-})|j \in H_2\}],$$
$$[\{\underset{i}{Max}(F_{ij}^{-})|j \in H_1\}; \{\underset{i}{Min}(F_{ij}^{-})|j \in H_2\}] >, j = 1, 2, \ldots, n;$$

where $H_1$ and $H_2$ represent benefit and cost type attributes, respectively.

**Step 3.** The weighted cross entropy between an alternative $B_i$, $i = 1, 2, \ldots, m$, and the ideal alternative $p^*$ is determined by

$$C_B(B_i, p*)_w = \sum_{i=1}^{n} w_i \left[ \begin{array}{l} \sqrt{\frac{T_{ij}^{+} + T_j^{*+}}{2}} - \left(\frac{\sqrt{T_{ij}^{+}} + \sqrt{T_j^{*+}}}{2}\right) + \sqrt{\frac{I_{ij}^{+} + I_j^{*+}}{2}} - \left(\frac{\sqrt{I_{ij}^{+}} + \sqrt{I_j^{*+}}}{2}\right) + \sqrt{\frac{[1 - I_{ij}^{+}] + [1 - I_j^{*+}]}{2}} - \\ \left(\frac{\sqrt{1 - I_j^{+}} + \sqrt{[1 - I_j^{*+}]}}{2}\right) + \sqrt{\frac{F_{ij}^{+} + F_j^{*+}}{2}} - \left(\frac{\sqrt{F_{ij}^{+}} + \sqrt{F_j^{*+}}}{2}\right) + \sqrt{\frac{-(T_{ij}^{-} + T_j^{*-})}{2}} - \\ \left(\frac{\sqrt{(-T_{ij}^{-})} + \sqrt{(-T_j^{*-})}}{2}\right) + \sqrt{\frac{-(I_{ij}^{-} + I_j^{*-})}{2}} - \left(\frac{\sqrt{(-I_{ij}^{-})} + \sqrt{(-I_j^{*-})}}{2}\right) + \sqrt{\frac{[1 + I_{ij}^{-}] + [1 + I_j^{*-}]}{2}} \\ - \left(\frac{\sqrt{1 + I_{ij}^{-}} + \sqrt{[1 + I_j^{*-}]}}{2}\right) + \sqrt{\frac{-(F_{ij}^{-} + F_j^{*-})}{2}} - \left(\frac{\sqrt{(-F_{ij}^{-})} + \sqrt{(-F_j^{*-})}}{2}\right) \end{array} \right]. \quad (5)$$

**Step 4.** A smaller value of $C_B(B_i, p^*)_w$, $i = 1, 2, \ldots, m$ represents that an alternative $B_i$, $i = 1, 2, \ldots, m$ is closer to the PIS $p^*$. Therefore, the alternative with the smallest weighted cross entropy measure is the best alternative.

*5.2. MADM Strategy Based on Weighted Cross Entropy Measures of IBNSs*

The steps for solving MADM problems with interval bipolar neutrosophic information are presented as follows.

**Step 1.** In an interval bipolar neutrosophic environment, the rating of the performance value of alternative $B_i$ ($i = 1, 2, \ldots, m$) with respect to the predefined attribute $C_j$ ($j = 1, 2, \ldots, n$) can be represented as follows:

$$B_i = \{C_j, < [\inf T_{B_i}^{+}(C_j), \sup T_{B_i}^{+}(C_j)], [\inf I_{B_i}^{+}(C_j), \sup I_{B_i}^{+}(C_j)], [\inf F_{B_i}^{+}(C_j), \sup F_{B_i}^{+}(C_j)],$$
$$[\inf T_{B_i}^{-}(C_j), \sup T_{B_i}^{-}(C_j)], [\inf I_{B_i}^{-}(C_j), \sup I_{B_i}^{-}(C_j)], [\inf F_{B_i}^{-}(C_j), \sup F_{B_i}^{-}(C_j)] > |C_j \in C_j,$$
$$j = 1, 2, \ldots, n\}$$

where $0 \leq \sup T_{B_i}^{+}(C_j) + \sup I_{B_i}^{+}(C_j) + \sup F_{B_i}^{+}(C_j) \leq 3$ and $-3 \leq \sup T_{B_i}^{-}(C_j) + \sup I_{B_i}^{-}(C_j) + \sup F_{B_i}^{-}(C_j) \leq 0$; $j = 1, 2, \ldots, n$. Let $\tilde{g}_{ij} = <[{}^{L}T_{ij}^{+}, {}^{U}T_{ij}^{+}], [{}^{L}I_{ij}^{+}, {}^{U}I_{ij}^{+}], [{}^{L}F_{ij}^{+}, {}^{U}F_{ij}^{+}], [{}^{L}T_{ij}^{-}, {}^{U}T_{ij}^{-}], [{}^{L}I_{ij}^{-}, {}^{U}I_{ij}^{-}], [{}^{L}F_{ij}^{-}, {}^{U}F_{ij}^{-}]>$ be the bipolar neutrosophic decision matrix whose entries are the rating values of the alternatives with respect to the attributes provided by the expert or decision-maker. The interval bipolar neutrosophic decision matrix $[\tilde{g}_{ij}]_{m \times n}$ can be presented as follows:

$$[\tilde{g}_{ij}]_{m \times n} = \begin{array}{c} \\ B_1 \\ B_2 \\ . \\ . \\ B_m \end{array} \begin{array}{c} C_1 \; C_2 \; \ldots \; C_n \\ \left( \begin{array}{cccc} g_{11} & g_{12} & \cdots & g_{1n} \\ g_{21} & g_{22} & \cdots & g_{2n} \\ . & . & \cdots & . \\ . & . & \cdots & . \\ g_{m1} & g_{m2} & \cdots & g_{mn} \end{array} \right) \end{array}.$$

**Step 2.** The PIS $<q^* = (g_1^*, g_2^*, \ldots, g_n^*)>$ of the interval bipolar neutrosophic information is obtained as follows:

$$
\begin{aligned}
q_j^* &= <[{}^{L}T_{ij}^{*+}, {}^{U}T_{ij}^{*+}], [{}^{L}I_{ij}^{*+}, {}^{U}I_{ij}^{*+}], [{}^{L}F_{ij}^{*+}, {}^{U}F_{ij}^{*+}], [{}^{L}T_{ij}^{*-}, {}^{U}T_{ij}^{*-}], [{}^{L}I_{ij}^{*-}, {}^{U}I_{ij}^{*-}], [{}^{L}F_{ij}^{*-}, {}^{U}F_{ij}^{*-}]>, \\
&= <[\{\underset{i}{Max}\,({}^{L}T_{ij}^{+})|j \in H_1\}; \{\underset{i}{Min}\,({}^{L}T_{ij}^{+})|j \in H_2\}, \{\underset{i}{Max}\,({}^{U}T_{ij}^{+})|j \in H_1\}; \{\underset{i}{Min}\,({}^{U}T_{ij}^{+})|j \in H_2\}], \\
&[\{\underset{i}{Min}\,({}^{L}I_{ij}^{+})|j \in H_1\}; \{\underset{i}{Max}\,({}^{L}I_{ij}^{+})|j \in H_2\}, \{\underset{i}{Min}\,({}^{U}I_{ij}^{+})|j \in H_1\}; \{\underset{i}{Max}\,({}^{U}I_{ij}^{+})|j \in H_2\}], \\
&[\{\underset{i}{Min}\,({}^{L}F_{ij}^{+})|j \in H_1\}; \{\underset{i}{Max}\,({}^{L}F_{ij}^{+})|j \in H_2\}, \{\underset{i}{Min}\,({}^{U}F_{ij}^{+})|j \in H_1\}; \{\underset{i}{Max}\,({}^{U}F_{ij}^{+})|j \in H_2\}], \\
&[\{\underset{i}{Min}\,({}^{L}T_{ij}^{-})|j \in H_1\}; \{\underset{i}{Max}\,({}^{L}T_{ij}^{-})|j \in H_2\}, \{\underset{i}{Min}\,({}^{U}T_{ij}^{-})|j \in H_1\}; \{\underset{i}{Max}\,({}^{U}T_{ij}^{-})|j \in H_2\}], \\
&[\{\underset{i}{Max}\,({}^{L}I_{ij}^{-})|j \in H_1\}; \{\underset{i}{Min}\,({}^{L}I_{ij}^{-})|j \in H_2\}, \{\underset{i}{Max}\,({}^{U}I_{ij}^{-})|j \in H_1\}; \{\underset{i}{Min}\,({}^{U}I_{ij}^{-})|j \in H_2\}], \\
&[\{\underset{i}{Max}\,({}^{L}F_{ij}^{-})|j \in H_1\}; \{\underset{i}{Min}\,({}^{L}F_{ij}^{-})|j \in H_2\}, \{\underset{i}{Max}\,({}^{U}F_{ij}^{-})|j \in H_1\}; \{\underset{i}{Min}\,({}^{U}F_{ij}^{-})|j \in H_2\}]>, \\
j &= 1, 2, \ldots, n
\end{aligned}
$$

where $H_1$ and $H_2$ stand for benefit and cost type attributes, respectively.

**Step 3.** The weighted cross entropy between an alternative $B_i$, $i = 1, 2, \ldots, m$, and the ideal alternative $q^*$ under an interval bipolar neutrosophic setting is computed as follows:

$$
C_{IB}(B_i, q^*)_w = \frac{1}{2} \times \sum_{i=1}^{n} w_i
\left[
\begin{array}{l}
\sqrt{\frac{{}^{L}T_{ij}^{+} + {}^{L}T_{j}^{*+}}{2}} - \left(\frac{\sqrt{{}^{L}T_{ij}^{+}} + \sqrt{{}^{L}T_{j}^{*+}}}{2}\right) + \sqrt{\frac{{}^{U}T_{ij}^{+} + {}^{U}T_{j}^{*+}}{2}} - \left(\frac{\sqrt{{}^{U}T_{ij}^{+}} + \sqrt{{}^{U}T_{j}^{*+}}}{2}\right) + \sqrt{\frac{{}^{L}I_{ij}^{+} + {}^{L}I_{j}^{*+}}{2}} - \left(\frac{\sqrt{{}^{L}I_{ij}^{+}} + \sqrt{{}^{L}I_{j}^{*+}}}{2}\right) + \\
\sqrt{\frac{[1 - {}^{L}I_{ij}^{+}] + [1 - {}^{L}I_{j}^{*+}]}{2}} - \left(\frac{\sqrt{1 - {}^{L}I_{j}^{+}} + \sqrt{[1 - {}^{L}I_{j}^{*+}]}}{2}\right) + \sqrt{\frac{{}^{U}I_{ij}^{+} + {}^{U}I_{j}^{*+}}{2}} - \left(\frac{\sqrt{{}^{U}I_{ij}^{+}} + \sqrt{{}^{U}I_{j}^{*+}}}{2}\right) + \sqrt{\frac{[1 - {}^{U}I_{ij}^{+}] + [1 - {}^{U}I_{j}^{*+}]}{2}} - \\
\left(\frac{\sqrt{1 - {}^{U}I_{j}^{+}} + \sqrt{[1 - {}^{U}I_{j}^{*+}]}}{2}\right) + \sqrt{\frac{{}^{L}F_{ij}^{+} + {}^{L}F_{j}^{*+}}{2}} - \left(\frac{\sqrt{{}^{L}F_{ij}^{+}} + \sqrt{{}^{L}F_{j}^{*+}}}{2}\right) + \sqrt{\frac{{}^{U}F_{ij}^{+} + {}^{U}F_{j}^{*+}}{2}} - \left(\frac{\sqrt{{}^{U}F_{ij}^{+}} + \sqrt{{}^{U}F_{j}^{*+}}}{2}\right) + \sqrt{\frac{-({}^{L}T_{ij}^{-} + {}^{L}T_{j}^{*-})}{2}} - \\
\left(\frac{\sqrt{(-{}^{L}T_{ij}^{-})} + \sqrt{(-{}^{L}T_{j}^{*-})}}{2}\right) + \sqrt{\frac{-({}^{U}T_{ij}^{-} + {}^{U}T_{j}^{*-})}{2}} - \left(\frac{\sqrt{(-{}^{U}T_{ij}^{-})} + \sqrt{(-{}^{U}T_{j}^{*-})}}{2}\right) + \sqrt{\frac{-({}^{L}I_{ij}^{-} + {}^{L}I_{j}^{*-})}{2}} - \left(\frac{\sqrt{(-{}^{L}I_{ij}^{-})} + \sqrt{(-{}^{L}I_{j}^{*-})}}{2}\right) + \\
\sqrt{\frac{-({}^{U}I_{ij}^{-} + {}^{U}I_{j}^{*-})}{2}} - \left(\frac{\sqrt{(-{}^{U}I_{ij}^{-})} + \sqrt{(-{}^{U}I_{j}^{*-})}}{2}\right) + \sqrt{\frac{[1 + {}^{L}I_{ij}^{-}] + [1 + {}^{L}I_{j}^{*-}]}{2}} - \left(\frac{\sqrt{1 + {}^{L}I_{ij}^{-}} + \sqrt{[1 + {}^{L}I_{j}^{*-}]}}{2}\right) + \sqrt{\frac{[1 + {}^{U}I_{ij}^{-}] + [1 + {}^{U}I_{j}^{*-}]}{2}} - \\
\left(\frac{\sqrt{1 + {}^{U}I_{ij}^{-}} + \sqrt{[1 + {}^{U}I_{j}^{*-}]}}{2}\right) + \sqrt{\frac{-({}^{L}F_{ij}^{-} + {}^{L}F_{j}^{*-})}{2}} - \left(\frac{\sqrt{(-{}^{L}F_{ij}^{-})} + \sqrt{(-{}^{L}F_{j}^{*-})}}{2}\right) + \sqrt{\frac{-({}^{U}F_{ij}^{-} + {}^{U}F_{j}^{*-})}{2}} - \left(\frac{\sqrt{(-{}^{U}F_{ij}^{-})} + \sqrt{(-{}^{U}F_{j}^{*-})}}{2}\right)
\end{array}
\right]. \quad (6)
$$

**Step 4.** A smaller value of $C_{IB}(B_i, p^*)_w$, $i = 1, 2, \ldots, m$ indicates that an alternative $B_i$, $i = 1, 2, \ldots, m$ is closer to the PIS $q^*$. Hence, the alternative with the smallest weighted cross entropy measure will be identified as the best alternative.

A conceptual model of the proposed strategy is shown in Figure 1.

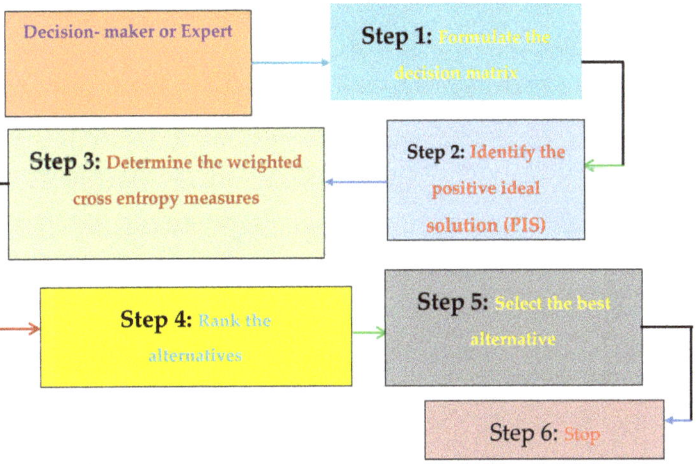


**Figure 1.** Conceptual model of the proposed strategy.

## 6. Illustrative Example

In this section we solve two numerical MADM problems and a comparison with other existing strategies is presented to verify the applicability and effectiveness of the proposed strategies in bipolar neutrosophic and interval bipolar neutrosophic environments.

### 6.1. Car Selection Problem with Bipolar Neutrosophic Information

Consider the problem discussed in [81,86–88] where a buyer wants to purchase a car based on some predefined attributes. Suppose that four types of cars (alternatives) $B_i$, ($i = 1, 2, 3, 4$) are available in the market. Four attributes are taken into consideration in the decision-making environment, namely, Fuel economy ($C_1$), Aerod ($C_2$), Comfort ($C_3$), Safety ($C_4$), to select the most desirable car. Assume that the weight vector for the four attributes is known and given by $w = (w_1, w_2, w_3, w_4) = (0.5, 0.25, 0.125, 0.125)$. Therefore, the bipolar neutrosophic decision matrix $\langle d_{ij} \rangle_{4\times 4}$ can be obtained as given below.

The bipolar neutrosophic decision matrix $[\tilde{d}_{ij}]_{4\times 4}$ =

| | $C_1$ | $C_2$ | $C_3$ | $C_4$ |
|---|---|---|---|---|
| $B_1$ | $<0.5, 0.7, 0.2, -0.7, -0.3, -0.6>$ | $<0.4, 0.4, 0.5, -0.7, -0.8, -0.4>$ | $<0.7, 0.7, 0.5, -0.8, -0.7, -0.6>$ | $<0.1, 0.5, 0.7, -0.5, -0.2, -0.8>$ |
| $B_2$ | $<0.9, 0.7, 0.5, -0.7, -0.7, -0.1>$ | $<0.7, 0.6, 0.8, -0.7, -0.5, -0.1>$ | $<0.9, 0.4, 0.6, -0.1, -0.7, -0.5>$ | $<0.5, 0.2, 0.7, -0.5, -0.1, -0.9>$ |
| $B_3$ | $<0.3, 0.4, 0.2, -0.6, -0.3, -0.7>$ | $<0.2, 0.2, 0.2, -0.4, -0.7, -0.4>$ | $<0.9, 0.5, 0.5, -0.6, -0.5, -0.2>$ | $<0.7, 0.5, 0.3, -0.4, -0.2, -0.2>$ |
| $B_4$ | $<0.9, 0.7, 0.2, -0.8, -0.6, -0.1>$ | $<0.3, 0.5, 0.2, -0.5, -0.5, -0.2>$ | $<0.5, 0.4, 0.5, -0.1, -0.7, -0.2>$ | $<0.2, 0.4, 0.8, -0.5, -0.5, -0.6>$ |

The positive ideal bipolar neutrosophic solutions are computed from $[\tilde{d}_{ij}]_{4\times 4}$ as follows:

$$p^* = [<0.9, 0.4, 0.2, -0.8, -0.3, -0.1>, <0.7, 0.2, 0.2, -0.7, -0.5, -0.1>,$$
$$<0.9, 0.4, 0.5, -0.8, -0.5, -0.2>, <0.7, 0.2, 0.3, -0.5, -0.1, -0.2>].$$

Using Equation (5), the weighted cross entropy measure $C_B(B_i, p^*)_w$ is obtained as follows:

$$C_B(B_1, p^*)_w = 0.0734,\ C_B(B_2, p^*)_w = 0.0688,\ C_B(B_3, p^*)_w = 0.0642,\ C_B(B_4, p^*)_w = 0.0516. \quad (7)$$

According to the weighted cross entropy measure $C_B(B_i, p^*)_w$, the order of the four alternatives is $B_4 \prec B_3 \prec B_2 \prec B_1$; therefore, $B_4$ is the best car.

We compare our obtained result with the results of other existing strategies (see Table 1), where the known weight of the attributes is given by $w = (w_1, w_2, w_3, w_4) = (0.5, 0.25, 0.125, 0.125)$. It is to be noted that the ranking results obtained from the other existing strategies are different from the result of the proposed strategies in some cases. The reason is that the different authors adopted different decision-making strategies and thereby obtained different ranking results. However, the proposed strategies are simple and straightforward and can effectively solve decision-making problems with bipolar neutrosophic information.

**Table 1.** The results of the car selection problem obtained from different methods.

| Methods | Ranking Results | Best Option |
|---|---|---|
| The proposed weighted cross entropy measure | $B_4 \prec B_3 \prec B_2 \prec B_1$ | $B_4$ |
| Dey et al.'s TOPSIS strategy [87] | $B_1 \prec B_3 \prec B_2 \prec B_4$ | $B_4$ |
| Deli et al.'s strategy [81] | $B_1 \prec B_2 \prec B_4 \prec B_3$ | $B_3$ |
| Projection measure [88] | $B_3 \prec B_4 \prec B_1 \prec B_2$ | $B_2$ |
| Bidirectional projection measure [88] | $B_2 \prec B_1 \prec B_4 \prec B_3$ | $B_3$ |
| Hybrid projection measure [88] with $\rho = 0.25$ | $B_2 \prec B_1 \prec B_3 \prec B_4$ | $B_4$ |
| Hybrid projection measure [88] with $\rho = 0.50$ | $B_3 \prec B_2 \prec B_1 \prec B_4$ | $B_4$ |
| Hybrid projection measure [88] with $\rho = 0.75$ | $B_1 \prec B_3 \prec B_4 \prec B_2$ | $B_2$ |
| Hybrid projection measure [88] with $\rho = 0.90$ | $B_3 \prec B_4 \prec B_2 \prec B_1$ | $B_1$ |
| Hybrid similarity measure [88] with $\rho = 0.25$ | $B_2 \prec B_4 \prec B_1 \prec B_3$ | $B_3$ |
| Hybrid similarity measure [88] with $\rho = 0.30$ | $B_2 \prec B_4 \prec B_1 \prec B_3$ | $B_3$ |
| Hybrid similarity measure [88] with $\rho = 0.60$ | $B_2 \prec B_4 \prec B_1 \prec B_3$ | $B_3$ |
| Hybrid similarity measure [88] with $\rho = 0.90$ | $B_2 \prec B_4 \prec B_3 \prec B_1$ | $B_1$ |

### 6.2. Interval Bipolar Neutrosophic MADM Investment Problem

Consider an interval bipolar neutrosophic MADM problem studied in [91] with four possible alternatives with the aim to invest a sum of money in the best choice. The four alternatives are:

> a food company ($B_1$),
> a car company ($B_2$),
> an arms company ($B_3$), and
> a computer company ($B_4$).

The investment company selects the best option based on three predefined attributes, namely, growth analysis ($C_1$), risk analysis ($C_2$), and environment analysis ($C_3$). We consider $C_1$ and $C_2$ to be benefit type attributes and $C_3$ to be a cost type attribute based on Ye [93]. Assume that the weight vector [91] of $C_1$, $C_2$, and $C_3$ is given by $w = (w_1, w_2, w_3) = (0.35, 0.25, 0.4)$. The interval bipolar neutrosophic decision matrix $[\tilde{g}_{ij}]_{4\times 3}$ presented by the decision-maker or expert is as follows.

Interval bipolar neutrosophic decision matrix $[\tilde{g}_{ij}]_{4\times 3}$ =

$$C_1$$

$$\begin{pmatrix} B_1 & [[0.4, 0.5], [0.2, 0.3], [0.3, 0.4], [-0.3, -0.2], [-0.4, -0.3], [-0.5, -0.4]] \\ B_2 & [[0.6, 0.7], [0.1, 0.2], [0.2, 0.3], [-0.2, -0.1], [-0.3, -0.2], [-0.7, -0.6]] \\ B_3 & [[0.3, 0.6], [0.2, 0.3], [0.3, 0.4], [-0.3, -0.2], [-0.4, -0.3], [-0.6, -0.3]] \\ B_4 & [[0.7, 0.8], [0.0, 0.1], [0.1, 0.2], [-0.1, -0.0], [-0.2, -0.1], [-0.8, -0.7]] \end{pmatrix}$$

$$C_2$$

$$\begin{pmatrix} B_1 & [[0.4, 0.6], [0.1, 0.3], [0.2, 0.4], [-0.3, -0.1], [-0.4, -0.2], [-0.6, -0.4]] \\ B_2 & [[0.6, 0.7], [0.1, 0.2], [0.2, 0.3], [-0.2, -0.1], [-0.3, -0.2], [-0.7, -0.6]] \\ B_3 & [[0.5, 0.6], [0.2, 0.3], [0.3, 0.4], [-0.3, -0.2], [-0.4, -0.3], [-0.6, -0.5]] \\ B_4 & [[0.6, 0.7], [0.1, 0.2], [0.1, 0.3], [-0.2 -0.1], [-0.3, -0.1], [-0.7, -0.6]] \end{pmatrix}$$

$$C_3$$

$$\begin{pmatrix} B_1 & [[0.7, 0.9], [0.2, 0.3], [0.4, 0.5], [-0.3, -0.2], [-0.5, -0.4], [-0.9, -0.7]] \\ B_2 & [[0.3, 0.6], [0.3, 0.5], [0.8, 0.9], [-0.5, -0.3], [-0.9, -0.8], [-0.6, -0.3]] \\ B_3 & [[0.4, 0.5], [0.2, 0.4], [0.7, 0.9], [-0.4, -0.2], [-0.9, -0.7], [-0.5, -0.4]] \\ B_4 & [[0.6, 0.7], [0.3, 0.4], [0.8, 0.9], [-0.4, -0.3], [-0.9, -0.8], [-0.7, -0.6]] \end{pmatrix}$$

From the matrix $[\tilde{g}_{ij}]_{4\times 3}$, we determine the positive ideal interval bipolar neutrosophic solution ($q^*$) by using Equation (6) as follows:

$$\begin{aligned} q^* = &<[0.7, 0.8], [0.0, 0.1], [0.1, 0.2], [-0.3, -0.2], [-0.2, -0.1], [-0.5, -0.3]>; \\ &<[0.6, 0.7], [0.1, 0.2], [0.1, 0.3], [-0.3, -0.2], [-0.3, -0.1], [-0.6, -0.4]>; \\ &<[0.3, 0.5], [0.3, 0.5], [0.8, 0.9], [-0.3, -0.2], [-0.9, -0.8], [-0.9, -0.7]>. \end{aligned}$$

The weighted cross entropy between an alternative $B_i$, $i = 1, 2, \ldots, m$, and the ideal alternative $q^*$ can be obtained as given below:

$$C_{IB}(B_1, q^*)_w = 0.0606,\ C_{IB}(B_2, q^*)_w = 0.0286,\ C_{IB}(B_3, q^*)_w = 0.0426,\ C_{IB}(B_4, q^*)_w = 0.0423.$$

On the basis of the weighted cross entropy measure $C_{IB}(B_i, q^*)_w$, the order of the four alternatives is $B_2 \prec B_4 \prec B_3 \prec B_1$; therefore, $B_2$ is the best choice.

Next, the comparison of the results obtained from different methods is presented in Table 2 where the weight vector of the attribute is given by $w = (w_1, w_2, w_3) = (0.35, 0.25, 0.4)$. We observe that $B_2$ is the best option obtained using the proposed method and $B_4$ is the best option obtained using the method of Mahmood et al. [91]. The reason for this may be that we use the interval bipolar neutrosophic cross entropy method whereas Mahmood et al. [91] derived the most desirable alternative based on a weighted arithmetic average operator in an interval bipolar neutrosophic setting.

**Table 2.** The results of the investment problem obtained from different methods.

| Methods | Ranking Results | Best Option |
|---|---|---|
| The proposed weighted cross entropy measure | $B_2 \prec B_4 \prec B_3 \prec B_1$ | $B_2$ |
| Mahmood et al.'s strategy [91] | $B_2 \prec B_3 \prec B_1 \prec B_4$ | $B_4$ |

## 7. Conclusions

In this paper we defined cross entropy and weighted cross entropy measures for bipolar neutrosophic sets and proved their basic properties. We also extended the proposed concept to the interval bipolar neutrosophic environment and proved its basic properties. The proposed cross entropy measures were then employed to develop two new multi-attribute decision-making strategies. Two illustrative numerical examples were solved and comparisons with existing strategies were provided to demonstrate the feasibility, applicability, and efficiency of the proposed strategies. We hope that the proposed cross entropy measures can be effective in dealing with group decision-making, data analysis, medical diagnosis, selection of a suitable company to build power plants [94], teacher selection [95], quality brick selection [96], weaver selection [97,98], etc. In future, the cross entropy measures can be extended to other neutrosophic hybrid environments, such as bipolar neutrosophic soft expert sets, bipolar neutrosophic refined sets, etc.

**Author Contributions:** Surapati Pramanik and Partha Pratim Dey conceived and designed the experiments; Surapati Pramanik performed the experiments; Jun Ye and Florentin Smarandache analyzed the data; Surapati Pramanik and Partha Pratim Dey wrote the paper.

**Conflicts of Interest:** The authors declare no conflict of interest.

## Appendix A

### Proof of Theorem 2

(1) From the inequality stated in Theorem 1, we can easily obtain $C_B(M, N)_w \geqslant 0$.

(2) $C_B(M, N)_w = 0$ if, and only if, $M = N$, i.e., $T_M^+(x_i) = T_N^+(x_i)$, $I_M^+(x_i) = I_N^+(x_i)$, $F_M^+(x_i) = F_N^+(x_i)$, $T_M^-(x_i) = T_N^-(x_i)$, $I_M^-(x_i) = I_N^-(x_i)$, $F_M^-(x_i) = F_N^-(x_i)\ \forall\, x \in U$.

(3)
$$C_B(M,N)_w = \sum_{i=1}^{n} w_i\, i \left[ \begin{array}{l} \sqrt{\frac{T_M^+(x_i)+T_N^+(x_i)}{2}} - \left(\frac{\sqrt{T_M^+(x_i)}+\sqrt{T_N^+(x_i)}}{2}\right) + \sqrt{\frac{I_M^+(x_i)+I_N^+(x_i)}{2}} - \left(\frac{\sqrt{I_M^+(x_i)}+\sqrt{I_N^+(x_i)}}{2}\right) + \\ \sqrt{\frac{(1-I_M^+(x_i))+(1-I_N^+(x_i))}{2}} - \left(\frac{\sqrt{1-I_M^+(x_i)}+\sqrt{1-I_N^+(x_i)}}{2}\right) + \sqrt{\frac{F_M^+(x_i)+F_N^+(x_i)}{2}} - \left(\frac{\sqrt{F_M^+(x_i)}+\sqrt{F_N^+(x_i)}}{2}\right) + \\ \sqrt{\frac{-(T_M^-(x_i)+T_N^-(x_i))}{2}} - \left(\frac{\sqrt{(-T_M^-(x_i))}+\sqrt{(-T_N^-(x_i))}}{2}\right) + \sqrt{\frac{-(I_M^-(x_i)+I_N^-(x_i))}{2}} - \left(\frac{\sqrt{(-I_M^-(x_i))}+\sqrt{(-I_N^-(x_i))}}{2}\right) + \\ \sqrt{\frac{(1+I_M^-(x_i))+(1+I_N^-(x_i))}{2}} - \left(\frac{\sqrt{1+I_M^-(x_i)}+\sqrt{1+I_N^-(x_i)}}{2}\right) + \sqrt{\frac{-(F_M^-(x_i)+F_N^-(x_i))}{2}} - \left(\frac{\sqrt{(-F_M^-(x_i))}+\sqrt{(-F_N^-(x_i))}}{2}\right) \end{array} \right]$$

$$= \sum_{i=1}^{n} w_i \left[ \begin{array}{l} \sqrt{\frac{T_N^+(x_i)+T_M^+(x_i)}{2}} - \left(\frac{\sqrt{T_N^+(x_i)}+\sqrt{T_M^+(x_i)}}{2}\right) + \sqrt{\frac{I_N^+(x_i)+I_M^+(x_i)}{2}} - \left(\frac{\sqrt{I_N^+(x_i)}+\sqrt{I_M^+(x_i)}}{2}\right) + \\ \sqrt{\frac{(1-I_N^+(x_i))+(1-I_M^+(x_i))}{2}} - \left(\frac{\sqrt{1-I_N^+(x_i)}+\sqrt{1-I_M^+(x_i)}}{2}\right) + \sqrt{\frac{F_N^+(x_i)+F_M^+(x_i)}{2}} - \left(\frac{\sqrt{F_N^+(x_i)}+\sqrt{F_M^+(x_i)}}{2}\right) + \\ \sqrt{\frac{-(T_N^-(x_i)+T_M^-(x_i))}{2}} - \left(\frac{\sqrt{(-T_N^-(x_i))}+\sqrt{(-T_M^-(x_i))}}{2}\right) + \sqrt{\frac{-(I_N^-(x_i)+I_M^-(x_i))}{2}} - \left(\frac{\sqrt{(-I_N^-(x_i))}+\sqrt{(-I_M^-(x_i))}}{2}\right) + \\ \sqrt{\frac{(1+I_N^-(x_i))+(1+I_M^-(x_i))}{2}} - \left(\frac{\sqrt{1+I_N^-(x_i)}+\sqrt{1+I_M^-(x_i)}}{2}\right) + \sqrt{\frac{-(F_N^-(x_i)+F_M^-(x_i))}{2}} - \left(\frac{\sqrt{(-F_N^-(x_i))}+\sqrt{(-F_M^-(x_i))}}{2}\right) \end{array} \right]$$

$= C_B(N, M)_w$.

(4) $C_B(M^C, N^C)_w =$

$$\sum_{i=1}^{n}\left[\begin{array}{l}\sqrt{\frac{F_M^+(x_i)+F_N^+(x_i)}{2}}-\left(\frac{\sqrt{F_M^+(x_i)}+\sqrt{F_N^+(x_i)}}{2}\right)+\sqrt{\frac{\left(1-I_M^+(x_i)\right)+\left(1-I_N^+(x_i)\right)}{2}}-\left(\frac{\sqrt{\left(1-I_M^+(x_i)\right)}+\sqrt{1-I_N^+(x_i)\big)}}{2}\right)+\\ \sqrt{\frac{1-\left(1-I_M^+(x_i)\right)+1-\left(1-I_N^+(x_i)\right)}{2}}-\left(\frac{\sqrt{1-\left(1-I_M^+(x_i)\right)}+\sqrt{1-\left(1-I_N^+(x_i)\right)}}{2}\right)+\sqrt{\frac{T_M^+(x_i)+T_N^+(x_i)}{2}}-\left(\frac{\sqrt{T_M^+(x_i)}+\sqrt{T_N^+(x_i)}}{2}\right)+\\ \sqrt{\frac{-\left(F_M^-(x_i)+F_N^-(x_i)\right)}{2}}-\left(\frac{\sqrt{\left(-F_M^-(x_i)\right)}+\sqrt{\left(-F_N^-(x_i)\right)}}{2}\right)+\sqrt{\frac{-\left(-1-I_M^-(x_i)\right)-\left(-1-I_N^-(x_i)\right)}{2}}-\left(\frac{\sqrt{-\left(-1-I_M^-(x_i)\right)}+\sqrt{-\left(-1-I_N^-(x_i)\right)}}{2}\right)+\\ \sqrt{\frac{1+\left(-1-I_M^-(x_i)\right)+1+\left(-1-I_N^-(x_i)\right)}{2}}-\left(\frac{\sqrt{1+\left(-1-I_M^-(x_i)\right)}+\sqrt{1+\left(-1-I_N^-(x_i)\right)}}{2}\right)+\sqrt{\frac{-\left(T_M^-(x_i)+T_N^-(x_i)\right)}{2}}-\left(\frac{\sqrt{\left(-T_M^-(x_i)\right)}+\sqrt{\left(-T_N^-(x_i)\right)}}{2}\right)\end{array}\right]$$

$$=\sum_{i=1}^{n} w_i\, i\left[\begin{array}{l}\sqrt{\frac{T_M^+(x_i)+T_N^+(x_i)}{2}}-\left(\frac{\sqrt{T_M^+(x_i)}+\sqrt{T_N^+(x_i)}}{2}\right)+\sqrt{\frac{I_M^+(x_i)+I_N^+(x_i)}{2}}-\left(\frac{\sqrt{I_M^+(x_i)}+\sqrt{I_N^+(x_i)}}{2}\right)+\\ \sqrt{\frac{\left(1-I_M^+(x_i)\right)+\left(1-I_N^+(x_i)\right)}{2}}-\left(\frac{\sqrt{1-I_M^+(x_i)}+\sqrt{1-I_N^+(x_i)}}{2}\right)+\sqrt{\frac{F_M^+(x_i)+F_N^+(x_i)}{2}}-\left(\frac{\sqrt{F_M^+(x_i)}+\sqrt{F_N^+(x_i)}}{2}\right)+\\ \sqrt{\frac{-\left(T_M^-(x_i)+T_N^-(x_i)\right)}{2}}-\left(\frac{\sqrt{\left(-T_M^-(x_i)\right)}+\sqrt{\left(-T_N^-(x_i)\right)}}{2}\right)+\sqrt{\frac{-\left(I_M^-(x_i)+I_N^-(x_i)\right)}{2}}-\left(\frac{\sqrt{\left(-I_M^-(x_i)\right)}+\sqrt{\left(-I_N^-(x_i)\right)}}{2}\right)+\\ \sqrt{\frac{\left(1+I_M^-(x_i)\right)+\left(1+I_N^-(x_i)\right)}{2}}-\left(\frac{\sqrt{1+I_M^-(x_i)}+\sqrt{1+I_N^-(x_i)}}{2}\right)+\sqrt{\frac{-\left(F_M^-(x_i)+F_N^-(x_i)\right)}{2}}-\left(\frac{\sqrt{\left(-F_M^-(x_i)\right)}+\sqrt{\left(-F_N^-(x_i)\right)}}{2}\right)\end{array}\right]$$

$= C_B(M^C, N^C)_w$. □

## Appendix B

### Proof of Theorem 4

(1) Obviously, we can easily get $C_{IB}(R, S)_w \geqslant 0$.

(2) If $C_{IB}(R, S)_w = 0$ then $R = S$, and if $\inf T_R^+(x_i) = \inf T_S^+(x_i)$, $\sup T_R^+(x_i) = \sup T_S^+(x_i)$, $\inf I_R^+(x_i) = \inf I_S^+(x_i)$, $\sup I_R^+(x_i) = \sup I_S^+(x_i)$, $\inf F_R^+(x_i) = \inf F_S^+(x_i)$, $\sup F_R^+(x_i) = \sup F_S^+(x_i)$, $\inf T_R^-(x_i) = \inf T_S^-(x_i)$, $\sup T_R^-(x_i) = \sup T_S^-(x_i)$, $\inf I_R^-(x_i) = \inf I_S^-(x_i)$, $\sup I_R^-(x_i) = \sup I_S^-(x_i)$, $\inf F_R^-(x_i) = \inf F_S^-(x_i)$, $\sup F_R^-(x_i) = \sup F_S^-(x_i)$ $\forall\, x \in U$, then we obtain $C_{IB}(R, S)_w = 0$.

(3)

$$C_{IB}(R,S)_w=\frac{1}{2}\sum_{i=1}^{n} w_i\left(\left[\begin{array}{l}\sqrt{\frac{\inf T_R^+(x_i)+\inf T_S^+(x_i)}{2}}-\left(\frac{\sqrt{\inf T_R^+(x_i)}+\sqrt{\inf T_S^+(x_i)}}{2}\right)+\sqrt{\frac{\sup T_R^+(x_i)+\sup T_S^+(x_i)}{2}}-\left(\frac{\sqrt{\sup T_R^+(x_i)}+\sqrt{\sup T_S^+(x_i)}}{2}\right)+\\ \sqrt{\frac{\inf I_R^+(x_i)+\inf I_S^+(x_i)}{2}}-\left(\frac{\sqrt{\inf I_R^+(x_i)}+\sqrt{\inf I_S^+(x_i)}}{2}\right)+\sqrt{\frac{\sup I_R^+(x_i)+\sup I_S^+(x_i)}{2}}-\left(\frac{\sqrt{\sup I_R^+(x_i)}+\sqrt{\sup I_S^+(x_i)}}{2}\right)+\\ \sqrt{\frac{\left(1-\inf I_R^+(x_i)\right)+\left(1-\inf I_S^+(x_i)\right)}{2}}-\left(\frac{\sqrt{1-\inf I_R^+(x_i)}+\sqrt{1-\inf I_S^+(x_i)}}{2}\right)+\sqrt{\frac{\left(1-\sup I_R^+(x_i)\right)+\left(1-\sup I_S^+(x_i)\right)}{2}}-\\ \left(\frac{\sqrt{1-\sup I_R^+(x_i)}+\sqrt{1-\sup I_S^+(x_i)}}{2}\right)+\sqrt{\frac{\inf F_R^+(x_i)+\inf F_S^+(x_i)}{2}}-\left(\frac{\sqrt{\inf F_R^+(x_i)}+\sqrt{\inf F_S^+(x_i)}}{2}\right)+\\ \sqrt{\frac{\sup F_R^+(x_i)+\sup F_S^+(x_i)}{2}}-\left(\frac{\sqrt{\sup F_R^+(x_i)}+\sqrt{\sup F_S^+(x_i)}}{2}\right)+\sqrt{\frac{-\left(\inf T_R^-(x_i)+\inf T_S^+(x_i)\right)}{2}}-\\ \left(\frac{\sqrt{\left(-\inf T_R^-(x_i)\right)}+\sqrt{\left(-\inf T_S^-(x_i)\right)}}{2}\right)+\sqrt{\frac{-\left(\sup T_R^-(x_i)+\sup T_S^+(x_i)\right)}{2}}-\left(\frac{\sqrt{\left(-\sup T_R^-(x_i)\right)}+\sqrt{\left(-\sup T_S^-(x_i)\right)}}{2}\right)+\\ \sqrt{\frac{-\left(\inf I_R^-(x_i)+\inf I_S^-(x_i)\right)}{2}}-\left(\frac{\sqrt{\left(-\inf I_R^-(x_i)\right)}+\sqrt{\left(-\inf I_S^-(x_i)\right)}}{2}\right)+\sqrt{\frac{-\left(\sup I_R^-(x_i)+\sup I_S^-(x_i)\right)}{2}}-\\ \left(\frac{\sqrt{\left(-\sup I_R^-(x_i)\right)}+\sqrt{\left(-\sup I_S^-(x_i)\right)}}{2}\right)+\sqrt{\frac{\left(1+\inf I_R^-(x_i)\right)+\left(1+\inf I_S^-(x_i)\right)}{2}}-\left(\frac{\sqrt{1+\inf I_R^-(x_i)}+\sqrt{1+\inf I_S^-(x_i)}}{2}\right)+\\ \sqrt{\frac{\left(1+\sup I_R^-(x_i)\right)+\left(1+\sup I_S^-(x_i)\right)}{2}}-\left(\frac{\sqrt{1+\sup I_R^-(x_i)}+\sqrt{1+\sup I_S^-(x_i)}}{2}\right)+\sqrt{\frac{-\left(\inf F_R^-(x_i)+\inf F_S^-(x_i)\right)}{2}}-\\ \left(\frac{\sqrt{\left(-\inf F_R^-(x_i)\right)}+\sqrt{\left(-\inf F_S^-(x_i)\right)}}{2}\right)+\sqrt{\frac{-\left(\sup F_R^-(x_i)+\sup F_S^-(x_i)\right)}{2}}-\left(\frac{\sqrt{\left(-\sup F_R^-(x_i)\right)}+\sqrt{\left(-\sup F_S^-(x_i)\right)}}{2}\right)\end{array}\right]\right)$$

$$
\begin{aligned}
&= \frac{1}{2}\sum_{i=1}^{n} w_i \left( \left[ \begin{array}{l}
\sqrt{\frac{\inf T_S^{+}(x_i)+\inf T_R^{+}(x_i)}{2}} - \left(\frac{\sqrt{\inf T_S^{+}(x_i)}+\sqrt{\inf T_R^{+}(x_i)}}{2}\right) + \sqrt{\frac{\sup T_S^{+}(x_i)+\sup T_R^{+}(x_i)}{2}} - \left(\frac{\sqrt{\sup T_S^{+}(x_i)}+\sqrt{\sup T_R^{+}(x_i)}}{2}\right) + \\
\sqrt{\frac{\inf I_S^{+}(x_i)+\inf I_R^{+}(x_i)}{2}} - \left(\frac{\sqrt{\inf I_S^{+}(x_i)}+\sqrt{\inf I_R^{+}(x_i)}}{2}\right) + \sqrt{\frac{\sup I_S^{+}(x_i)+\sup I_R^{+}(x_i)}{2}} - \left(\frac{\sqrt{\sup I_S^{+}(x_i)}+\sqrt{\sup I_R^{+}(x_i)}}{2}\right) + \\
\sqrt{\frac{\left(1-\inf I_S^{+}(x_i)\right)+\left(1-\inf I_R^{+}(x_i)\right)}{2}} - \left(\frac{\sqrt{1-\inf I_S^{+}(x_i)}+\sqrt{1-\inf I_R^{+}(x_i)}}{2}\right) + \sqrt{\frac{\left(1-\sup I_S^{+}(x_i)\right)+\left(1-\sup I_R^{+}(x_i)\right)}{2}} - \\
\left(\frac{\sqrt{1-\sup I_S^{+}(x_i)}+\sqrt{[1-\sup I_R^{+}(x_i)]}}{2}\right) + \sqrt{\frac{\inf F_S^{+}(x_i)+\inf F_R^{+}(x_i)}{2}} - \left(\frac{\sqrt{\inf F_S^{+}(x_i)}+\sqrt{\inf F_R^{+}(x_i)}}{2}\right) + \\
\sqrt{\frac{\sup F_S^{+}(x_i)+\sup F_R^{+}(x_i)}{2}} - \left(\frac{\sqrt{\sup F_S^{+}(x_i)}+\sqrt{\sup F_R^{+}(x_i)}}{2}\right) + \sqrt{\frac{-\left(\inf T_S^{-}(x_i)+\inf T_R^{-}(x_i)\right)}{2}} - \\
\left(\frac{\sqrt{\left(-\inf T_S^{-}(x_i)\right)}+\sqrt{\left(-\inf T_R^{-}(x_i)\right)}}{2}\right) + \sqrt{\frac{-\left(\sup T_S^{-}(x_i)+\sup T_R^{-}(x_i)\right)}{2}} - \left(\frac{\sqrt{\left(-\sup T_S^{-}(x_i)\right)}+\sqrt{\left(-\sup T_R^{-}(x_i)\right)}}{2}\right) + \\
\sqrt{\frac{-\left(\inf I_S^{-}(x_i)+\inf I_R^{-}(x_i)\right)}{2}} - \left(\frac{\sqrt{\left(-\inf I_S^{-}(x_i)\right)}+\sqrt{\left(-\inf I_R^{-}(x_i)\right)}}{2}\right) + \sqrt{\frac{-\left(\sup I_S^{-}(x_i)+\sup I_R^{-}(x_i)\right)}{2}} - \\
\left(\frac{\sqrt{\left(-\sup I_S^{-}(x_i)\right)}+\sqrt{\left(-\sup I_R^{-}(x_i)\right)}}{2}\right) + \sqrt{\frac{\left(1+\inf I_S^{-}(x_i)\right)+\left(1+\inf I_R^{-}(x_i)\right)}{2}} - \left(\frac{\sqrt{1+\inf I_S^{-}(x_i)}+\sqrt{1+\inf I_R^{-}(x_i)}}{2}\right) + \\
\sqrt{\frac{\left(1+\sup I_S^{-}(x_i)\right)+\left(1+\sup I_R^{-}(x_i)\right)}{2}} - \left(\frac{\sqrt{1+\sup I_S^{-}(x_i)}+\sqrt{[1+\sup I_R^{-}(x_i)]}}{2}\right) + \sqrt{\frac{-\left(\inf F_S^{-}(x_i)+\inf F_R^{-}(x_i)\right)}{2}} - \\
\left(\frac{\sqrt{\left(-\inf F_S^{-}(x_i)\right)}+\sqrt{\left(-\inf F_R^{-}(x_i)\right)}}{2}\right) + \sqrt{\frac{-\left(\sup F_S^{-}(x_i)+\sup F_R^{-}(x_i)\right)}{2}} - \left(\frac{\sqrt{\left(-\sup F_S^{-}(x_i)\right)}+\sqrt{\left(-\sup F_R^{-}(x_i)\right)}}{2}\right)
\end{array} \right] \right) \\
&= C_{IB}(S,R)_w.
\end{aligned}
$$

(4) $C_{IB}(R^C, S^C)_w =$

$$
\frac{1}{2}\sum_{i=1}^{n} w_i \left( \left[ \begin{array}{l}
\sqrt{\frac{\inf F_R^{+}(x_i)+\inf F_S^{+}(x_i)}{2}} - \left(\frac{\sqrt{\inf F_R^{+}(x_i)}+\sqrt{\inf F_S^{+}(x_i)}}{2}\right) + \sqrt{\frac{\sup F_R^{+}(x_i)+\sup F_S^{+}(x_i)}{2}} - \left(\frac{\sqrt{\sup F_R^{+}(x_i)}+\sqrt{\sup F_S^{+}(x_i)}}{2}\right) + \\
\sqrt{\frac{\left(1-\inf I_R^{+}(x_i)\right)+\left(1-\inf I_S^{+}(x_i)\right)}{2}} - \left(\frac{\sqrt{\left(1-\inf I_R^{+}(x_i)\right)}+\sqrt{\left(1-\inf I_S^{+}(x_i)\right)}}{2}\right) + \sqrt{\frac{\left(1-\sup I_R^{+}(x_i)\right)+\left(1-\sup I_S^{+}(x_i)\right),}{2}} - \left(\frac{\sqrt{\left(1-\sup I_R^{+}(x_i)\right)}+\sqrt{\left(1-\sup I_S^{+}(x_i)\right)}}{2}\right) + \\
\sqrt{\frac{1-\left(1-\inf I_R^{+}(x_i)\right)+1-\left(1-\inf I_S^{+}(x_i)\right)}{2}} - \left(\frac{\sqrt{1-\left(1-\inf I_R^{+}(x_i)\right)}+\sqrt{1-\left(1-\inf I_S^{+}(x_i)\right)}}{2}\right) + \sqrt{\frac{1-\left(1-\sup I_R^{+}(x_i)\right)+1-\left(1-\sup I_S^{+}(x_i)\right)}{2}} - \\
\left(\frac{\sqrt{1-\left(1-\sup I_R^{+}(x_i)\right)}+\sqrt{1-\left(1-\sup I_S^{+}(x_i)\right)}}{2}\right) + \sqrt{\frac{\inf T_R^{+}(x_i)+\inf T_S^{+}(x_i)}{2}} - \left(\frac{\sqrt{\inf T_R^{+}(x_i)}+\sqrt{\inf T_S^{+}(x_i)}}{2}\right) + \sqrt{\frac{\sup T_R^{+}(x_i)+\sup T_S^{+}(x_i)}{2}} - \left(\frac{\sqrt{\sup T_R^{+}(x_i)}+\sqrt{\sup T_S^{+}(x_i)}}{2}\right) \\
+\sqrt{\frac{-\left(\inf F_R^{-}(x_i)+\inf F_S^{-}(x_i)\right)}{2}} - \left(\frac{\sqrt{\left(-\inf F_R^{-}(x_i)\right)}+\sqrt{\left(-\inf F_S^{-}(x_i)\right)}}{2}\right) + \sqrt{\frac{-\left(\sup F_R^{-}(x_i)+\sup F_S^{-}(x_i)\right)}{2}} - \left(\frac{\sqrt{\left(-\sup F_R^{-}(x_i)\right)}+\sqrt{\left(-\sup F_S^{-}(x_i)\right)}}{2}\right) + \\
\sqrt{\frac{-\left(\left(-1-\inf I_R^{-}(x_i)\right)+\left(-1-\inf I_S^{-}(x_i)\right)\right)}{2}} - \left(\frac{\sqrt{-\left(-1-\inf I_R^{-}(x_i)\right)}+\sqrt{-\left(-1-\inf I_S^{-}(x_i)\right)}}{2}\right) + \sqrt{\frac{-\left(\left(-1-\sup I_R^{-}(x_i)\right)+\left(-1-\sup I_S^{-}(x_i)\right)\right)}{2}} - \\
\left(\frac{\sqrt{-\left(-1-\sup I_R^{-}(x_i)\right)}+\sqrt{-\left(-1-\sup I_S^{-}(x_i)\right)}}{2}\right) + \sqrt{\frac{1+\left(-1-\inf I_R^{-}(x_i)\right)+1+\left(-1-\inf I_S^{-}(x_i)\right)}{2}} - \left(\frac{\sqrt{1+\left(-1-\inf I_R^{-}(x_i)\right)}+\sqrt{1+\left(-1-\inf I_S^{-}(x_i)\right)}}{2}\right) + \\
\sqrt{\frac{1+\left(-1-\sup I_R^{-}(x_i)\right)+1+\left(-1-\sup I_S^{-}(x_i)\right)}{2}} - \left(\frac{\sqrt{1+\left(-1-\sup I_R^{-}(x_i)\right)}+\sqrt{1+\left(-1-\sup I_S^{-}(x_i)\right)}}{2}\right) + \sqrt{\frac{-\left(\inf T_R^{-}(x_i)+\inf T_S^{-}(x_i)\right)}{2}} - \\
\left(\frac{\sqrt{\left(-\inf T_S^{-}(x_i)\right)}+\sqrt{\left(-\inf T_R^{-}(x_i)\right)}}{2}\right) + \sqrt{\frac{-\left(\sup T_S^{-}(x_i)+\sup T_R^{-}(x_i)\right)}{2}} - \left(\frac{\sqrt{\left(-\sup T_S^{-}(x_i)\right)}+\sqrt{\left(-\sup T_R^{-}(x_i)\right)}}{2}\right)
\end{array} \right] \right)
$$

$$
= \frac{1}{2}\sum_{i=1}^{n} w_i \left\| \left[ \begin{array}{l}
\sqrt{\frac{\inf T_R^+(x_i)+\inf T_S^+(x_i)}{2}} - \left(\frac{\sqrt{\inf T_R^+(x_i)}+\sqrt{\inf T_S^+(x_i)}}{2}\right) + \sqrt{\frac{\sup T_R^+(x_i)+\sup T_S^+(x_i)}{2}} - \left(\frac{\sqrt{\sup T_R^+(x_i)}+\sqrt{\sup T_S^+(x_i)}}{2}\right) + \\
\sqrt{\frac{\inf I_R^+(x_i)+\inf I_S^+(x_i)}{2}} - \left(\frac{\sqrt{\inf I_R^+(x_i)}+\sqrt{\inf I_S^+(x_i)}}{2}\right) + \sqrt{\frac{\sup I_R^+(x_i)+\sup I_S^+(x_i)}{2}} - \left(\frac{\sqrt{\sup I_R^+(x_i)}+\sqrt{\sup I_S^+(x_i)}}{2}\right) + \\
\sqrt{\frac{\left(1-\inf I_R^+(x_i)\right)+\left(1-\inf I_S^+(x_i)\right)}{2}} - \left(\frac{\sqrt{1-\inf I_R^+(x_i)}+\sqrt{1-\inf I_S^+(x_i)}}{2}\right) + \sqrt{\frac{\left(1-\sup I_R^+(x_i)\right)+\left(1-\sup I_S^+(x_i)\right)}{2}} - \\
\left(\frac{\sqrt{1-\sup I_R^+(x_i)}+\sqrt{1-\sup I_S^+(x_i)}}{2}\right) + \sqrt{\frac{\inf F_R^+(x_i)+\inf F_S^+(x_i)}{2}} - \left(\frac{\sqrt{\inf F_R^+(x_i)}+\sqrt{\inf F_S^+(x_i)}}{2}\right) + \\
\sqrt{\frac{\sup F_R^+(x_i)+\sup F_S^+(x_i)}{2}} - \left(\frac{\sqrt{\sup F_R^+(x_i)}+\sqrt{\sup F_S^+(x_i)}}{2}\right) + \sqrt{\frac{-\left(\inf T_R^-(x_i)+\inf T_S^-(x_i)\right)}{2}} - \\
\left(\frac{\sqrt{\left(-\inf T_R^-(x_i)\right)}+\sqrt{\left(-\inf T_S^-(x_i)\right)}}{2}\right) + \sqrt{\frac{-\left(\sup T_R^-(x_i)+\sup T_S^-(x_i)\right)}{2}} - \left(\frac{\sqrt{\left(-\sup T_R^-(x_i)\right)}+\sqrt{\left(-\sup T_S^-(x_i)\right)}}{2}\right) + \\
\sqrt{\frac{-\left(\inf I_R^-(x_i)+\inf I_S^-(x_i)\right)}{2}} - \left(\frac{\sqrt{\left(-\inf I_R^-(x_i)\right)}+\sqrt{\left(-\inf I_S^-(x_i)\right)}}{2}\right) + \sqrt{\frac{-\left(\sup I_R^-(x_i)+\sup I_S^-(x_i)\right)}{2}} - \\
\left(\frac{\sqrt{\left(-\sup I_R^-(x_i)\right)}+\sqrt{\left(-\sup I_S^-(x_i)\right)}}{2}\right) + \sqrt{\frac{\left(1+\inf I_R^-(x_i)\right)+\left(1+\inf I_S^-(x_i)\right)}{2}} - \left(\frac{\sqrt{1+\inf I_R^-(x_i)}+\sqrt{1+\inf I_S^-(x_i)}}{2}\right) + \\
\sqrt{\frac{\left(1+\sup I_R^-(x_i)\right)+\left(1+\sup I_S^-(x_i)\right)}{2}} - \left(\frac{\sqrt{1+\sup I_R^-(x_i)}+\sqrt{1+\sup I_S^-(x_i)}}{2}\right) + \sqrt{\frac{-\left(\inf F_R^-(x_i)+\inf F_S^-(x_i)\right)}{2}} - \\
\left(\frac{\sqrt{\left(-\inf F_R^-(x_i)\right)}+\sqrt{\left(-\inf F_S^-(x_i)\right)}}{2}\right) + \sqrt{\frac{-\left(\sup F_R^-(x_i)+\sup F_S^-(x_i)\right)}{2}} - \left(\frac{\sqrt{\left(-\sup F_R^-(x_i)\right)}+\sqrt{\left(-\sup F_S^-(x_i)\right)}}{2}\right)
\end{array} \right] \right\|
$$

$$= C_{IB}(R, S)_w.$$

This completes the proof. □

## References


1. Shannon, C.E.; Weaver, W. *The Mathematical Theory of Communications*; The University of Illinois Press: Urbana, IL, USA, 1949.
2. Shannon, C.E. A mathematical theory of communications. *Bell Syst. Tech. J.* **1948**, *27*, 379–423. [CrossRef]
3. Criado, F.; Gachechiladze, T. Entropy of fuzzy events. *Fuzzy Sets Syst.* **1997**, *88*, 99–106. [CrossRef]
4. Herencia, J.; Lamta, M. Entropy measure associated with fuzzy basic probability assignment. In Proceedings of the IEEE International Conference on Fuzzy Systems, Barcelona, Spain, 5 July 1997; Volume 2, pp. 863–868.
5. Rudas, I.; Kaynak, M. Entropy basedoperations on fuzzy sets. *IEEE Trans. Fuzzy Syst.* **1998**, *6*, 33–39. [CrossRef]
6. Zadeh, L.A. Probality measures of fuzzy events. *J. Math. Anal. Appl.* **1968**, *23*, 421–427. [CrossRef]
7. Luca, A.D.; Termini, S. A definition of non-probabilistic entropy in the setting of fuzzy set theory. *Inf. Control* **1972**, *20*, 301–312. [CrossRef]
8. Sander, W. On measure of fuzziness. *Fuzzy Sets Syst.* **1989**, *29*, 49–55. [CrossRef]
9. Xie, W.; Bedrosian, S. An information measure for fuzzy sets. *IEEE Trans. Syst. Man Cybern.* **1984**, *14*, 151–156. [CrossRef]
10. Pal, N.; Pal, S. Higher order fuzzy entropy and hybridentropy of a fuzzy set. *Inf. Sci.* **1992**, *61*, 211–221. [CrossRef]
11. Kaufmann, A.; Gupta, M. *Introduction of Fuzzy Arithmetic: Theory and Applications*; Van Nostrand Reinhold Co.: New York, NY, USA, 1985.
12. Yager, R. On the measure of fuzziness and negation. Part I: Membership in the unit interval. *Int. J. Gen. Syst.* **1979**, *5*, 221–229. [CrossRef]
13. Yager, R. On the measure of fuzziness and negation. Part II: Lattice. *Inf. Control* **1980**, *44*, 236–260. [CrossRef]
14. Kosko, B. Fuzzy entropy and conditioning. *Inf. Sci.* **1986**, *40*, 165–174. [CrossRef]
15. Kosko, B. Concepts of fuzzy information measure on continuous domains. *Int. J. Gen. Syst.* **1990**, *17*, 211–240. [CrossRef]
16. Prakash, O.; Sharma, P.K.; Mahajan, R. New measures of weighted fuzzy entropy and their applications for the study of maximum weighted fuzzy entropy principle. *Inf. Sci.* **2008**, *178*, 2389–2395. [CrossRef]

17. Burillo, P.; Bustince, H. Entropy on intuitionistic fuzzy sets and on interval–valued fuzzy sets. *Fuzzy Sets Syst.* **1996**, *78*, 305–316. [CrossRef]
18. Szmidt, E.; Kacprzyk, J. Entropy for intuitionistic fuzzy sets. *Fuzzy Sets Syst.* **2001**, *118*, 467–477. [CrossRef]
19. Wei, C.P.; Wang, P.; Zhang, Y.Z. Entropy, similarity measure of interval–valued intuitionistic fuzzy sets and their applications. *Inf. Sci.* **2011**, *181*, 4273–4286. [CrossRef]
20. Li, X.Y. Interval–valued intuitionistic fuzzy continuous cross entropy and its application in multi-attribute decision-making. *Comput. Eng. Appl.* **2013**, *49*, 234–237.
21. Shang, X.G.; Jiang, W.S. A note on fuzzy information measures. *Pattern Recogit. Lett.* **1997**, *18*, 425–432. [CrossRef]
22. Vlachos, I.K.; Sergiadis, G.D. Intuitionistic fuzzy information applications to pattern recognition. *Pattern Recogit. Lett.* **2007**, *28*, 197–206. [CrossRef]
23. Ye, J. Fuzzy cross entropy of interval–valued intuitionistic fuzzy sets and its optimal decision-making method based on the weights of the alternatives. *Expert Syst. Appl.* **2011**, *38*, 6179–6183. [CrossRef]
24. Xia, M.M.; Xu, Z.S. Entropy/cross entropy–based group decision making under intuitionistic fuzzy sets. *Inf. Fusion* **2012**, *13*, 31–47. [CrossRef]
25. Tong, X.; Yu, L.A. A novel MADM approach based on fuzzy cross entropy with interval-valued intuitionistic fuzzy sets. *Math. Probl. Eng.* **2015**, *2015*, 1–9. [CrossRef]
26. Smarandache, F. *A Unifying Field of Logics. Neutrosophy: Neutrosophic Probability, Set and Logic*; American Research Press: Rehoboth, DE, USA, 1998.
27. Zadeh, L.A. Fuzzy sets. *Inf. Control* **1965**, *8*, 338–353. [CrossRef]
28. Atanassov, K.T. Intuitionistic fuzzy sets. *Fuzzy Sets Syst.* **1986**, *20*, 87–96. [CrossRef]
29. Wang, H.; Smarandache, F.; Zhang, Y.Q.; Sunderraman, R. Single valued neutrosophic sets. *Multispace Multistruct.* **2010**, *4*, 410–413.
30. Pramanik, S.; Biswas, P.; Giri, B.C. Hybrid vector similarity measures and their applications to multi-attribute decision making under neutrosophic environment. *Neural Comput. Appl.* **2017**, *28*, 1163–1176. [CrossRef]
31. Biswas, P.; Pramanik, S.; Giri, B.C. Entropy based grey relational analysis method for multi-attribute decision making under single valued neutrosophic assessments. *Neutrosophic Sets Syst.* **2014**, *2*, 102–110.
32. Biswas, P.; Pramanik, S.; Giri, B.C. A new methodology for neutrosophic multi-attribute decision making with unknown weight information. *Neutrosophic Sets Syst.* **2014**, *3*, 42–52.
33. Biswas, P.; Pramanik, S.; Giri, B.C. TOPSIS method for multi-attribute group decision-making under single valued neutrosophic environment. *Neural Comput. Appl.* **2015**. [CrossRef]
34. Biswas, P.; Pramanik, S.; Giri, B.C. Aggregation of triangular fuzzy neutrosophic set information and its application to multi-attribute decision making. *Neutrosophic Sets Syst.* **2016**, *12*, 20–40.
35. Biswas, P.; Pramanik, S.; Giri, B.C. Value and ambiguity index based ranking method of single-valued trapezoidal neutrosophic numbers and its application to multi-attribute decision making. *Neutrosophic Sets Syst.* **2016**, *12*, 127–138.
36. Biswas, P.; Pramanik, S.; Giri, B.C. Multi-attribute group decision making based on expected value of neutrosophic trapezoidal numbers. In *New Trends in Neutrosophic Theory and Applications*; Smarandache, F., Pramanik, S., Eds.; Pons Editions: Brussels, Belgium, 2017; Volume II, in press.
37. Biswas, P.; Pramanik, S.; Giri, B.C. Non-linear programming approach for single-valued neutrosophic TOPSIS method. *New Math. Nat. Comput.* **2017**, in press.
38. Deli, I.; Subas, Y. A ranking method of single valued neutrosophic numbers and its applications to multi-attribute decision making problems. *Int. J. Mach. Learn. Cybern.* **2016**. [CrossRef]
39. Ji, P.; Wang, J.Q.; Zhang, H.Y. Frank prioritized Bonferroni mean operator with single-valued neutrosophic sets and its application in selecting third-party logistics providers. *Neural Comput. Appl.* **2016**. [CrossRef]
40. Kharal, A. A neutrosophic multi-criteria decision making method. *New Math. Nat. Comput.* **2014**, *10*, 143–162. [CrossRef]
41. Liang, R.X.; Wang, J.Q.; Li, L. Multi-criteria group decision making method based on interdependent inputs of single valued trapezoidal neutrosophic information. *Neural Comput. Appl.* **2016**. [CrossRef]
42. Liang, R.X.; Wang, J.Q.; Zhang, H.Y. A multi-criteria decision-making method based on single-valued trapezoidal neutrosophic preference relations with complete weight information. *Neural Comput. Appl.* **2017**. [CrossRef]

43. Liu, P.; Chu, Y.; Li, Y.; Chen, Y. Some generalized neutrosophic number Hamacher aggregation operators and their application to group decision making. *Int. J. Fuzzy Syst.* **2014**, *16*, 242–255.
44. Liu, P.D.; Li, H.G. Multiple attribute decision-making method based on some normal neutrosophic Bonferroni mean operators. *Neural Comput. Appl.* **2017**, *28*, 179–194. [CrossRef]
45. Liu, P.; Wang, Y. Multiple attribute decision-making method based on single-valued neutrosophic normalized weighted Bonferroni mean. *Neural Comput. Appl.* **2014**, *25*, 2001–2010. [CrossRef]
46. Peng, J.J.; Wang, J.Q.; Wang, J.; Zhang, H.Y.; Chen, X.H. Simplified neutrosophic sets and their applications in multi-criteria group decision-making problems. *Int. J. Syst. Sci.* **2016**, *47*, 2342–2358. [CrossRef]
47. Peng, J.; Wang, J.; Zhang, H.; Chen, X. An outranking approach for multi-criteria decision-making problems with simplified neutrosophic sets. *Appl. Soft Comput.* **2014**, *25*, 336–346. [CrossRef]
48. Pramanik, S.; Banerjee, D.; Giri, B.C. Multi–criteria group decision making model in neutrosophic refined set and its application. *Glob. J. Eng. Sci. Res. Manag.* **2016**, *3*, 12–18.
49. Pramanik, S.; Dalapati, S.; Roy, T.K. Logistics center location selection approach based on neutrosophic multi-criteria decision making. In *New Trends in Neutrosophic Theory and Applications*; Smarandache, F., Pramanik, S., Eds.; Pons Asbl: Brussels, Belgium, 2016; Volume 1, pp. 161–174, ISBN 978-1-59973-498-9.
50. Sahin, R.; Karabacak, M. A multi attribute decision making method based on inclusion measure for interval neutrosophic sets. *Int. J. Eng. Appl. Sci.* **2014**, *2*, 13–15.
51. Sahin, R.; Kucuk, A. Subsethood measure for single valued neutrosophic sets. *J. Intell. Fuzzy Syst.* **2014**. [CrossRef]
52. Sahin, R.; Liu, P. Maximizing deviation method for neutrosophic multiple attribute decision making with incomplete weight information. *Neural Comput. Appl.* **2016**, *27*, 2017–2029. [CrossRef]
53. Sodenkamp, M. Models, Strategies and Applications of Group Multiple-Criteria Decision Analysis in Complex and Uncertain Systems. Ph.D. Thesis, University of Paderborn, Paderborn, Germany, 2013.
54. Ye, J. Multicriteria decision-making method using the correlation coefficient under single-valued neutrosophic environment. *Int. J. Gen. Syst.* **2013**, *42*, 386–394. [CrossRef]
55. Ye, J. Another form of correlation coefficient between single valued neutrosophic sets and its multiple attribute decision making method. *Neutrosophic Sets Syst.* **2013**, *1*, 8–12.
56. Ye, J. A multi criteria decision-making method using aggregation operators for simplified neutrosophic sets. *J. Intell. Fuzzy Syst.* **2014**, *26*, 2459–2466.
57. Ye, J. Trapezoidal neutrosophic set and its application to multiple attribute decision-making. *Neural Comput. Appl.* **2015**, *26*, 1157–1166. [CrossRef]
58. Ye, J. Bidirectional projection method for multiple attribute group decision making with neutrosophic number. *Neural Comput. Appl.* **2015**. [CrossRef]
59. Ye, J. Projection and bidirectional projection measures of single valued neutrosophic sets and their decision—Making method for mechanical design scheme. *J. Exp. Theor. Artif. Intell.* **2016**. [CrossRef]
60. Nancy, G.H. Novel single-valued neutrosophic decision making operators under Frank norm operations and its application. *Int. J. Uncertain. Quant.* **2016**, *6*, 361–375. [CrossRef]
61. Nancy, G.H. Some new biparametric distance measures on single-valued neutrosophic sets with applications to pattern recognition and medical diagnosis. *Information* **2017**, *8*, 162. [CrossRef]
62. Pramanik, S.; Roy, T.K. Neutrosophic game theoretic approach to Indo-Pak conflict over Jammu-Kashmir. *Neutrosophic Sets Syst.* **2014**, *2*, 82–101.
63. Mondal, K.; Pramanik, S. Multi-criteria group decision making approach for teacher recruitment in higher education under simplified Neutrosophic environment. *Neutrosophic Sets Syst.* **2014**, *6*, 28–34.
64. Mondal, K.; Pramanik, S. Neutrosophic decision making model of school choice. *Neutrosophic Sets Syst.* **2015**, *7*, 62–68.
65. Cheng, H.D.; Guo, Y. A new neutrosophic approach to image thresholding. *New Math. Nat. Comput.* **2008**, *4*, 291–308. [CrossRef]
66. Guo, Y.; Cheng, H.D. New neutrosophic approach to image segmentation. *Pattern Recogit.* **2009**, *42*, 587–595. [CrossRef]
67. Guo, Y.; Sengur, A.; Ye, J. A novel image thresholding algorithm based on neutrosophic similarity score. *Measurement* **2014**, *58*, 175–186. [CrossRef]
68. Ye, J. Single valued neutrosophic minimum spanning tree and its clustering method. *J. Intell. Syst.* **2014**, *23*, 311–324. [CrossRef]

69. Ye, J. Clustering strategies using distance-based similarity measures of single-valued neutrosophic sets. *J. Intell. Syst.* **2014**, *23*, 379–389.
70. Mondal, K.; Pramanik, S. A study on problems of Hijras in West Bengal based on neutrosophic cognitive maps. *Neutrosophic Sets Syst.* **2014**, *5*, 21–26.
71. Pramanik, S.; Chakrabarti, S. A study on problems of construction workers in West Bengal based on neutrosophic cognitive maps. *Int. J. Innov. Res. Sci. Eng. Technol.* **2013**, *2*, 6387–6394.
72. Majumdar, P.; Samanta, S.K. On similarity and entropy of neutrosophic sets. *J. Intell. Fuzzy Syst.* **2014**, *26*, 1245–1252.
73. Ye, J. Single valued neutrosophic cross-entropy for multi criteria decision making problems. *Appl. Math. Model.* **2014**, *38*, 1170–1175. [CrossRef]
74. Ye, J. Improved cross entropy measures of single valued neutrosophic sets and interval neutrosophic sets and their multi criteria decision making strategies. *Cybern. Inf. Technol.* **2015**, *15*, 13–26. [CrossRef]
75. Wang, H.; Smarandache, F.; Zhang, Y.Q.; Sunderraman, R. *Interval Neutrosophic Sets and Logic: Theory and Applications in Computing*; Hexis: Phoenix, AZ, USA, 2005.
76. Pramanik, S.; Dalapati, S.; Alam, S.; Smarandache, F.; Roy, T.K. NS-cross entropy-based MAGDM under single-valued neutrosophic set environment. *Information* **2018**, *9*, 37. [CrossRef]
77. Sahin, R. Cross-entropy measure on interval neutrosophic sets and its applications in multi criteria decision making. *Neural Comput. Appl.* **2015**. [CrossRef]
78. Tian, Z.P.; Zhang, H.Y.; Wang, J.; Wang, J.Q.; Chen, X.H. Multi-criteria decision-making method based on a cross-entropy with interval neutrosophic sets. *Int. J. Syst. Sci.* **2015**. [CrossRef]
79. Hwang, C.L.; Yoon, K. *Multiple Attribute Decision Making: Methods and Applications*; Springer: New York, NY, USA, 1981.
80. Dalapati, S.; Pramanik, S.; Alam, S.; Smarandache, S.; Roy, T.K. IN-cross entropy based magdm strategy under interval neutrosophic set environment. *Neutrosophic Sets Syst.* **2017**, *18*, 43–57. [CrossRef]
81. Deli, I.; Ali, M.; Smarandache, F. Bipolar neutrosophic sets and their application based on multi-criteria decision making problems. In Proceedings of the 2015 International Conference on Advanced Mechatronic Systems (ICAMechS), Beijing, China, 22–24 August 2015; pp. 249–254.
82. Zhang, W.R. Bipolar fuzzy sets. In Proceedings of the IEEE World Congress on Computational Science (FuzzIEEE), Anchorage, AK, USA, 4–9 May 1998; pp. 835–840. [CrossRef]
83. Zhang, W.R. Bipolar fuzzy sets and relations: A computational framework for cognitive modeling and multiagent decision analysis. In Proceedings of the IEEE Industrial Fuzzy Control and Intelligent Systems Conference, and the NASA Joint Technology Workshop on Neural Networks and Fuzzy Logic, Fuzzy Information Processing Society Biannual Conference, San Antonio, TX, USA, 18–21 December 1994; pp. 305–309. [CrossRef]
84. Deli, I.; Subas, Y.A. Multiple criteria decision making method on single valued bipolar neutrosophic set based on correlation coefficient similarity measure. In Proceedings of the International Conference on Mathematics and Mathematics Education (ICMME-2016), Elazg, Turkey, 12–14 May 2016.
85. Şahin, M.; Deli, I.; Uluçay, V. Jaccard vector similarity measure of bipolar neutrosophic set based on multi-criteria decision making. In Proceedings of the International Conference on Natural Science and Engineering (ICNASE'16), Killis, Turkey, 19–20 March 2016.
86. Uluçay, V.; Deli, I.; Şahin, M. Similarity measures of bipolar neutrosophic sets and their application to multiple criteria decision making. *Neural Comput. Appl.* **2016**. [CrossRef]
87. Dey, P.P.; Pramanik, S.; Giri, B.C. TOPSIS for solving multi-attribute decision making problems under bi-polar neutrosophic environment. In *New Trends in Neutrosophic Theory and Applications*; Smarandache, F., Pramanik, S., Eds.; Pons Asbl: Brussells, Belgium, 2016; pp. 65–77, ISBN 978-1-59973-498-9.
88. Pramanik, S.; Dey, P.P.; Giri, B.C.; Smarandache, F. Bipolar neutrosophic projection based models for solving multi-attribute decision making problems. *Neutrosophic Sets Syst.* **2017**, *15*, 74–83. [CrossRef]
89. Wang, L.; Zhang, H.; Wang, J. Frank Choquet Bonferroni operators of bipolar neutrosophic sets and their applications to multi-criteria decision-making problems. *Int. J. Fuzzy Syst.* **2017**. [CrossRef]
90. Pramanik, S.; Dalapati, S.; Alam, S.; Smarandache, F.; Roy, T.K. TODIM Method for Group Decision Making under Bipolar Neutrosophic Set Environment. In *New Trends in Neutrosophic Theory and Applications*; Smarandache, F., Pramanik, S., Eds.; Pons Editions: Brussels, Belgium, 2016; Volume II, in press.

91. Mahmood, T.; Ye, J.; Khan, Q. Bipolar Interval Neutrosophic Set and Its Application in Multicriteria Decision Making. Available online: https://archive.org/details/BipolarIntervalNeutrosophicSet (accessed on 9 October 2017).
92. Deli, I.; Şubaş, Y.; Smarandache, F.; Ali, M. Interval Valued Bipolar Neutrosophic Sets and Their Application in Pattern Recognition. Available online: https://www.researchgate.net/publication/289587637 (accessed on 9 October 2017).
93. Ye, J. Similarity measures between interval neutrosophic sets and their applications in multicriteria decision-making. *J. Intell. Fuzzy Syst.* **2014**, *26*, 165–172. [CrossRef]
94. Garg, H. Non-linear programming method for multi-criteria decision making problems under interval neutrosophic set environment. *Appl. Intell.* **2017**, 1–15. [CrossRef]
95. Pramanik, S.; Mukhopadhyaya, D. Grey relational analysis-based intuitionistic fuzzy multi-criteria group decision-making approach for teacher selection in higher education. *Int. J. Comput. Appl.* **2011**, *34*, 21–29. [CrossRef]
96. Mondal, K.; Pramanik, S. Intuitionistic fuzzy multi criteria group decision making approach to quality-brick selection problem. *J. Appl. Quant. Methods* **2014**, *9*, 35–50.
97. Dey, P.P.; Pramanik, S.; Giri, B.C. Multi-criteria group decision making in intuitionistic fuzzy environment based on grey relational analysis for weaver selection in Khadi institution. *J. Appl. Quant. Methods* **2015**, *10*, 1–14.
98. Dey, P.P.; Pramanik, S.; Giri, B.C. An extended grey relational analysis based interval neutrosophic multi-attribute decision making for weaver selection. *J. New Theory* **2015**, *9*, 82–93.

axioms

*Article*

# Multi-Attribute Decision-Making Method Based on Neutrosophic Soft Rough Information

**Muhammad Akram [1,*], Sundas Shahzadi [1] and Florentin Smarandache [2]**

1 Department of Mathematics, University of the Punjab, New Campus, Lahore 54590, Pakistan; sundas1011@gmail.com
2 Mathematics & Science Department, University of New Mexico, 705 Gurley Ave., Gallup, NM 87301, USA; fsmarandache@gmail.com
* Correspondence: m.akram@pucit.edu.pk

Received: 17 February 2018; Accepted: 19 March 2018; Published: 20 March 2018

**Abstract:** Soft sets (SSs), neutrosophic sets (NSs), and rough sets (RSs) are different mathematical models for handling uncertainties, but they are mutually related. In this research paper, we introduce the notions of soft rough neutrosophic sets (SRNSs) and neutrosophic soft rough sets (NSRSs) as hybrid models for soft computing. We describe a mathematical approach to handle decision-making problems in view of NSRSs. We also present an efficient algorithm of our proposed hybrid model to solve decision-making problems.

**Keywords:** soft rough neutrosophic sets; neutrosophic soft rough sets; decision-making; algorithm

**MSC:** 03E72; 68R10; 68R05

---

## 1. Introduction

Smarandache [1] initiated the concept of neutrosophic set (NS). Smarandache's NS is characterized by three parts: truth, indeterminacy, and falsity. Truth, indeterminacy and falsity membership values behave independently and deal with the problems of having uncertain, indeterminant and imprecise data. Wang et al. [2] gave a new concept of single valued neutrosophic set (SVNS) and defined the set of theoretic operators in an instance of NS called SVNS. Ye [3–5] studied the correlation coefficient and improved correlation coefficient of NSs, and also determined that, in NSs, the cosine similarity measure is a special case of the correlation coefficient. Peng et al. [6] discussed the operations of simplified neutrosophic numbers and introduced an outranking idea of simplified neutrosophic numbers.

Molodtsov [7] introduced the notion of soft set as a novel mathematical approach for handling uncertainties. Molodtsov's soft sets give us new technique for dealing with uncertainty from the viewpoint of parameters. Maji et al. [8–10] introduced neutrosophic soft sets (NSSs), intuitionistic fuzzy soft sets (IFSSs) and fuzzy soft sets (FSSs). Babitha and Sunil gave the idea of soft set relations [11]. In [12], Sahin and Kucuk presented NSS in the form of neutrosophic relation.

Rough set theory was initiated by Pawlak [13] in 1982. Rough set theory is used to study the intelligence systems containing incomplete, uncertain or inexact information. The lower and upper approximation operators of RSs are used for managing hidden information in a system. Therefore, many hybrid models have been built such as soft rough sets (SRSs), rough fuzzy sets (RFSs), fuzzy rough sets (FRSs), soft fuzzy rough sets (SFRSs), soft rough fuzzy sets (SRFSs), intuitionistic fuzzy soft rough sets (IFSRS), neutrosophic rough sets (NRSs), and rough neutrosophic sets (RNSs) for handling uncertainty and incomplete information effectively. Soft set theory and RS theory are two different mathematical tools to deal with uncertainty. Evidently, there is no direct relationship between these two mathematical tools, but efforts have been made to define some kind of relation [14,15]. Feng et al. [15] took a significant step to introduce parametrization tools in RSs. They introduced SRSs, in which parameterized subsets of universal sets are elementary building blocks for approximation operators of a subset. Shabir et al. [16] introduced another approach to study roughness through SSs, and this approach is known as modified SRSs (MSR-sets). In MSR-sets, some results proved to be valid that failed to hold in SRSs. Feng et al. [17] introduced a modification of Pawlak approximation space known as soft approximation space (SAS) in which SAS SRSs were proposed. Moreover, they introduced soft rough fuzzy approximation operators in SAS and initiated a idea of SRFSs, which is an extension of RFSs introduced by Dubois and Prade [18] . Meng et al. [19] provide further discussion of the combination of SSs, RSs and FSs. In various decision-making problems, RSs have been used. The existing results of RSs and other extended RSs such as RFSs, generalized RFSs, SFRSs and IFRSs based decision-making models have their advantages and limitations [20,21]. In a different way, RS approximations have been constructed into the IF environment and are known as IFRSs, RIFSs and generalized IFRSs [22–24]. Zhang et al. [25,26] presented the notions of SRSs, SRIFSs, and IFSRSs, its application in decision-making, and also introduced generalized IFSRSs. Broumi et al. [27,28] developed a hybrid structure by combining RSs and NSs, called RNSs. They also presented interval valued neutrosophic soft rough sets by combining interval valued neutrosophic soft sets and RSs. Yang et al. [29] proposed single valued neutrosophic rough sets (SVNRSs) by combining SVNSs and RSs, and established an algorithm for decision-making problems based on SVNRSs in two universes. For some papers related to NSs and multi-criteria decision-making (MCDM), the readers are referred to [30–38]. The notion of SRNSs is a extension of SRSs, SRIFSs, IFSRSs, introduced by Zhang et al. motivated by the idea of single valued neutrosophic rough sets (SVNRSs) introduced, we extend the single valued neutrosophic rough sets' lower and upper approximations to the case of a neutrosophic soft rough set. The concept of a neutrosophic soft rough set is introduced by coupling both the neutrosophic soft sets and rough sets. In this research paper, we introduce the notions of SRNSs and NSRSs as hybrid models for soft computing. Approximation operators of SRNSs and NSRSs are described and their relevant properties are investigated in detail. We describe a mathematical approach to handle decision-making problems in view of NSRSs. We also present an efficient algorithm of our proposed hybrid model to solve decision-making problems.

## 2. Construction of Soft Rough Neutrosophic Sets

In this section, we introduce the notions of SRNSs by combining soft sets with RNSs and soft rough neutrosophic relations (SRNRs). Soft rough neutrosophic sets consist of two basic components, namely neutrosophic sets and soft relations, which are the mathematical basis of SRNSs. The basic idea of soft rough neutrosophic sets is based on the approximation of sets by a couple of sets known as the lower soft rough neutrosophic approximation and the upper soft rough neutrosophic approximation of a set. Here, the lower and upper approximation operators are based on an arbitrary soft relation. The concept of soft rough neutrosophic sets is an extension of the crisp set, rough set for the study of intelligent systems characterized by inexact, uncertain or insufficient information. It is a useful tool for dealing with uncertainty or imprecision information. The concept of neutrosophic soft sets is powerful logic to handle indeterminate and inconsistent situations, and the theory of rough neutrosophic sets is also powerful mathematical logic to handle incompleteness. We introduce the notions of soft rough neutrosophic sets (SRNSs) and neutrosophic soft rough sets (NSRSs) as hybrid models for soft computing. The rating of all alternatives is expressed with the upper soft rough neutrosophic approximation and lower soft rough neutrosophic approximation operator and the pair of neutrosophic sets that are characterized by truth-membership degree, indeterminacy-membership degree, and falsity-membership degree from the view point of parameters.

**Definition 1.** *Let $Y$ be an initial universal set and $M$ a universal set of parameters. For an arbitrary soft relation $P$ over $Y \times M$, let $P_s : Y \to \mathcal{N}(M)$ be a set-valued function defined as $P_s(u) = \{k \in M \mid (u,k) \in P\}, u \in Y$.*

*Let $(Y, M, P)$ be an SAS. For any NS $C = \{(k, T_C(k), I_C(k), F_C(k)) \mid k \in M\} \in \mathcal{N}(M)$, where $\mathcal{N}(M)$ is a neutrosophic power set of parameter set $M$, the lower soft rough neutrosophic approximation (LSRNA) and* *the upper soft rough neutrosophic approximation (USRNA) operators of C w.r.t* $(Y, M, P)$ *denoted by* $\underline{P}(C)$ *and* $\overline{P}(C)$*, are, respectively, defined as follows:*

$$\overline{P}(C) = \{(u, T_{\overline{P}(C)}(u), I_{\overline{P}(C)}(u), F_{\overline{P}(C)}(u)) \mid u \in Y\},$$

$$\underline{P}(C) = \{(u, T_{\underline{P}(C)}(u), I_{\underline{P}(C)}(u), F_{\underline{P}(C)}(u)) \mid u \in Y\},$$

*where*

$$T_{\overline{P}(C)}(u) = \bigvee_{k \in P_s(u)} T_C(k), \quad I_{\overline{P}(C)}(u) = \bigwedge_{k \in P_s(u)} I_C(k), \quad F_{\overline{P}(C)}(u) = \bigwedge_{k \in P_s(u)} F_C(k),$$

$$T_{\underline{P}(C)}(u) = \bigwedge_{k \in P_s(u)} T_C(k), \quad I_{\underline{P}(C)}(u) = \bigvee_{k \in P_s(u)} I_C(k), \quad F_{\underline{P}(C)}(u) = \bigvee_{k \in P_s(u)} F_C(k).$$

*It is observed that* $\overline{P}(C)$ *and* $\underline{P}(C)$ *are two NSs on* $Y$, $\underline{P}(C), \overline{P}(C) : \mathcal{N}(M) \to \mathcal{P}(Y)$ *are referred to as the LSRNA and the USRNA operators, respectively. The pair* $(\underline{P}(C), \overline{P}(C))$ *is called SRNS of C* w.r.t $(Y, M, P)$.

**Remark 1.** *Let* $(Y, M, P)$ *be an SAS. If* $C \in IF(M)$ *and* $C \in \mathcal{P}(M)$*, where* $IF(M)$ *and* $\mathcal{P}(M)$ *are intuitionistic fuzzy power set and crisp power set of M, respectively. Then, the above SRNA operators* $\underline{P}(C)$ *and* $\overline{P}(C)$ *degenerate to SRIFA and SRA operators, respectively. Hence, SRNA operators are an extension of SRIFA and SRA operators.*

**Example 1.** *Suppose that* $Y = \{w_1, w_2, w_3, w_4, w_5\}$ *is the set of five careers under observation, and Mr. X wants to select best suitable career. Let* $M = \{k_1, k_2, k_3, k_4\}$ *be a set of decision parameters. The parameters* $k_1, k_2, k_3$ *and* $k_4$ *stand for "aptitude", "work value", "skill" and "recent advancement", respectively. Mr. X describes the "most suitable career" by defining a soft relation P from Y to M, which is a crisp soft set as shown in Table 1.*

**Table 1.** Crisp soft relation $P$.

| **P** | $w_1$ | $w_2$ | $w_3$ | $w_4$ | $w_5$ |
|---|---|---|---|---|---|
| $k_1$ | 1 | 1 | 0 | 1 | 0 |
| $k_2$ | 0 | 1 | 1 | 0 | 1 |
| $k_3$ | 0 | 1 | 0 | 0 | 0 |
| $k_4$ | 1 | 1 | 1 | 0 | 1 |

$P_s : Y \to \mathcal{N}(M)$ *is a set valued function, and we have* $P_s(w_1) = \{k_1, k_4\}, P_s(w_2) = \{k_1, k_2, k_3, k_4\}, P_s(w_3) = \{k_2, k_4\}, P_s(w_4) = \{k_1\}$ *and* $P_s(w_5) = \{k_2, k_4\}$. *Mr. X gives most the favorable parameter object C, which is an NS defined as follows:*

$$C = \{(k_1, 0.2, 0.5, 0.6), (k_2, 0.4, 0.3, 0.2), (k_3, 0.2, 0.4, 0.5), (k_4, 0.6, 0.2, 0.1)\}.$$

*From the Definition 1, we have*

$$T_{\overline{P}(C)}(w_1) = \bigvee_{k \in P_s(w_1)} T_C(k) = \bigvee\{0.2, 0.6\} = 0.6,$$

$$I_{\overline{P}(C)}(w_1) = \bigwedge_{k \in P_s(w_1)} I_C(k) = \bigwedge\{0.5, 0.2\} = 0.2,$$

$$F_{\overline{P}(C)}(w_1) = \bigwedge_{k \in P_s(w_1)} F_C(k) = \bigwedge\{0.6, 0.1\} = 0.1,$$

$$T_{\overline{P}(C)}(w_2) = 0.6, \quad I_{\overline{P}(C)}(w_2) = 0.2, \quad F_{\overline{P}(C)}(w_2) = 0.1,$$

$$T_{\overline{P}(C)}(w_3) = 0.6, \quad I_{\overline{P}(C)}(w_3) = 0.2, \quad F_{\overline{P}(C)}(w_3) = 0.1,$$

$$T_{\overline{P}(C)}(w_4) = 0.2, \quad I_{\overline{P}(C)}(w_4) = 0.5, \quad F_{\overline{P}(C)}(w_4) = 0.6,$$
$$T_{\overline{P}(C)}(w_5) = 0.6, \quad I_{\overline{P}(C)}(w_5) = 0.2, \quad F_{\overline{P}(C)}(w_5) = 0.1.$$

*Similarly,*

$$\begin{aligned}
T_{\underline{P}(C)}(w_1) &= \bigwedge_{k \in P_s(w_1)} T_C(k) = \bigwedge\{0.2, 0.6\} = 0.2, \\
I_{\underline{P}(C)}(w_1) &= \bigvee_{k \in P_s(w_1)} I_C(k) = \bigvee\{0.5, 0.2\} = 0.5, \\
F_{\underline{P}(C)}(w_1) &= \bigvee_{k \in P_s(w_1)} F_C(k) = \bigvee\{0.6, 0.1\} = 0.6,
\end{aligned}$$

$$T_{\underline{P}(C)}(w_2) = 0.2, \quad I_{\underline{P}(C)}(w_2) = 0.5, \quad F_{\underline{P}(C)}(w_2) = 0.6,$$
$$T_{\underline{P}(C)}(w_3) = 0.4, \quad I_{\underline{P}(C)}(w_3) = 0.3, \quad F_{\underline{P}(C)}(w_3) = 0.2,$$
$$T_{\underline{P}(C)}(w_4) = 0.2, \quad I_{\underline{P}(C)}(w_4) = 0.5, \quad F_{\underline{P}(C)}(w_4) = 0.6,$$
$$T_{\underline{P}(C)}(w_5) = 0.4, \quad I_{\underline{P}(C)}(w_5) = 0.3, \quad F_{\underline{P}(C)}(w_5) = 0.2.$$

*Thus, we obtain*

$$\begin{aligned}
\overline{P}(C) &= \{(w_1, 0.6, 0.2, 0.1), (w_2, 0.6, 0.2, 0.1), (w_3, 0.6, 0.2, 0.1), (w_4, 0.2, 0.5, 0.6), (w_5, 0.6, 0.2, 0.1)\}, \\
\underline{P}(C) &= \{(w_1, 0.2, 0.5, 0.6), (w_2, 0.2, 0.5, 0.6), (w_3, 0.4, 0.3, 0.2), (w_4, 0.2, 0.5, 0.6), (w_5, 0.4, 0.3, 0.2)\}.
\end{aligned}$$

*Hence,* $(\underline{P}(C), \overline{P}(C))$ *is an SRNS of* $C$.

**Theorem 1.** *Let* $(Y, M, P)$ *be an SAS. Then, the LSRNA and the USRNA operators* $\underline{P}(C)$ *and* $\overline{P}(C)$ *satisfy the following properties for all* $C, D \in \mathcal{N}(M)$:

(i) $\overline{P}(C) = \sim \underline{P}(\sim C)$,
(ii) $\underline{P}(C \cap D) = \underline{P}(C) \cap \underline{P}(D)$,
(iii) $C \subseteq D \Rightarrow \underline{P}(C) \subseteq \underline{P}(D)$,
(iv) $\underline{P}(C \cup D) \supseteq \underline{P}(C) \cup \underline{P}(D)$,
(v) $\underline{P}(C) = \sim \overline{P}(\sim C)$,
(vi) $\overline{P}(C \cup D) = \overline{P}(C) \cup \overline{P}(D)$,
(vii) $C \subseteq D \Rightarrow \overline{P}(C) \subseteq \overline{P}(D)$,
(viii) $\overline{P}(C \cap D) \subseteq \overline{P}(C) \cap \overline{P}(D)$,

*where* $\sim C$ *is the complement of* $C$.

**Proof.** (i) By definition of SRNS, we have

$$\begin{aligned}
\sim C &= \{(k, F_C(k), 1 - I_C(k), T_C(k))\}, \\
\underline{P}(\sim C) &= \{(u, T_{\underline{P}(\sim C)}(u), I_{\underline{P}(\sim C)}(u), F_{\underline{P}(\sim C)}(u)) \mid u \in Y\}, \\
\sim \underline{P}(\sim C) &= \{(u, F_{\underline{P}(\sim C)}(u), 1 - I_{\underline{P}(\sim C)}(u), T_{\underline{P}(\sim C)}(u)) \mid u \in Y\},
\end{aligned}$$

where

$$F_{\underline{P}(\sim C)}(u) = \bigvee_{k \in P_s(u)} T_C(k), \; I_{\underline{P}(\sim C)}(u) = \bigvee_{k \in P_s(u)} (1 - I_C(k)), \; T_{\underline{P}(\sim C)}(u) = \bigwedge_{k \in P_s(u)} F_C(k).$$

Hence, $\sim \underline{P}(\sim C) = \overline{P}(C)$.

(ii)

$$\begin{aligned}
\underline{P}(C\cap D) &= \{(u, T_{\underline{P}(C\cap D)}(u), I_{\underline{P}(C\cap D)}(u), F_{\underline{P}(C\cap D)}(u)) \mid u \in Y\} \\
&= \{(u, \bigwedge_{k\in P_s(u)} T_{(C\cap D)}(k), \bigvee_{k\in P_s(u)} I_{(C\cap D)}(k), \bigvee_{k\in P_s(u)} F_{(C\cap D)}(k)) \mid u \in Y\} \\
&= \{(u, \bigwedge_{k\in P_s(u)} (T_C(k)\wedge T_D(k)), \bigvee_{k\in P_s(u)} (I_C(k)\vee I_D(k)), \\
&\quad \bigvee_{k\in P_s(u)} (F_C(k)\vee F_D(k)) \mid u \in Y\} \\
&= \{(u, T_{\underline{P}(C)}(u)\wedge T_{\underline{P}(D)}(u), I_{\underline{P}(C)}(u)\vee I_{\underline{P}(D)}(u), F_{\underline{P}(C)}(u)\vee F_{\underline{P}(D)}(u)) \mid u \in Y\} \\
&= \underline{P}(C)\cap \underline{P}(D).
\end{aligned}$$

(iii) It can be easily proved by Definition 1.

(iv)

$$\begin{aligned}
T_{\underline{P}(C\cup D)}(u) &= \bigwedge_{k\in P_s(u)} T_{C\cup D}(k) \\
&= \bigwedge_{k\in P_s(u)} (T_C(k)\vee T_D(k)) \\
&\geq (\bigwedge_{k\in P_s(u)} T_C(k) \vee \bigwedge_{k\in P_s(u)} T_D(k)) \\
&\geq (T_{\underline{P}(C)}(u)\vee T_{\underline{P}(D)}(u)), \\
T_{\underline{P}(C\cup D)}(u) &\geq T_{\underline{P}(C)}(u)\cup T_{\underline{P}(D)}(u).
\end{aligned}$$

Similarly, we can prove that

$$\begin{aligned}
I_{\underline{P}(C\cup D)}(u) &\leq I_{\underline{P}(C)}(u)\cup I_{\underline{P}(D)}(u), \\
F_{\underline{P}(C\cup D)}(u) &\leq F_{\underline{P}(C)}(u)\cup F_{\underline{P}(D)}(u).
\end{aligned}$$

Thus, $\underline{P}(C\cup D)\supseteq \underline{P}(C)\cup\underline{P}(D)$.

The properties (v)–(viii) of the USRNA $\overline{P}(C)$ can be easily proved similarly. □

**Example 2.** *Considering Example 1, we have*

$$\begin{aligned}
\sim C &= \{(k_1,0.6,0.5,0.2),(k_2,0.2,0.7,0.4),(k_3,0.5,0.6,0.2),(k_4,0.1,0.8,0.6)\}, \\
\overline{P}(\sim C) &= \{(w_1,0.6,0.5,0.2),(w_2,0.6,0.5,0.2),(w_3,0.2,0.7,0.4),(w_4,0.6,0.5,0.2), \\
&\quad (w_5,0.2,0.7,0.4)\}, \\
\sim \overline{P}(\sim C) &= \{(w_1,0.2,0.5,0.6),(w_2,0.2,0.5,0.6),(w_3,0.4,0.3,0.2),(w_4,0.2,0.5,0.6), \\
&\quad (w_5,0.4,0.3,0.2)\}, \\
&= \underline{P}(C). \\
\textit{Let}\quad D &= \{(k_1,0.4,0.2,0.6),(k_2,0.5,0.3,0.2),(k_3,0.5,0.5,0.1),(k_4,0.6,0.4,0.7)\}, \\
\underline{P}(D) &= \{(w_1,0.4,0.4,0.7),(w_2,0.4,0.5,0.6),(w_3,0.5,0.4,0.7),(w_4,0.4,0.2,0.6), \\
&\quad (w_5,0.5,0.4,0.7)\}, \\
C\cap D &= \{(k_1,0.2,0.5,0.6),(k_2,0.4,0.3,0.2),(k_3,0.2,0.5,0.5),(k_4,0.6,0.4,0.7)\},
\end{aligned}$$

$$
\begin{aligned}
\underline{P}(C \cap D) &= \{(w_1, 0.2, 0.5, 0.7), (w_2, 0.2, 0.5, 0.6), (w_3, 0.4, 0.4, 0.7), (w_4, 0.2, 0.5, 0.6), \\
&\quad (w_5, 0.4, 0.4, 0.7)\}, \\
\underline{P}(C) \cap \underline{P}(D) &= \{(w_1, 0.2, 0.5, 0.7), (w_2, 0.2, 0.5, 0.6), (w_3, 0.4, 0.4, 0.7), (w_4, 0.2, 0.5, 0.6), \\
&\quad (w_5, 0.4, 0.4, 0.7)\}, \\
\underline{P}(C \cap D) &= \underline{P}(C) \cap \underline{P}(D), \\
C \cup D &= \{(k_1, 0.4, 0.2, 0.6), (k_2, 0.5, 0.3, 0.2), (k_3, 0.5, 0.4, 0.1), (k_4, 0.6, 0.2, 0.1)\}, \\
\underline{P}(C \cup D) &= \{(w_1, 0.4, 0.2, 0.6), (w_2, 0.4, 0.4, 0.6), (w_3, 0.5, 0.3, 0.2), (w_4, 0.4, 0.2, 0.6), \\
&\quad (w_5, 0.5, 0.3, 0.2)\}, \\
\underline{P}(C) \cup \underline{P}(D) &= \{(w_1, 0.4, 0.4, 0.6), (w_2, 0.4, 0.5, 0.6), (w_3, 0.5, 0.3, 0.2), (w_4, 0.4, 0.2, 0.6), \\
&\quad (w_5, 0.5, 0.3, 0.2)\}.
\end{aligned}
$$

*Clearly, $\underline{P}(C \cup D) \supseteq \underline{P}(C) \cup \underline{P}(D)$. Hence, properties of the LSRNA operator hold, and we can easily verify the properties of the USRNA operator.*

The conventional soft set is a mapping from a parameter to the subset of universe and let $(P, M)$ be a crisp soft set. In [11], Babitha and Sunil introduced the concept of soft set relation. Now, we present the constructive definition of SRNR by using a soft relation $R$ from $M \times M = \acute{M}$ to $\mathcal{P}(Y \times Y = \acute{Y})$, where $Y$ is a universal set and $M$ is a set of parameter.

**Definition 2.** *A SRNR $(\underline{R}(D), \overline{R}(D))$ on $Y$ is a SRNS, $R : \acute{M} \to \mathcal{P}(\acute{Y})$ is a soft relation on $Y$ defined by*

$$R(k_i k_j) = \{u_i u_j \mid \exists u_i \in P(k_i), u_j \in P(k_j)\}, \ u_i u_j \in \acute{Y}.$$

*Let $R_s : \acute{Y} \to \mathcal{P}(\acute{M})$ be a set-valued function by*

$$R_s(u_i u_j) = \{k_i k_j \in \acute{M} \mid (u_i u_j, k_i k_j) \in R\}, \ u_i u_j \in \acute{Y}.$$

*For any $D \in \mathcal{N}(\acute{M})$, the USRNA and the LSRNA operators of $D$ w.r.t $(\acute{Y}, \acute{M}, R)$ defined as follows:*

$$\overline{R}(D) = \{(u_i u_j, T_{\overline{R}(D)}(u_i u_j), I_{\overline{R}(D)}(u_i u_j), F_{\overline{R}(D)}(u_i u_j)) \mid u_i u_j \in \acute{Y}\},$$

$$\underline{R}(D) = \{(u_i u_j, T_{\underline{R}(D)}(u_i u_j), I_{\underline{R}(D)}(u_i u_j), F_{\underline{R}(D)}(u_i u_j)) \mid u_i u_j \in \acute{Y}\},$$

*where*

$$
\begin{aligned}
T_{\overline{R}(D)}(u_i u_j) &= \bigvee_{k_i k_j \in R_s(u_i u_j)} T_D(k_i k_j), \quad I_{\overline{R}(D)}(u_i u_j) = \bigwedge_{k_i k_j \in R_s(u_i u_j)} I_D(k_i k_j), \\
F_{\overline{R}(D)}(u_i u_j) &= \bigwedge_{k_i k_j \in R_s(u_i u_j)} F_D(k_i k_j),
\end{aligned}
$$

$$
\begin{aligned}
T_{\underline{R}(D)}(u_i u_j) &= \bigwedge_{k_i k_j \in R_s(u_i u_j)} T_D(k_i k_j), \quad I_{\underline{R}(D)}(u_i u_j) = \bigvee_{k_i k_j \in R_s(u_i u_j)} I_D(k_i k_j), \\
F_{\underline{R}(D)}(u_i u_j) &= \bigvee_{k_i k_j \in R_s(u_i u_j)} F_D(k_i k_j).
\end{aligned}
$$

*The pair $(\underline{R}(D), \overline{R}(D))$ is called SRNR and $\underline{R}, \overline{R} : \mathcal{N}(\acute{M}) \to \mathcal{P}(\acute{Y})$ are called the LSRNA and the USRNA operators, respectively.*

**Remark 2.** *For an NS D on $\acute{M}$ and an NS C on M,*

$$\begin{aligned} T_D(k_ik_j) &\leq \min_{k_i \in M}\{T_C(k_i)\}, \\ I_D(k_ik_j) &\leq \min_{k_i \in M}\{I_C(k_i)\}, \\ F_D(k_ik_j) &\leq \min_{k_i \in M}\{F_C(k_i)\}. \end{aligned}$$

*According to the definition of SRNR, we get*

$$\begin{aligned} T_{\overline{R}(D)}(u_iu_j) &\leq \min\{T_{\overline{R}(C)}(u_i), T_{\overline{R}(C)}(u_j)\}, \\ I_{\overline{R}(D)}(u_iu_j) &\leq \max\{I_{\overline{R}(C)}(u_i), I_{\overline{R}(C)}(u_j)\}, \\ F_{\overline{R}(D)}(u_iu_j) &\leq \max\{F_{\overline{R}(C)}(u_i), F_{\overline{R}(C)}(u_j)\}. \end{aligned}$$

*Similarly, for the LSRNA operator $\underline{R}(D)$,*

$$\begin{aligned} T_{\underline{R}(D)}(u_iu_j) &\leq \min\{T_{\underline{R}(C)}(u_i), T_{\underline{R}(C)}(u_j)\}, \\ I_{\underline{R}(D)}(u_iu_j) &\leq \max\{I_{\underline{R}(C)}(u_i), I_{\underline{R}(C)}(u_j)\}, \\ F_{\underline{R}(D)}(u_iu_j) &\leq \max\{F_{\underline{R}(C)}(u_i), F_{\underline{R}(C)}(u_j)\}. \end{aligned}$$

**Example 3.** *Let $Y = \{u_1, u_2, u_3\}$ be a universal set and $M = \{k_1, k_2, k_3\}$ be a set of parameters. A soft set $(P, M)$ on Y can be defined in tabular form (see Table 2) as follows:*

**Table 2.** Soft set $(P, M)$.

| **P** | $u_1$ | $u_2$ | $u_3$ |
|---|---|---|---|
| $k_1$ | 1 | 1 | 0 |
| $k_2$ | 0 | 0 | 1 |
| $k_3$ | 1 | 1 | 1 |

*Let $E = \{u_1u_2, u_2u_3, u_2u_2, u_3u_2\} \subseteq \acute{Y}$ and $L = \{k_1k_3, k_2k_1, k_3k_2\} \subseteq \acute{M}$. Then, a soft relation R on E (from L to E) can be defined in tabular form (see Table 3) as follows:*

**Table 3.** Soft relation $R$.

| **R** | $u_1u_2$ | $u_2u_3$ | $u_2u_2$ | $u_3u_2$ |
|---|---|---|---|---|
| $k_1k_3$ | 1 | 1 | 1 | 0 |
| $k_2k_1$ | 0 | 0 | 0 | 1 |
| $k_3k_2$ | 0 | 1 | 0 | 0 |

*Now, we can define set-valued function $R_s$ such that*

$$R_s(u_1u_2) = \{k_1k_3\},\ R_s(u_2u_3) = \{k_1k_3, k_3k_2\},\ R_s(u_2u_2) = \{k_1k_3\},\ R_s(u_3u_2) = \{k_2k_1\}.$$

*Let $C = \{(k_1, 0.2, 0.4, 0.6), (k_2, 0.4, 0.5, 0.2), (k_3, 0.1, 0.2, 0.4)\}$ be an NS on M, then*
$\overline{R}(C) = \{(u_1, 0.2, 0.2, 0.4), (u_2, 0.2, 0.4, 0.4), (u_3, 0.4, 0.2, 0.2)\}$,
$\underline{R}(C) = \{(u_1, 0.1, 0.4, 0.6), (u_2, 0.1, 0.4, 0.6), (u_3, 0.1, 0.5, 0.4)\}$,
*Let $D = \{(k_1k_3, 0.1, 0.2, 0.2), (k_2k_1, 0.1, 0.1, 0.2), (k_3k_2, 0.1, 0.2, 0.1)\}$ be an NS on L, then*
$\overline{R}(D) = \{(u_1u_2, 0.1, 0.2, 0.2), (u_2u_3, 0.1, 0.2, 0.1), (u_2u_2, 0.1, 0.2, 0.2), (u_3u_2, 0.1, 0.1, 0.2)\}$,
$\underline{R}(D) = \{(u_1u_2, 0.1, 0.2, 0.2), (u_2u_3, 0.1, 0.2, 0.1), (u_2u_2, 0.1, 0.2, 0.2), (u_3u_2, 0.1, 0.1, 0.2)\}$.

*Hence, $R(D) = (\underline{R}(D), \overline{R}(D))$ is SRNR.*

## 3. Construction of Neutrosophic Soft Rough Sets

In this section, we will introduce the notions of NSRSs, neutrosophic soft rough relations (NSRRs).

**Definition 3.** *Let $Y$ be an initial universal set and $M$ a universal set of parameters. For an arbitrary neutrosophic soft relation $\tilde{P}$ from $Y$ to $M$, $(Y, M, \tilde{P})$ is called neutrosophic soft approximation space (NSAS). For any NS $C \in \mathcal{N}(M)$, we define the upper neutrosophic soft approximation (UNSA) and the lower neutrosophic soft approximation (LNSA) operators of $C$ with respect to $(Y, M, \tilde{P})$ denoted by $\overline{\tilde{P}}(C)$ and $\underline{\tilde{P}}(C)$, respectively as follows:*

$$\begin{aligned}\overline{\tilde{P}}(C) &= \{(u, T_{\overline{\tilde{P}}(C)}(u), I_{\overline{\tilde{P}}(C)}(u), F_{\overline{\tilde{P}}(C)}(u)) \mid u \in Y\},\\ \underline{\tilde{P}}(C) &= \{(u, T_{\underline{\tilde{P}}(C)}(u), I_{\underline{\tilde{P}}(C)}(u), F_{\underline{\tilde{P}}(C)}(u)) \mid u \in Y\},\end{aligned}$$

*where*

$$\begin{aligned}T_{\overline{\tilde{P}}(C)}(u) &= \bigvee_{k \in M} (T_{\tilde{P}(C)}(u,k) \wedge T_C(k)), \quad I_{\overline{\tilde{P}}(C)}(u) = \bigwedge_{k \in M} (I_{\tilde{P}(C)}(u,k) \vee I_C(k)),\\ F_{\overline{\tilde{P}}(C)}(u) &= \bigwedge_{k \in M} (F_{\tilde{P}(C)}(u,k) \vee F_C(k)),\\ T_{\underline{\tilde{P}}(C)}(u) &= \bigwedge_{k \in M} (F_{\tilde{P}(C)}(u,k) \vee T_C(k)), \quad I_{\underline{\tilde{P}}(C)}(u) = \bigvee_{k \in M} ((1 - I_{\tilde{P}(C)}(u,k)) \wedge I_C(k)),\\ F_{\underline{\tilde{P}}(C)}(u) &= \bigvee_{k \in M} (T_{\tilde{P}(C)}(u,k) \wedge F_C(k)).\end{aligned}$$

*The pair $(\underline{\tilde{P}}(C), \overline{\tilde{P}}(C))$ is called NSRS of $C$ w.r.t $(Y, M, \tilde{P})$, and $\underline{\tilde{P}}$ and $\overline{\tilde{P}}$ are referred to as the LNSRA and the UNSRA operators, respectively.*

**Remark 3.** *A neutrosophic soft relation over $Y \times M$ is actually a neutrosophic soft set on $Y$. The NSRA operators are defined over two distinct universes $Y$ and $M$. As we know, universal set $Y$ and parameter set $M$ are two different universes of discourse but have solid relations. These universes can not be considered as identical universes; therefore, the reflexive, symmetric and transitive properties of neutrosophic soft relations from $Y$ to $M$ do not exist.*

*Let $\tilde{P}$ be a neutrosophic soft relation from $Y$ to $M$, if, for each $u \in Y$, there exists $k \in M$ such that $T_{\tilde{P}}(u,k) = 1, I_{\tilde{P}}(u,k) = 0, F_{\tilde{P}}(u,k) = 0$. Then, $\tilde{P}$ is referred to as a* serial *neutrosophic soft relation from $Y$ to parameter set $M$.*

**Example 4.** *Suppose that $Y = \{w_1, w_2, w_3, w_4\}$ is the set of careers under consideration, and Mr. X wants to select the most suitable career. $M = \{k_1, k_2, k_3\}$ is a set of decision parameters. Mr. X describes the "most suitable career" by defining a neutrosophic soft set $(\tilde{P}, M)$ on $Y$ that is a neutrosophic relation from $Y$ to $M$ as shown in Table 4.*

**Table 4.** Neutrosophic soft relation $\tilde{P}$.

| $\tilde{P}$ | $w_1$ | $w_2$ | $w_3$ | $w_4$ |
|---|---|---|---|---|
| $k_1$ | $(0.3, 0.4, 0.5)$ | $(0.4, 0.2, 0.3)$ | $(0.1, 0.5, 0.4)$ | $(0.2, 0.3, 0.4)$ |
| $k_2$ | $(0.1, 0.5, 0.4)$ | $(0.3, 0.4, 0.6)$ | $(0.4, 0.4, 0.3)$ | $(0.5, 0.3, 0.8)$ |
| $k_3$ | $(0.3, 0.4, 0.4)$ | $(0.4, 0.6, 0.7)$ | $(0.3, 0.5, 0.4)$ | $(0.5, 0.4, 0.6)$ |

*Now, Mr. X gives the most favorable decision object C, which is an NS on M defined as follows:* $C = \{(k_1, 0.5, 0.2, 0.4), (k_2, 0.2, 0.3, 0.1), (k_3, 0.2, 0.4, 0.6)\}$. *By Definition 3, we have*

$$\begin{aligned}
T_{\overline{\tilde{P}}(C)}(w_1) &= \bigvee_{k\in M} (T_{\tilde{P}(C)}(w_1, k) \wedge T_C(k)) = \bigvee\{0.3, 0.1, 0.2\} = 0.3,\\
I_{\overline{\tilde{P}}(C)}(w_1) &= \bigwedge_{k\in M} (I_{\tilde{P}(C)}(w_1, k) \vee I_C(k)) = \bigwedge\{0.4, 0.5, 0.4\} = 0.4,\\
F_{\overline{\tilde{P}}(C)}(w_1) &= \bigwedge_{k\in M} (F_{\tilde{P}(C)}(w_1, k) \vee F_C(k)) = \bigwedge\{0.5, 0.4, 0.6\} = 0.4,
\end{aligned}$$

$$T_{\overline{\tilde{P}}(C)}(w_2) = 0.4,\quad I_{\overline{\tilde{P}}(C)}(w_2) = 0.2,\quad F_{\overline{\tilde{P}}(C)}(w_2) = 0.4,$$
$$T_{\overline{\tilde{P}}(C)}(w_3) = 0.2,\quad I_{\overline{\tilde{P}}(C)}(w_3) = 0.4,\quad F_{\overline{\tilde{P}}(C)}(w_3) = 0.3,$$
$$T_{\overline{\tilde{P}}(C)}(w_4) = 0.2,\quad I_{\overline{\tilde{P}}(C)}(w_4) = 0.3,\quad F_{\overline{\tilde{P}}(C)}(w_4) = 0.4.$$

*Similarly,*

$$\begin{aligned}
T_{\underline{\tilde{P}}(C)}(w_1) &= \bigwedge_{k\in M} (F_{\tilde{P}(C)}(w_1, k) \vee T_C(k)) = \bigwedge\{0.5, 0.4, 0.4\} = 0.4,\\
I_{\underline{\tilde{P}}(C)}(w_1) &= \bigvee_{k\in M} ((1 - I_{\tilde{P}(C)}(w_1, k)) \wedge I_C(k)) = \bigvee\{0.2, 0.3, 0.4\} = 0.4,\\
F_{\underline{\tilde{P}}(C)}(w_1) &= \bigvee_{k\in M} (T_{\tilde{P}(C)}(w_1, k) \wedge F_C(k)) = \bigvee\{0.3, 0.1, 0.3\} = 0.3,
\end{aligned}$$

$$T_{\underline{\tilde{P}}(C)}(w_2) = 0.5,\quad I_{\underline{\tilde{P}}(C)}(w_2) = 0.4,\quad F_{\underline{\tilde{P}}(C)}(w_2) = 0.4,$$
$$T_{\underline{\tilde{P}}(C)}(w_3) = 0.4,\quad I_{\underline{\tilde{P}}(C)}(w_3) = 0.4,\quad F_{\underline{\tilde{P}}(C)}(w_3) = 0.3,$$
$$T_{\underline{\tilde{P}}(C)}(w_4) = 0.5,\quad I_{\underline{\tilde{P}}(C)}(w_4) = 0.4,\quad F_{\underline{\tilde{P}}(C)}(w_4) = 0.5.$$

*Thus, we obtain*

$$\begin{aligned}
\overline{\tilde{P}}(C) &= \{(w_1, 0.3, 0.4, 0.4), (w_2, 0.4, 0.2, 0.4), (w_3, 0.2, 0.4, 0.3), (w_4, 0.2, 0.3, 0.4)\},\\
\underline{\tilde{P}}(C) &= \{(w_1, 0.4, 0.4, 0.3), (w_2, 0.5, 0.4, 0.4), (w_3, 0.4, 0.4, 0.3), (w_4, 0.5, 0.4, 0.5)\}.
\end{aligned}$$

*Hence,* $(\underline{\tilde{P}}(C), \overline{\tilde{P}}(C))$ *is an NSRS of C.*

**Theorem 2.** *Let* $(Y, M, \tilde{P})$ *be an NSAS. Then, the UNSRA and the LNSRA operators* $\overline{\tilde{P}}(C)$ *and* $\underline{\tilde{P}}(C)$ *satisfy the following properties for all* $C, D \in \mathcal{N}(M)$:

(i) $\underline{\tilde{P}}(C) = \sim \overline{\tilde{P}}(\sim A)$,
(ii) $\underline{\tilde{P}}(C \cap D) = \underline{\tilde{P}}(C) \cap \underline{\tilde{P}}(D)$,
(iii) $C \subseteq D \Rightarrow \underline{\tilde{P}}(C) \subseteq \underline{\tilde{P}}(D)$,
(iv) $\underline{\tilde{P}}(C \cup D) \supseteq \underline{\tilde{P}}(C) \cup \underline{\tilde{P}}(D)$,
(v) $\overline{\tilde{P}}(C) = \sim \underline{\tilde{P}}(\sim C)$,
(vi) $\overline{\tilde{P}}(C \cup D) = \overline{\tilde{P}}(C) \cup \overline{\tilde{P}}(D)$,
(vii) $C \subseteq D \Rightarrow \overline{\tilde{P}}(C) \subseteq \overline{\tilde{P}}(D)$,
(viii) $\overline{\tilde{P}}(C \cap D) \subseteq \overline{\tilde{P}}(C) \cap \overline{\tilde{P}}(D)$.

**Proof.** (i)

$$\sim C = \{(k, F_C(k), 1 - I_C(k), T_C(k)) \mid k \in M\}.$$

By definition of NSRS, we have

$$
\begin{aligned}
\tilde{P}(\sim C) &= \{(u, T_{\overline{\tilde{P}}(\sim C)}(u), I_{\overline{\tilde{P}}(\sim C)}(u), F_{\overline{\tilde{P}}(\sim C)}(u)) \mid u \in Y\},\\
\sim \tilde{P}(\sim C) &= \{(u, F_{\overline{\tilde{P}}(\sim C)}(u), 1 - I_{\overline{\tilde{P}}(\sim C)}(u), T_{\overline{\tilde{P}}(\sim C)}(u)) \mid u \in Y\},\\
F_{\overline{\tilde{P}}(\sim C)}(u) &= \bigwedge_{k \in M} (F_{\tilde{P}}(u,k) \vee T_C(k))\\
&= T_{\underline{\tilde{P}}(C)}(u),\\
1 - I_{\overline{\tilde{P}}(\sim C)}(u) &= 1 - \Big( \bigwedge_{k \in M} [I_{\tilde{P}}(u,k) \vee I_{\sim C}(k)] \Big)\\
&= \bigvee_{k \in M} \Big( (1 - I_{\tilde{P}}(u,k)) \wedge (1 - I_{\sim C}(k)) \Big)\\
&= \bigvee_{k \in M} \Big( (1 - I_{\tilde{P}}(u,k)) \wedge (1 - (1 - I_C(k))) \Big)\\
&= \bigvee_{k \in M} \Big( (1 - I_{\tilde{P}}(u,k)) \wedge I_C(k) \Big)\\
&= I_{\underline{\tilde{P}}(C)}(u),\\
T_{\overline{\tilde{P}}(\sim C)}(u) &= \bigvee_{k \in M} (T_{\tilde{P}}(u,k) \wedge T_{\sim C}(k))\\
&= \bigvee_{k \in M} (T_{\tilde{P}}(u,k) \wedge F_C(k))\\
&= F_{\underline{\tilde{P}}(C)}(u).\\
\text{Thus, } \underline{\tilde{P}}(C) &= \sim \overline{\tilde{P}}(\sim C).
\end{aligned}
$$

(ii)

$$
\begin{aligned}
\underline{\tilde{P}}(C \cap D) &= \{(u, T_{\underline{\tilde{P}}(C \cap D)}(u), I_{\underline{\tilde{P}}(C \cap D)}(u), F_{\underline{\tilde{P}}(C \cap D)}(u))\},\\
\underline{\tilde{P}}(C) \cap \underline{\tilde{P}}(D) &= \{(u, T_{\underline{\tilde{P}}(C)}(u) \wedge T_{\underline{\tilde{P}}(D)}(u), I_{\underline{\tilde{P}}(C)}(u) \vee I_{\underline{\tilde{P}}(D)}(u), F_{\underline{\tilde{P}}(C)}(u) \vee F_{\underline{\tilde{P}}(D)}(u))\}.
\end{aligned}
$$

Now, consider

$$
\begin{aligned}
T_{\underline{\tilde{P}}(C \cap D)}(u) &= \bigwedge_{k \in M} (F_{\underline{\tilde{P}}}(u,k) \vee T_{C \cap D}(k))\\
&= \bigwedge_{k \in M} (F_{\underline{\tilde{P}}}(u,k) \vee (T_C(k) \wedge T_D(k)))\\
&= \bigwedge_{k \in M} (F_{\underline{\tilde{P}}}(u,k) \vee T_C(k)) \wedge \bigwedge_{k \in M} (F_{\underline{\tilde{P}}}(u,k) \vee T_D(k))\\
&= T_{\underline{\tilde{P}}(C)}(u) \wedge T_{\underline{\tilde{P}}(D)}(u),
\end{aligned}
$$

$$\begin{aligned}
I_{\underline{\tilde{P}}(C\cap D)}(u) &= \bigvee_{k\in M}((1-I_{\underline{\tilde{P}}}(u,k))\wedge I_{C\cap D}(k))\\
&= \bigvee_{k\in M}((1-I_{\underline{\tilde{P}}}(u,k))\wedge (I_C(k)\vee I_D(k)))\\
&= \bigvee_{k\in M}((1-I_{\underline{\tilde{P}}}(u,k)))\wedge I_C(k))\vee \bigvee_{k\in M}((1-I_{\underline{\tilde{P}}}(u,k))\vee I_D(k))\\
&= I_{\underline{\tilde{P}}(C)}(u)\vee I_{\underline{\tilde{P}}(D)}(u),\\
F_{\underline{\tilde{P}}(C\cap D)}(u) &= \bigvee_{k\in M}(T_{\underline{\tilde{P}}}(u,k)\wedge F_{C\cap D}(k))\\
&= \bigvee_{k\in M}(T_{\underline{\tilde{P}}}(u,k)\wedge (F_C(k)\vee F_D(k)))\\
&= \bigvee_{k\in M}(T_{\underline{\tilde{P}}}(u,k)\wedge F_C(k))\vee \bigvee_{k\in M}(T_{\underline{\tilde{P}}}(u,k)\wedge F_D(k))\\
&= F_{\underline{\tilde{P}}(C)}(u)\vee F_{\underline{\tilde{P}}(D)}(u).\\
\text{Thus, } \underline{\tilde{P}}(C\cap D) &= \underline{\tilde{P}}(C)\cap \underline{\tilde{P}}(D).
\end{aligned}$$

(iii) It can be easily proven by Definition 3.

(iv)

$$\begin{aligned}
\underline{\tilde{P}}(C\cup D) &= \{(u, T_{\underline{\tilde{P}}(C\cup D)}(u), I_{\underline{\tilde{P}}(C\cup D)}(u), F_{\underline{\tilde{P}}(C\cup D)}(u))\},\\
\underline{\tilde{P}}(C)\cup \underline{\tilde{P}}(D) &= \{(u, T_{\underline{\tilde{P}}(C)}(u)\vee T_{\underline{\tilde{P}}(D)}(u), I_{\underline{\tilde{P}}(C)}(u)\wedge I_{\underline{\tilde{P}}(D)}(u), F_{\underline{\tilde{P}}(C)}(u)\wedge F_{\underline{\tilde{P}}(D)}(u))\},\\
T_{\underline{\tilde{P}}(C\cup D)}(u) &= \bigwedge_{k\in M}(F_{\tilde{P}}(u,k)\vee T_{C\cup D}(k))\\
&= \bigwedge_{k\in M}\Big(F_{\tilde{P}}(u,k)\vee [T_C(k)\vee T_D(k)]\Big)\\
&= \bigwedge_{k\in M}\Big([F_{\tilde{P}}(u,k)\vee T_C(k)]\vee [F_{\tilde{P}}(u,k)\vee T_D(k)]\Big)\\
&\geq \bigwedge_{k\in M}\Big(F_{\tilde{P}}(u,k)\vee T_C(k)\Big)\vee \bigwedge_{k\in M}\Big(F_{\tilde{P}}(u,k)\vee T_D(k)\Big)\\
&= T_{\underline{\tilde{P}}(C)}(u)\vee T_{\underline{\tilde{P}}(D)}(u),\\
I_{\underline{\tilde{P}}(C\cup D)}(u) &= \bigvee_{k\in M}\Big((1-I_{\underline{\tilde{P}}}(u,k))\wedge I_{C\cup D}(k)\Big)\\
&= \bigvee_{k\in M}\Big((1-I_{\underline{\tilde{P}}}(u,k))\wedge [I_C(k)\wedge I_D(k)]\Big)\\
&= \bigvee_{k\in M}\Big([1-I_{\underline{\tilde{P}}}(u,k))\wedge I_C(k)]\wedge [(1-I_{\underline{\tilde{P}}}(u,k))\wedge I_D(k)]\Big)\\
&\leq \bigvee_{k\in M}((1-I_{\tilde{P}}(u,k))\wedge I_C(k))\wedge \bigvee_{k\in M}((1-I_{\underline{\tilde{P}}}(u,k))\wedge I_D(k))\\
&= I_{\underline{\tilde{P}}(C)}(u)\wedge I_{\underline{\tilde{P}}(D)}(u),\\
F_{\underline{\tilde{P}}(C\cup D)}(u) &= \bigvee_{k\in M}(T_{\underline{\tilde{P}}}(u,k)\wedge F_{C\cup D}(k))\\
&= \bigvee_{k\in M}(T_{\underline{\tilde{P}}}(u,k)\wedge [F_C(k)\wedge F_D(k)])\\
&= \bigvee_{k\in M}\Big([T_{\underline{\tilde{P}}}(u,k)\wedge F_C(k)]\wedge [T_{\underline{\tilde{P}}}(u,k)\wedge F_D(k)]\Big)\\
&\leq \bigvee_{k\in M}(T_{\underline{\tilde{P}}}(u,k)\wedge F_C(k))\wedge \bigvee_{k\in M}(T_{\underline{\tilde{P}}}(u,k)\wedge F_D(k))\\
&= F_{\underline{\tilde{P}}(C)}(u)\wedge F_{\underline{\tilde{P}}(D)}(u).
\end{aligned}$$

(vii)

$$\begin{aligned}
\overline{\tilde{P}}(C\cap D) &= \{(u, T_{\overline{\tilde{P}}(C\cap D)}(u), I_{\overline{\tilde{P}}(C\cap D)}(u), F_{\overline{\tilde{P}}(C\cap D)}(u))\},\\
\overline{\tilde{P}}(C)\cap\overline{\tilde{P}}(D) &= \{(u, T_{\overline{\tilde{P}}(C)}(u)\wedge T_{\overline{\tilde{P}}(D)}(u), I_{\overline{\tilde{P}}(C)}(u)\vee I_{\overline{\tilde{P}}(D)}(u), F_{\overline{\tilde{P}}(C)}(u)\vee F_{\overline{\tilde{P}}(D)}(u))\},\\
T_{\overline{\tilde{P}}(C\cap D)}(u) &= \bigvee_{k\in M}(T_{\tilde{P}}(u,k)\wedge T_{C\cap D}(k))\\
&= \bigvee_{k\in M}\Big(T_{\tilde{P}}(u,k)\wedge [T_C(k)\wedge T_D(k)]\Big)\\
&= \bigvee_{k\in M}\Big([T_{\tilde{P}}(u,k)\wedge T_C(k)]\wedge [T_{\tilde{P}}(u,k)\wedge T_D(k)]\Big)\\
&\le \bigvee_{k\in M}\Big(T_{\tilde{P}}(u,k)\wedge T_C(k)\Big)\wedge\bigvee_{k\in M}\Big(T_{\tilde{P}}(u,k)\wedge T_D(k)\Big)\\
&= T_{\overline{\tilde{P}}(C)}(u)\wedge T_{\overline{\tilde{P}}(D)}(u),\\
I_{\overline{\tilde{P}}(C\cap D)}(u) &= \bigwedge_{k\in M}\Big(I_{\overline{\tilde{P}}}(u,k)\vee I_{C\cap D}(k)\Big)\\
&= \bigwedge_{k\in M}\Big(I_{\overline{\tilde{P}}}(u,k)\vee [I_C(k)\vee I_D(k)]\Big)\\
&= \bigwedge_{k\in M}\Big([I_{\overline{\tilde{P}}}(u,k)\vee I_C(k)]\vee [I_{\overline{\tilde{P}}}(u,k)\vee I_D(k)]\Big)\\
&\ge \bigwedge_{k\in M}((I_{\overline{\tilde{P}}}(u,k))\vee I_C(k))\vee\bigwedge_{k\in M}((I_{\overline{\tilde{P}}}(u,k))\vee I_D(k))\\
&= I_{\overline{\tilde{P}}(C)}(u)\vee I_{\overline{\tilde{P}}(D)}(u),\\
F_{\overline{\tilde{P}}(C\cap D)}(u) &= \bigwedge_{k\in M}(F_{\overline{\tilde{P}}}(u,k)\vee F_{C\cap D}(k))\\
&= \bigwedge_{k\in M}(F_{\overline{\tilde{P}}}(u,k)\vee [F_C(k)\vee F_D(k)])\\
&= \bigwedge_{k\in M}\Big([F_{\overline{\tilde{P}}}(u,k)\vee F_C(k)]\vee [F_{\overline{\tilde{P}}}(u,k)\vee F_D(k)]\Big)\\
&\ge \bigwedge_{k\in M}(F_{\overline{\tilde{P}}}(u,k)\vee F_C(k))\vee\bigwedge_{k\in M}(F_{\overline{\tilde{P}}}(u,k)\vee F_D(k))\\
&= F_{\overline{\tilde{P}}(C)}(u)\vee F_{\overline{\tilde{P}}(D)}(u).\\
\text{Thus, }\overline{\tilde{P}}(C\cap D) &\subseteq \overline{\tilde{P}}(C)\cap\overline{\tilde{P}}(D).
\end{aligned}$$

The properties (v)–(vii) of the UNSRA operator $\overline{\tilde{P}}(C)$ can be easily proved similarly. □

**Theorem 3.** *Let $(Y, M, \tilde{P})$ be an NSAS. The UNSRA and the LNSRA operators $\overline{\tilde{P}}$ and $\underline{\tilde{P}}$ satisfy the following properties for all $C, D \in \mathcal{N}(M)$:*

(i) $\underline{\tilde{P}}(C-D)\supseteq\underline{\tilde{P}}(C)-\overline{\tilde{P}}(D)$,
(ii) $\overline{\tilde{P}}(C-D)\subseteq\overline{\tilde{P}}(C)-\underline{\tilde{P}}(D)$.

**Proof.** (i) By Definition 3 and definition of difference of two NSs, for all $u \in Y$,

$$\begin{aligned}
T_{\underline{\tilde{P}}(C-D)}(u) &= \bigwedge_{k\in M} (F_{\tilde{P}}(u,k) \vee T_{C-D}(k)) \\
&= \bigwedge_{k\in M} (F_{\tilde{P}}(u,k) \vee (T_C(k) \wedge F_D(k))) \\
&= \bigwedge_{k\in M} \left([F_{\tilde{P}}(u,k) \vee T_C(k)] \wedge [F_{\tilde{P}}(u,k) \vee F_D(k)]\right) \\
&= \bigwedge_{k\in M} (F_{\tilde{P}}(u,k) \vee T_C(k)) \wedge \bigwedge_{k\in M} (F_{\tilde{P}}(u,k) \vee F_D(k)) \\
&= T_{\underline{\tilde{P}}(C)}(u) \wedge F_{\overline{\tilde{P}}(D)}(u) \\
&= T_{\underline{\tilde{P}}(C)-\overline{\tilde{P}}(D)}(u), \\
I_{\underline{\tilde{P}}(C-D)}(u) &= \bigvee_{k\in M} ((1 - I_{\tilde{P}}(u,k)) \wedge I_{C-D}(k)) \\
&= \bigvee_{k\in M} ((1 - I_{\tilde{P}}(u,k)) \wedge (I_C(k) \wedge (1 - I_D(k)))) \\
&= \bigvee_{k\in M} \left([(1 - I_{\tilde{P}}(u,k)) \wedge I_C(k)] \wedge [(1 - I_{\tilde{P}}(u,k)) \wedge (1 - I_D(k))]\right) \\
&= \bigvee_{k\in M} \left([(1 - I_{\tilde{P}}(u,k)) \wedge I_C(k)] \wedge [1 - (I_{\tilde{P}}(u,k) \vee I_D(k))]\right) \\
&\leq \bigvee_{k\in M} ((1 - I_{\tilde{P}}(u,k)) \wedge I_C(k)) \wedge \bigvee_{k\in M} (1 - (I_{\tilde{P}}(u,k) \vee I_D(k))) \\
&\leq \bigvee_{k\in M} ((1 - I_{\tilde{P}}(u,k)) \wedge I_C(k)) \wedge \left(1 - \bigwedge_{k\in M} (I_{\tilde{P}}(u,k) \vee I_D(k))\right) \\
&= I_{\underline{\tilde{P}}(C)}(u) \wedge (1 - I_{\overline{\tilde{P}}(D)}(u)) \\
&= I_{\underline{\tilde{P}}(C)-\overline{\tilde{P}}(D)}(u), \\
F_{\underline{\tilde{P}}(C-D)}(u) &= \bigvee_{k\in M} (T_{\tilde{P}}(u,k) \wedge F_{C-D}(k)) \\
&= \bigvee_{k\in M} (T_{\tilde{P}}(u,k) \wedge (F_C(k) \wedge T_D(k))) \\
&= \bigvee_{k\in M} \left([T_{\tilde{P}}(u,k) \wedge F_C(k)] \wedge [T_{\tilde{P}}(u,k) \wedge T_D(k)]\right) \\
&\leq \bigvee_{k\in M} (T_{\tilde{P}}(u,k) \wedge F_C(k)) \wedge \bigvee_{k\in M} (T_{\tilde{P}}(u,k) \wedge T_D(k)) \\
&= F_{\underline{\tilde{P}}(C)}(u) \wedge T_{\overline{\tilde{P}}(D)}(u) \\
&= F_{\underline{\tilde{P}}(C)-\overline{\tilde{P}}(D)}(u). \\
\text{Thus, } \underline{\tilde{P}}(C-D) &\subseteq \underline{\tilde{P}}(C) - \overline{\tilde{P}}(D).
\end{aligned}$$

(ii) By Definition 3 and definition of difference of two NSs, for all $u \in Y$,

$$
\begin{aligned}
T_{\overline{\tilde{P}}(C-D)}(u) &= \bigvee_{k\in M} (T_{\tilde{P}}(u,k) \wedge T_{C-D}(k)) \\
&= \bigvee_{k\in M} (T_{\tilde{P}}(u,k) \wedge (T_C(k) \wedge F_D(k))) \\
&= \bigvee_{k\in M} \Big([T_{\tilde{P}}(u,k) \wedge T_C(k)] \wedge [T_{\tilde{P}}(u,k) \wedge F_D(k)]\Big) \\
&\leq \bigvee_{k\in M} (T_{\tilde{P}}(u,k) \wedge T_C(k)) \wedge \bigvee_{k\in M} (T_{\tilde{P}}(u,k) \wedge F_D(k)) \\
&= T_{\overline{\tilde{P}}(C)}(u) \wedge F_{\underline{\tilde{P}}(D)}(u) \\
&= T_{\overline{\tilde{P}}(C)-\underline{\tilde{P}}(D)}(u), \\
I_{\overline{\tilde{P}}(C-D)}(u) &= \bigwedge_{k\in M} (I_{\tilde{P}}(u,k) \vee I_{C-D}(k)) \\
&= \bigwedge_{k\in M} (I_{\tilde{P}}(u,k) \vee (I_C(k) \wedge (1 - I_D(k)))) \\
&= \bigwedge_{k\in M} \Big([I_{\tilde{P}}(u,k) \vee I_C(k)] \wedge [I_{\tilde{P}}(u,k) \vee (1 - I_D(k))]\Big) \\
&= \bigwedge_{k\in M} \Big([I_{\tilde{P}}(u,k) \vee I_C(k)] \wedge [1 - (1 - I_{\tilde{P}}(u,k)) \vee (1 - I_D(k))]\Big) \\
&= \bigwedge_{k\in M} (I_{\tilde{P}}(u,k) \vee I_C(k)) \wedge \Big(1 - \bigvee_{k\in M} ((1 - I_{\tilde{P}}(u,k)) \wedge I_D(k))\Big) \\
&= I_{\overline{\tilde{P}}(C)}(u) \wedge (1 - I_{\underline{\tilde{P}}(D)}(u)) \\
&= I_{\overline{\tilde{P}}(C)-\underline{\tilde{P}}(D)}(u), \\
F_{\overline{\tilde{P}}(C-D)}(u) &= \bigwedge_{k\in M} (F_{\tilde{P}}(u,k) \vee F_{C-D}(k)) \\
&= \bigwedge_{k\in M} (F_{\tilde{P}}(u,k) \vee (F_C(k) \wedge T_D(k))) \\
&= \bigwedge_{k\in M} \Big([F_{\tilde{P}}(u,k) \vee F_C(k)] \wedge [F_{\tilde{P}}(u,k) \vee T_D(k)]\Big) \\
&= \bigwedge_{k\in M} (F_{\tilde{P}}(u,k) \vee F_C(k)) \wedge \bigwedge_{k\in M} (F_{\tilde{P}}(u,k) \vee T_D(k)) \\
&= F_{\overline{\tilde{P}}(C)}(u) \wedge T_{\underline{\tilde{P}}(D)}(u) \\
&= F_{\overline{\tilde{P}}(C)-\underline{\tilde{P}}(D)}(u). \\
\text{Thus, } \overline{\tilde{P}}(C-D) &\subseteq \overline{\tilde{P}}(C) - \underline{\tilde{P}}(D).
\end{aligned}
$$

□

**Theorem 4.** *Let $(Y, M, \tilde{P})$ be an NSAS. If $\tilde{P}$ is serial, then the UNSA and the LNSA operators $\overline{\tilde{P}}$ and $\underline{\tilde{P}}$ satisfy the following properties for all $\emptyset, \mathbb{M}, C \in \mathcal{N}(M)$:*

(i) $\overline{\tilde{P}}(\emptyset) = \emptyset$, $\underline{\tilde{P}}(\mathbb{M}) = \mathbb{Y}$,

(ii) $\underline{\tilde{P}}(C) \subseteq \overline{\tilde{P}}(C)$.

**Proof.** (i)

$$\begin{aligned}
\overline{\tilde{P}}(\emptyset) &= \{(u, T_{\overline{\tilde{P}}(\emptyset)}(u), I_{\overline{\tilde{P}}(\emptyset)}(u), F_{\overline{\tilde{P}}(\emptyset)}(u)) \mid u \in Y\}, \\
T_{\overline{\tilde{P}}(\emptyset)}(u) &= \bigvee_{k\in M} (T_{\tilde{P}}(u,k) \wedge T_{\emptyset}(k)), \\
I_{\overline{\tilde{P}}(\emptyset)}(u) &= \bigwedge_{k\in M} (I_{\tilde{P}}(u,k) \vee I_{\emptyset}(k)), \\
F_{\overline{\tilde{P}}(\emptyset)}(u) &= \bigwedge_{k\in M} (F_{\tilde{P}}(u,k) \vee F_{\emptyset}(k)).
\end{aligned}$$

Since $\emptyset$ is a null NS on M, $T_{\emptyset}(k) = 0,\ I_{\emptyset}(k) = 1, F_{\emptyset}(k) = 1$, and this implies $T_{\overline{\tilde{P}}(\emptyset)}(u) = 0,\ I_{\overline{\tilde{P}}}(u) = 1,\ F_{\overline{\tilde{P}}}(u) = 1$. Thus, $\overline{\tilde{P}}(\emptyset) = \emptyset$.

Now,

$$\begin{aligned}
\underline{\tilde{P}}(\mathbb{M}) &= \{(u, T_{\underline{\tilde{P}}(\mathbb{M})}(u), I_{\underline{\tilde{P}}(\mathbb{M})}(u), F_{\underline{\tilde{P}}(\mathbb{M})}(u)) \mid u \in Y\}, \\
T_{\underline{\tilde{P}}(\mathbb{M})}(u) &= \bigwedge_{k\in M} (F_{\tilde{P}}(u,k) \vee T_{\mathbb{M}}(k)), \quad I_{\underline{\tilde{P}}(\mathbb{M})}(u) = \bigvee_{k\in M} ((1 - I_{\tilde{P}}(u,k)) \wedge I_{\mathbb{M}}(k)), \\
F_{\underline{\tilde{P}}(\mathbb{M})}(u) &= \bigvee_{k\in M} (T_{\tilde{P}}(u,k) \wedge F_{\mathbb{M}}(k)).
\end{aligned}$$

Since $\mathbb{M}$ is full NS on $M$, $T_{\mathbb{M}}(k) = 1, I_{\mathbb{M}}(k) = 0, F_{\mathbb{M}}(k) = 0$, for all $k \in M$, and this implies $T_{\underline{\tilde{P}}(\mathbb{M})}(u) = 1, I_{\underline{\tilde{P}}(\mathbb{M})}(u) = 0, F_{\underline{\tilde{P}}(\mathbb{M})}(u) = 0$. Thus, $\underline{\tilde{P}}(\mathbb{M}) = \mathbb{Y}$.

(ii) Since $(Y, M, \tilde{P})$ is an NSAS and $\tilde{P}$ is a serial neutrosophic soft relation, then, for each $u \in Y$, there exists $k \in M$, such that $T_{\tilde{P}}(u,k) = 1, I_{\tilde{P}}(u,k) = 0$, and $F_{\tilde{P}}(u,k) = 0$. The UNSRA and LNSRA operators $\overline{\tilde{P}}(C)$, and $\underline{\tilde{P}}(C)$ of an NS C can be defined as:

$$\begin{aligned}
T_{\overline{\tilde{P}}(C)}(u) &= \bigvee_{k\in M} T_C(k), \quad I_{\overline{\tilde{P}}(C)}(u) = \bigwedge_{k\in M} I_C(k), \\
F_{\overline{\tilde{P}}(C)}(u) &= \bigwedge_{k\in M} F_C(k),
\end{aligned}$$

$$\begin{aligned}
T_{\underline{\tilde{P}}(C)}(u) &= \bigwedge_{k\in M} T_C(k), \quad I_{\underline{\tilde{P}}(C)}(u) = \bigvee_{k\in M} I_C(k), \\
F_{\underline{\tilde{P}}(C)}(u) &= \bigvee_{k\in M} F_C(k).
\end{aligned}$$

Clearly, $T_{\underline{\tilde{P}}(C)}(u) \leq T_{\overline{\tilde{P}}(C)}(u)$, $I_{\underline{\tilde{P}}(C)}(u) \geq T_{\overline{\tilde{P}}(C)}(u)$, $F_{\underline{\tilde{P}}(C)}(u) \geq F_{\overline{\tilde{P}}(C)}(u)$ for all $u \in Y$. Thus, $\underline{\tilde{P}}(C) \subseteq \overline{\tilde{P}}(C)$.
□

The conventional NSS is a mapping from a parameter to the neutrosophic subset of universe and let $(\tilde{P}, M)$ be NSS. Now, we present the constructive definition of neutrosophic soft rough relation by using a neutrosophic soft relation $\tilde{R}$ from $M \times M = \acute{M}$ to $\mathcal{N}(Y \times Y = \acute{Y})$, where $Y$ is a universal set and $M$ is a set of parameters.

**Definition 4.** *A neutrosophic soft rough relation* $(\underline{\tilde{R}}(D), \overline{\tilde{R}}(D))$ *on* $Y$ *is an NSRS,* $\tilde{R} : \acute{M} \to \mathcal{N}(\acute{Y})$ *is a neutrosophic soft relation on* $Y$ *defined by*

$$\tilde{R}(k_i k_j) = \{u_i u_j \mid \exists u_i \in \tilde{P}(k_i), u_j \in \tilde{P}(k_j)\},\ u_i u_j \in \acute{Y},$$

*such that*

$$\begin{aligned} T_{\tilde{R}}(u_iu_j, k_ik_j) \leq & \min\{T_{\tilde{P}}(u_i, k_i), T_{\tilde{P}}(u_j, k_j)\}, \\ I_{\tilde{R}}(u_iu_j, k_ik_j) \leq & \max\{I_{\tilde{P}}(u_i, k_i), I_{\tilde{P}}(u_j, k_j)\}, \\ F_{\tilde{R}}(u_iu_j, k_ik_j) \leq & \max\{F_{\tilde{P}}(u_i, k_i), F_{\tilde{P}}(u_j, k_j)\}. \end{aligned}$$

*For any $D \in \mathcal{N}(\acute{M})$, the UNSA and the LNSA of B* w.r.t $(\acute{Y}, \acute{M}, \tilde{R})$ *are defined as follows:*

$$\overline{\tilde{R}}(D) = \{(u_iu_j, T_{\overline{\tilde{R}}(D)}(u_iu_j), I_{\overline{\tilde{R}}(D)}(u_iu_j), F_{\overline{\tilde{R}}(D)}(u_iu_j)) \mid u_iu_j \in \acute{Y}\},$$

$$\underline{\tilde{R}}(D) = \{(u_iu_j, T_{\underline{\tilde{R}}(D)}(u_iu_j), I_{\underline{\tilde{R}}(D)}(u_iu_j), F_{\underline{\tilde{R}}(D)}(u_iu_j)) \mid u_iu_j \in \acute{Y}\},$$

*where*

$$\begin{aligned} T_{\overline{\tilde{R}}(D)}(u_iu_j) &= \bigvee_{k_ik_j \in \acute{M}} (T_{\tilde{R}}(u_iu_j, k_ik_j) \wedge T_D(k_ik_j)), \\ I_{\overline{\tilde{R}}(D)}(u_iu_j) &= \bigwedge_{k_ik_j \in \tilde{M}} (I_{\tilde{R}}(u_iu_j, k_ik_j) \vee I_D(k_ik_j)), \\ F_{\overline{\tilde{R}}(D)}(u_iu_j) &= \bigwedge_{k_ik_j \in \tilde{M}} (F_{\tilde{R}}(u_iu_j, k_ik_j) \vee F_D(k_ik_j)), \end{aligned}$$

$$\begin{aligned} T_{\underline{\tilde{R}}(D)}(u_iu_j) &= \bigwedge_{k_ik_j \in \acute{M}} (F_{\tilde{R}}(u_iu_j, k_ik_j) \vee T_D(k_ik_j)), \\ I_{\underline{\tilde{R}}(D)}(u_iu_j) &= \bigvee_{k_ik_j \in \tilde{M}} ((1 - I_{\tilde{R}}(u_iu_j, k_ik_j)) \wedge I_D(k_ik_j)), \\ F_{\underline{\tilde{R}}(D)}(u_iu_j) &= \bigvee_{k_ik_j \in \tilde{M}} (T_{\tilde{R}}(u_iu_j, k_ik_j) \wedge F_D(k_ik_j)). \end{aligned}$$

*The pair $(\underline{\tilde{R}}(D), \overline{\tilde{R}}(D))$ is called NSRR and $\underline{\tilde{R}}, \overline{\tilde{R}} : \mathcal{N}(\acute{M}) \to \mathcal{N}(\acute{Y})$ are called the LNSRA and the UNSRA operators, respectively.*

**Remark 4.** *Consideer an NS D on $\acute{M}$ and an NS C on M,*

$$\begin{aligned} T_D(k_ik_j) &\leq \min\{T_C(k_i), T_C(k_j)\}, \\ I_D(k_ik_j) &\leq \max\{I_C(k_i), I_C(k_j)\}, \\ F_D(k_ik_j) &\leq \max\{F_C(k_i), F_C(k_j)\}. \end{aligned}$$

*According to the definition of NSRR, we get*

$$\begin{aligned} T_{\overline{\tilde{R}}(D)}(u_iu_j) &\leq \min\{T_{\overline{\tilde{R}}(C)}(u_i), T_{\overline{\tilde{R}}(C)}(u_j)\}, \\ I_{\overline{\tilde{R}}(D)}(u_iu_j) &\leq \max\{I_{\overline{\tilde{R}}(C)}(u_i), I_{\overline{\tilde{R}}(C)}(u_j)\}, \\ F_{\overline{\tilde{R}}(D)}(u_iu_j) &\leq \max\{F_{\overline{\tilde{R}}(C)}(u_i). F_{\overline{\tilde{R}}(C)}(u_j)\}. \end{aligned}$$

*Similarly, for LNSRA operator* $\underline{\tilde{R}}(D)$,

$$\begin{aligned} T_{\underline{\tilde{R}}(D)}(u_iu_j) &\leq \min\{T_{\underline{\tilde{R}}(C)}(u_i), T_{\underline{\tilde{R}}(C)}(u_j)\}, \\ I_{\underline{\tilde{R}}(D)}(u_iu_j) &\leq \max\{I_{\underline{\tilde{R}}(C)}(u_i), I_{\underline{\tilde{R}}(C)}(u_j)\}, \\ F_{\underline{\tilde{R}}(D)}(u_iu_j) &\leq \max\{F_{\underline{\tilde{R}}(C)}(u_i).F_{\underline{\tilde{R}}(C)}(u_j)\}. \end{aligned}$$

**Example 5.** *Let* $Y = \{u_1, u_2, u_3\}$ *be a universal set and* $M = \{k_1, k_2, k_3\}$ *a set of parameters. A neutrosophic soft set* $(\tilde{P}, M)$ *on* $Y$ *can be defined in tabular form (see Table 5) as follows:*

**Table 5.** Neutrosophic soft set $(\tilde{P}, M)$.

| $\tilde{P}$ | $u_1$ | $u_2$ | $u_3$ |
|---|---|---|---|
| $k_1$ | (0.4, 0.5, 0.6) | (0.7, 0.3, 0.2) | (0.6, 0.3, 0.4) |
| $k_2$ | (0.5, 0.3, 0.6) | (0.3, 0.4, 0.3) | (0.7, 0.2, 0.3) |
| $k_3$ | (0.7, 0.2, 0.3) | (0.6, 0.5, 0.4) | (0.7, 0.2, 0.4) |

*Let* $E = \{u_1u_2, u_2u_3, u_2u_2, u_3u_2\} \subseteq \acute{Y}$ *and* $L = \{k_1k_3, k_2k_1, k_3k_2\} \subseteq \acute{M}$.

*Then, a soft relation* $\tilde{R}$ *on* $E$ *(from* $L$ *to* $E$*) can be defined in tabular form (see Table 6) as follows:*

**Table 6.** Neutrosophic soft relation $\tilde{R}$.

| $\tilde{R}$ | $u_1u_2$ | $u_2u_3$ | $u_2u_2$ | $u_3u_2$ |
|---|---|---|---|---|
| $k_1k_3$ | (0.4, 0.4, 0.5) | (0.6, 0.3, 0.4) | (0.5, 0.4, 0.2) | (0.5, 0.4, 0.3) |
| $k_2k_1$ | (0.3, 0.3, 0.4) | (0.3, 0.2, 0.3) | (0.2, 0.3, 0.3) | (0.7, 0.2, 0.2) |
| $k_3k_2$ | (0.3, 0.3, 0.2) | (0.5, 0.3, 0.2) | (0.2, 0.4, 0.4) | (0.3, 0.4, 0.4) |

*Let* $C = \{(k_1, 0.2, 0.4, 0.6), (k_2, 0.4, 0.5, 0.2), (k_3, 0.1, 0.2, 0.4)\}$ *be an NS on* $M$, *then*
$\overline{\tilde{R}}(C) = \{(u_1, 0.4, 0.2, 0.4), (u_2, 0.3, 0.4, 0.3), (u_3, 0.4, 0.2, 0.3)\}$,
$\underline{\tilde{R}}(C) = \{(u_1, 0.3, 0.5, 0.4), (u_2, 0.2, 0.5, 0.6), (u_3, 0.4, 0.5, 0.6)\}$,
*Let* $B = \{(k_1k_3, 0.1, 0.3, 0.5), (k_2k_1, 0.2, 0.4, 0.3), (k_3k_2, 0.1, 0.2, 0.3)\}$ *be an NS on* $L$, *then*
$\overline{\tilde{R}}(D) = \{(u_1u_2, 0.2, 0.3, 0.3), (u_2u_3, 0.2, 0.3, 0.3), (u_2u_2, 0.2, 0.4, 0.3), (u_3u_2, 0.2, 0.4, 0.3)\}$,
$\underline{\tilde{R}}(D) = \{(u_1u_2, 0.2, 0.4, 0.4), (u_2u_3, 0.2, 0.4, 0.5), (u_2u_2, 0.3, 0.4, 0.5), (u_3u_2, 0.2, 0.4, 0.5)\}$.
*Hence,* $\tilde{R}(D) = (\overline{\tilde{R}}(D), \underline{\tilde{R}}(D))$ *is NSRR.*

**Theorem 5.** *Let* $\tilde{P}_1, \tilde{P}_2$ *be two NSRRs from universal* $Y$ *to a parameter set* $M$*; for all* $C \in \mathcal{N}(M)$, *we have*

(i) $\underline{\tilde{P}_1 \cup \tilde{P}_2}(C) = \underline{\tilde{P}_1}(C) \cap \underline{\tilde{P}_2}(C)$,
(ii) $\overline{\tilde{P}_1 \cup \tilde{P}_2}(C) = \overline{\tilde{P}_1}(C) \cup \overline{\tilde{P}_2}(C)$.

**Theorem 6.** *Let* $\tilde{P}_1, \tilde{P}_2$ *be two neutrosophic soft relations from universal* $Y$ *to a parameter set* $M$*; for all* $C \in \mathcal{N}(M)$, *we have*

(i) $\underline{\tilde{P}_1 \cap \tilde{P}_2}(C) \supseteq \underline{\tilde{P}_1}(C) \cup \underline{\tilde{P}_2}(C) \supseteq \underline{\tilde{P}_1}(C) \cap \underline{\tilde{P}_2}(C)$,
(ii) $\overline{\tilde{P}_1 \cap \tilde{P}_2}(C) \subseteq \overline{\tilde{P}_1}(C) \cap \overline{\tilde{P}_2}(C)$.

## 4. Application

In this section, we apply the concept of NSRSs to a decision-making problem. In recent times, the object recognition problem has gained considerable importance. The object recognition problem can be considered as a decision-making problem, in which final identification of object is founded on a given amount of information. A detailed description of the algorithm for the selection of the most suitable object based on an available set of alternatives is given, and the proposed decision-making method can be used to calculate lower and upper approximation operators to address deep concerns of the problem. The presented algorithms can be applied to avoid lengthy calculations when dealing with a large number of objects. This method can be applied in various domains for multi-criteria selection of objects. A multicriteria decision making (MCDM) can be modeled using neutrosophic soft rough sets and is ideally suited for solving problems.

In the pharmaceutical industry, different pharmaceutical companies develop, produce and discover pharmaceutical medicines (drugs) for use as medication. These pharmaceutical companies deal with "brand name medicine" and "generic medicine". Brand name medicine and generic medicine are bioequivalent, have a generic medicine rate and element of absorption. Brand name medicine and generic medicine have the same active ingredients, but the inactive ingredients may differ. The most important difference is cost. Generic medicine is less expensive as compared to brand names in comparison. Usually, generic drug manufacturers have competition to produce products that cost less. The product may possibly be slightly dissimilar in color, shape, or markings. The major difference is cost. We consider a brand name drug "$u$ = Claritin (loratadink)" with an ideal neutrosophic value number $n_u = (1, 0, 0)$ used for seasonal allergy medication. Consider

$$\begin{aligned} Y &= \{u_1 = \text{Nasacort Aq (Triamcinolone)}, u_2 = \text{Zyrtec D (Cetirizine/Pseudoephedrine)}, \\ &\quad u_3 = \text{Sudafed (Pseudoephedrine)}, u_4 = \text{Claritin-D (loratadine/pseudoephedrine)}, \\ &\quad u_5 = \text{Flonase (Fluticasone)}\} \end{aligned}$$

is a set of generic versions of "Clarition". We want to select the most suitable generic version of Claritin on the basis of parameters $e_1$ = Highly soluble, $e_2$ = Highly permeable, $e_3$ = Rapidly dissolving. $M = \{e_1, e_2, e_3\}$ be a set of paraments. Let $\tilde{P}$ be a neutrosophic soft relation from $Y$ to parameter set $M$ as shown in Table 7.

**Table 7.** Neutrosophic soft set $(\tilde{P}, M)$.

| $\tilde{P}$ | $e_1$ | $e_2$ | $e_3$ |
|---|---|---|---|
| $u_1$ | $(0.4, 0.5, 0.6)$ | $(0.7, 0.3, 0.2)$ | $(0.6, 0.3, 0.4)$ |
| $u_2$ | $(0.5, 0.3, 0.6)$ | $(0.3, 0.4, 0.3)$ | $(0.7, 0.2, 0.3)$ |
| $u_3$ | $(0.7, 0.2, 0.3)$ | $(0.6, 0.5, 0.4)$ | $(0.7, 0.2, 0.4)$ |
| $u_4$ | $(0.5, 0.7, 0.5)$ | $(0.8, 0.4, 0.6)$ | $(0.8, 0.7, 0.6)$ |
| $u_5$ | $(0.6, 0.5, 0.4)$ | $(0.7, 0.8, 0.5)$ | $(0.7, 0.3, 0.5)$ |

Suppose $C = \{(e_1, 0.2, 0.4, 0.5), (e_2, 0.5, 0.6, 0.4), (e_3, 0.7, 0.5, 0.4)\}$ is the most favorable object that is an NS on the parameter set $M$ under consideration. Then, $(\underline{\tilde{P}}(C), \overline{\tilde{P}}(C))$ is an NSRS in NSAS $(Y, M, \tilde{P})$, where

$$\begin{aligned} \overline{\tilde{P}}(C) &= \{(u_1, 0.6, 0.5, 0.4), (u_2, 0.7, 0.4, 0.4), (u_3, 0.7, 0.4, 0.4), (u_4, 0.7, 0.6, 0.5), (u_5, 0.7, 0.5, 0.5)\}, \\ \underline{\tilde{P}}(C) &= \{(u_1, 0.5, 0.6, 0.4), (u_2, 0.5, 0.6, 0.5), (u_3, 0.3, 0.3, 0.5), (u_4, 0.5, 0.6, 0.5), (u_5, 0.4, 0.5, 0.5)\}. \end{aligned}$$

In [6], the sum of two neutrosophic numbers is defined. The sum of LNSRA and the UNSRA operators $\overline{\tilde{P}}(C)$ and $\underline{\tilde{P}}(C)$ is an NS $\overline{\tilde{P}}(C) \oplus \underline{\tilde{P}}(C)$ defined by

$$\begin{aligned} \overline{\tilde{P}}(C) \oplus \underline{\tilde{P}}(C) &= \{(u_1, 0.8, 0.3, 0.16), (u_2, 0.85, 0.24, 0.2), (u_3, 0.79, 0.2, 0.2), (u_4, 0.85, 0.36, 0.25), \\ &\quad (u_5, 0.82, 0.25, 0.25)\}. \end{aligned}$$

Let $n_{u_i} = (T_{n_{u_i}}, I_{n_{u_i}}, F_{n_{u_i}})$ be a neutrosophic value number of generic versions medicine $u_i$. We can calculate the cosine similarity measure $S(n_{u_i}, n_u)$ between each neutrosophic value number $n_{u_i}$ of generic version $u_i$ and ideal value number $n_u$ of brand name drug $u$, and the grading of all generic version medicines of $Y$ can be determined. The cosine similarity measure is calculated as the inner product of two vectors divided by the product of their lengths. It is the cosine of the angle between the vector representations of two neutrosophic soft rough sets. The cosine similarity measure is a fundamental measure used in information technology. In [3], the cosine similarity is measured between neutrosophic numbers and demonstrates that the cosine similarity measure is a special case of the correlation coefficient in SVNS. Then, a decision-making method is proposed by the use of the cosine similarity measure of SVNSs, in which the evaluation information for alternatives with respect to criteria is carried out by truth-membership degree, indeterminacy-membership degree, and falsity-membership degree under single-valued neutrosophic environment. It defined as follows:

$$S(n_u, n_{u_i}) = \frac{T_{n_u} \cdot T_{n_{u_i}} + I_{n_u} \cdot I_{n_{u_i}} + F_{n_u} \cdot F_{n_{u_i}}}{\sqrt{T_{n_u}^2 + T_{n_u^2} + F_{n_u}^2} + \sqrt{T_{n_{u_i}}^2 + T_{n_{u_i}^2} + F_{n_{u_i}}^2}}. \quad (1)$$

Through the cosine similarity measure between each object and the ideal object, the ranking order of all objects can be determined and the best object can be easily identified as well. The advantage is that the proposed MCDM approach has some simple tools and concepts in the neutrosophic similarity measure approach among the existing ones. An illustrative application shows that the proposed method is simple and effective.

The generic version medicine $u_i$ with the larger similarity measure $S(n_{u_i}, n_u)$ is the most suitable version $u_i$ because it is close to the brand name drug $u$. By comparing the cosine similarity measure values, the grading of all generic medicines can be determined, and we can find the most suitable generic medicine after selection of suitable NS of parameters. By Equation (1), we can calculate the cosine similarity measure between neutrosophic value numbers $n_u$ of $u$ and $n_{u_i}$ of $u_i$ as follows:

$$\begin{aligned} S(n_u, n_{u_1}) &= 0.9203,\ S(n_u, n_{u_2}) = 0.9386,\ S(n_u, n_{u_3}) = 0.9415, \\ S(n_u, n_{u_4}) &= 0.8888\ S(n_u, n_{u_5}) = 0.9183. \end{aligned}$$

We get $S(n_u, n_{u_3}) > S(n_u, n_{u_2}) > S(n_u, n_{u_1}) > S(n_u, n_{u_5}) > S(n_u, n_{u_4})$. Thus, the optimal decision is $u_3$, and the most suitable generic version of Claritin is Sudafed (Pseudoephedrine). We have used software MATLAB (version 7, MathWorks, Natick, MA, USA) for calculations in the application. The flow chart of the algorithm is general for any number of objects with respect to certain parameters. The flow chart of our proposed method is given in Figure 1. The method is presented as an algorithm in Algorithm 1.

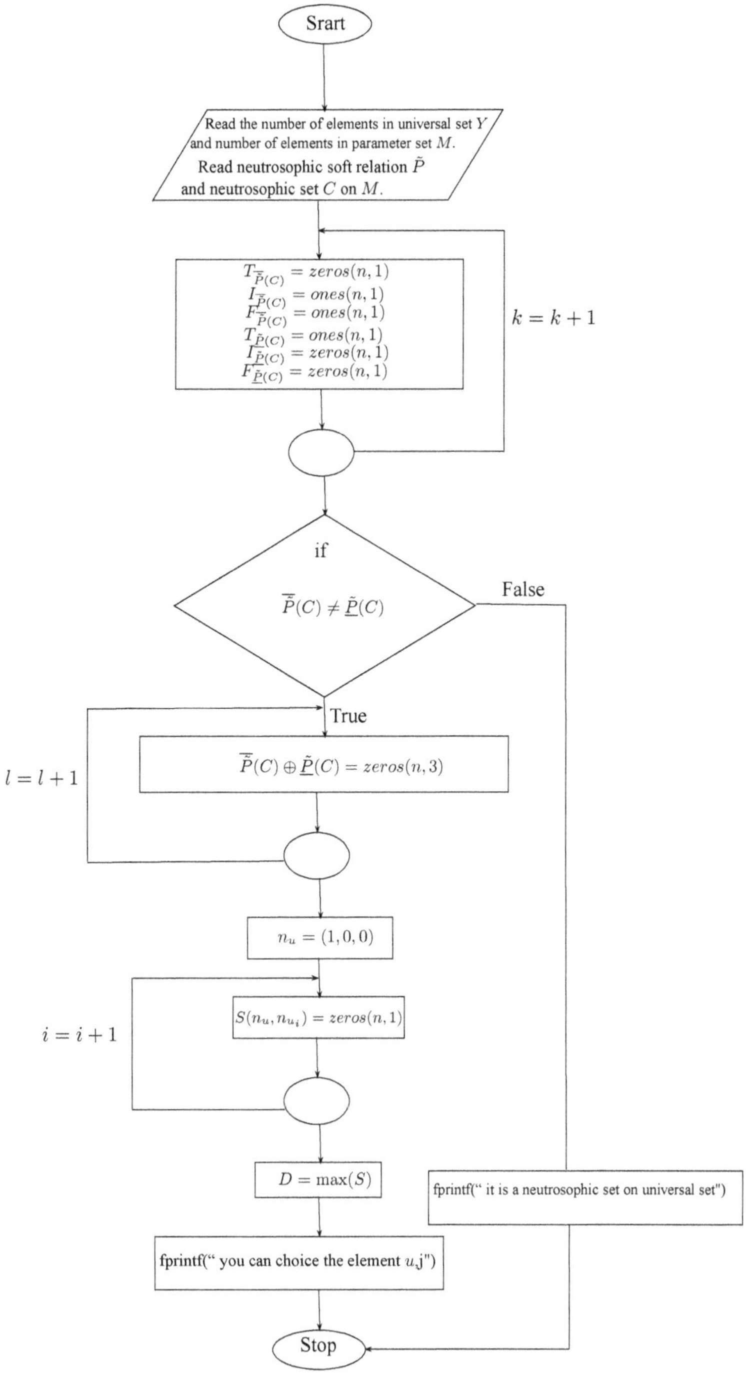


**Figure 1.** Flow chart for selection of most suitable objects.

---

**Algorithm 1:** Algorithm for selection of the most suitable objects

---

1. Begin
2.  Input the number of elements in universal set $Y = \{u_1, u_2, \ldots, u_n\}$.
3.  Input the number of elements in parameter set $M = \{e_1, e_2, \ldots, e_m\}$.
4.  Input a neutrosophic soft relation $\tilde{P}$ from $Y$ to $M$.
5.  Input an NS $C$ on $M$.
6.  if $size(\tilde{P}) \neq [n, 3 * m]$
7.    fprintf(\ size of neutrosophic soft relation from universal set to parameter
          set is not correct, it should be of order %dx%d; $', n, 3 * m$)
8.      error(\ Dimemsion of neutrosophic soft relation on vertex set is not correct. $'$)
9.  **end**
10. **if** $size(C) \neq [m, 3]$
11.   fprintf(\ size of NS on parameter set is not correct,
          it should be of order %dx3; $',m$)
12.   error('Dimemsion of NS on parameter set is not correct.')
13. **end**
14. $T_{\overline{\tilde{P}}(C)} = zeros(n, 1)$;
15. $I_{\overline{\tilde{P}}(C)} = ones(n, 1)$;
16. $F_{\overline{\tilde{P}}(C)} = ones(n, 1)$;
17. $T_{\underline{\tilde{P}}(C)} = ones(n, 1)$;
18. $I_{\underline{\tilde{P}}(C)} = zeros(n, 1)$;
19. $F_{\underline{\tilde{P}}(C)} = zeros(n, 1)$;
20.   **if** $size(\tilde{P}) == [n, 3 * m]$
21.    **if** $size(C) == [m, 3]$
22.     **if** $\tilde{P} >= 0$ && $\tilde{P} <= 1$
23.      **if** $C >= 0$ && $C <= 1$
24.       **for** $i = 1 : n$
25.        **for** $k = 1 : m$
26.         j=3*k-2;
27.         $T_{\overline{\tilde{P}}(C)}(i, 1) = \max(T_{\overline{\tilde{P}}(C)}(i, 1), \min(\tilde{P}(i, j), C(k, 1)))$;
28.         $I_{\overline{\tilde{P}}(C)}(i, 1) = \min(I_{\overline{\tilde{P}}(C)}(i, 1), \max(\tilde{P}(i, j + 1), C(k, 2)))$;
29.         $F_{\overline{\tilde{P}}(C)}(i, 1) = \min(F_{\overline{\tilde{P}}(C)}(i, 1), \max(\tilde{P}(i, j + 2), C(k, 3)))$;
30.         $T_{\underline{\tilde{P}}(C)}(i, 1) = \min(T_{\underline{\tilde{P}}(C)}(i, 1), \max(\tilde{P}(i, j + 2), C(k, 1)))$;
31.         $I_{\underline{\tilde{P}}(C)}(i, 1) = \max(I_{\underline{\tilde{P}}(C)}(i, 1), \min((1 - \tilde{P}(i, j + 1)), C(k, 2)))$;
32.         $F_{\underline{\tilde{P}}(C)}(i, 1) = \max(F_{\underline{\tilde{P}}(C)}(i, 1), \min(\tilde{P}(i, j), C(k, 3)))$;
33.        **end**
34.       **end**
35.      $\overline{\tilde{P}}(C) = (T_{\overline{\tilde{P}}(C)}, I_{\overline{\tilde{P}}(C)}, F_{\overline{\tilde{P}}(C)})$
36.      $\underline{\tilde{P}}(C) = (T_{\underline{\tilde{P}}(C)}, I_{\overline{\tilde{P}}(C)}, F_{\underline{\tilde{P}}(C)})$
37.       **if** $\overline{\tilde{P}}(C) == \underline{\tilde{P}}(C)$
38.         fprintf(\ it is a neutrosophic set on universal set. $'$)
39.       **else**
40.         fprintf(\it is an NSRS on universal set. $'$)
41.         $\overline{\tilde{P}}(C) \oplus \underline{\tilde{P}}(C) = zeros(n, 3)$;

---

**for** i=1:n

$$T_{\overline{\tilde{P}}(C)}(i) \oplus T_{\underline{\tilde{P}}(C)}(i) = T_{\overline{\tilde{P}}(C)}(i) + T_{\underline{\tilde{P}}(C)}(i) - T_{\overline{\tilde{P}}(C)}(i). * T_{\underline{\tilde{P}}(C)}(i);$$

$$I_{\overline{\tilde{P}}(C)}(i) \oplus I_{\underline{\tilde{P}}(C)}(i) = I_{\overline{\tilde{P}}(C)}(i). * I_{\underline{\tilde{P}}(C)}(i);$$

$$F_{\overline{\tilde{P}}}(C)(i) \oplus F_{\underline{\tilde{P}}(C)}(i) = F_{\overline{\tilde{P}}(C)}(i). * F_{\underline{\tilde{P}}(C)}(i);$$

**end**

$n_u = (1,0,0);$

$S(n_u, n_{u_i}) = zeros(n,1);$

**for** i=1:n

$$S(n_u, n_{u_i}) = \frac{T_{n_u} \cdot T_{n_{u_i}} + I_{n_u} \cdot I_{n_{u_i}} + F_{n_u} \cdot F_{n_{u_i}}}{\sqrt{T^2_{n_u} + T_{n^2_u} + F^2_{n_u}} + \sqrt{T^2_{n_{u_i}} + T_{n^2_{u_i}} + F^2_{n_{u_i}}}};$$

**end**

$S(n_u, n_{u_i})$

D=max(S);

l=0;

m=zeros(n,1);

D2=zeros(n,1);

**for** j=1:n

**if** S(j,1)==D

l=l+1;

D2(j,1)=S(j,1);

m(j)=j;

**end**

**end**

**for** $j = 1 : n$

**if** $m(j) = 0$

fprintf(' you can choice the element $u_{\%d}$ ',j)

**end**

**end**

**end**

**end**

**end**

**end**

**end**

End

## 5. Conclusions and Future Directions

Rough set theory can be considered as an extension of classical set theory. Rough set theory is a very useful mathematical model to handle vagueness. NS theory, RS theory and SS theory are three useful distinguished approaches to deal with vagueness. NS and RS models are used to handle uncertainty, and combining these two models with another remarkable model of SSs gives more precise results for decision-making problems. In this paper, we have first presented the notion of SRNSs. Furthermore, we have introduced NSRSs and investigated some properties of NSRSs in detail. The notion of NSRS can be utilized as a mathematical tool to deal with imprecise and unspecified information. In addition, a decision-making method based on NSRSs has been proposed. This research work can be extended to (1) rough bipolar neutrosophic soft sets; (2) bipolar neutrosophic soft rough sets; (3) interval-valued bipolar neutrosophic rough sets; and (4) neutrosophic soft rough graphs.

**Author Contributions:** Muhammad Akram and Sundas Shahzadi conceived and designed the experiments; Florentin Smarandache analyzed the data; Sundas Shahzadi wrote the paper.

**Conflicts of Interest:** The authors declare no conflict of interest.

## References


1. Smarandache, F. *Neutrosophy: Neutrosophic Probability, Set, and Logic*; American Research Press: Rehoboth, DE, USA, 1998; 105p.
2. Wang, H.; Smarandache, F.; Zhang, Y.; Sunderraman, R. Single-valued neutrosophic sets. *Multispace Multistruct.* **2010**, *4*, 410–413.
3. Ye, J. Multicriteria decision-making method using the correlation coefficient under single-valued neutrosophic environment. *Int. J. Gen. Syst.* **2013**, *42*, 386–394.
4. Ye, J. Improved correlation coefficients of single valued neutrosophic sets and interval neutrosophic sets for multiple attribute decision making. *J. Intell. Fuzzy Syst.* **2014**, *27*, 2453–2462.
5. Ye, J.; Fu, J. Multi-period medical diagnosis method using a single valued neutrosophic similarity measure based on tangent function. *Comput. Methods Prog. Biomed.* **2016**, *123*, 142–149.
6. Peng, J.J.; Wang, J.Q.; Zhang, H.Y.; Chen, X.H. An outranking approach for multi-criteria decision-making problems with simplified neutrosophic sets. *Appl. Soft Comput.* **2014**, *25*, 336–346.
7. Molodtsov, D.A. Soft set theory-first results. *Comput. Math. Appl.* **1999**, *37*, 19–31.
8. Maji, P.K.; Biswas, R.; Roy, A.R. Fuzzy soft sets. *J. Fuzzy Math.* **2001**, *9*, 589–602.
9. Maji, P.K.; Biswas, R.; Roy, A.R. Intuitionistic fuzzy soft sets. *J. Fuzzy Math.* **2001**, *9*, 677–692.
10. Maji, P.K. Neutrosophic soft set. *Ann. Fuzzy Math. Inform.* **2013**, 5, 157–168.
11. Babitha, K.V.; Sunil, J.J. Soft set relations and functions. *Comput. Math. Appl.* **2010**, *60*, 1840–1849.
12. Sahin, R.; Kucuk, A. On similarity and entropy of neutrosophic soft sets. *J. Intell. Fuzzy Syst. Appl. Eng. Technol.* **2014**, *27*, 2417–2430.
13. Pawlak, Z. Rough sets. *Int. J. Comput. Inf. Sci.* **1982**, *11*, 341–356.
14. Ali, M. A note on soft sets, rough sets and fuzzy soft sets. *Appl. Soft Comput.* **2011**, *11*, 3329–3332.
15. Feng, F.; Liu, X.; Leoreanu-Fotea, B.; Jun, Y.B. Soft sets and soft rough sets. *Inf. Sci.* **2011**, *181*, 1125–1137.
16. Shabir, M.; Ali, M.I.; Shaheen, T. Another approach to soft rough sets. *Knowl.-Based Syst.* **2013**, *40*, 72–80.
17. Feng, F.; Li, C.; Davvaz, B.; Ali, M.I. Soft sets combined with fuzzy sets and rough sets: A tentative approach. *Soft Comput.* **2010**, *14*, 899–911.
18. Dubois, D.; Prade, H. Rough fuzzy sets and fuzzy rough sets. *Int. J. Gen. Syst.* **1990**, *17*, 191–209.
19. Meng, D.; Zhang, X.; Qin, K. Soft rough fuzzy sets and soft fuzzy rough sets. *Comput. Math. Appl.* **2011**, *62*, 4635–4645.
20. Sun, B.Z.; Ma, W.; Liu, Q. An approach to decision making based on intuitionistic fuzzy rough sets over two universes. *J. Oper. Res. Soc.* **2013**, *64*, 1079–1089.
21. Sun, B.Z.; Ma, W. Soft fuzzy rough sets and its application in decision making. *Artif. Intell. Rev.* **2014**, *41*, 67–80.
22. Zhang, X.; Dai, J.; Yu, Y. On the union and intersection operations of rough sets based on various approximation spaces. *Inf. Sci.* **2015**, *292*, 214–229.
23. Zhang, H.; Shu, L. Generalized intuitionistic fuzzy rough set based on intuitionistic fuzzy covering. *Inf. Sci.* **2012**, *198*, 186–206.
24. Zhang, X.; Zhou, B.; Li, P. A general frame for intuitionistic fuzzy rough sets. *Inf. Sci.* **2012**, *216*, 34–49.
25. Zhang, H.; Shu, L.; Liao, S. Intuitionistic fuzzy soft rough set and its application in decision making. *Abstr. Appl. Anal.* **2014**, *2014*, 13.
26. Zhang, H.; Xiong, L.; Ma, W. Generalized intuitionistic fuzzy soft rough set and its application in decision making. *J. Comput. Anal. Appl.* **2016**, *20*, 750–766.
27. Broumi, S.; Smarandache, F. Interval-valued neutrosophic soft rough sets. *Int. J. Comput. Math.* **2015**, *2015*, 232919.
28. Broumi, S.; Smarandache, F.; Dhar, M. Rough Neutrosophic sets. *Neutrosophic Sets Syst.* **2014**, *3*, 62–67.
29. Yang, H.L.; Zhang, C.L.; Guo, Z.L.; Liu, Y.L.; Liao, X. A hybrid model of single valued neutrosophic sets and rough sets: Single valued neutrosophic rough set model. *Soft Comput.* **2016**, *21*, 6253–6267.

30. Faizi, S.; Salabun, W.; Rashid, T.; Watrbski, J.; Zafar, S. Group decision-making for hesitant fuzzy sets based on characteristic objects method. *Symmetry* **2017**, *9*, 136.
31. Faizi, S.; Rashid, T.; Salabun, W.; Zafar, S.; Watrbski, J. Decision making with uncertainty using hesitant fuzzy sets. *Int. J. Fuzzy Syst.* **2018**, *20*, 93–103.
32. Mardani, A.; Nilashi, M.; Antucheviciene, J.; Tavana, M.; Bausys, R.; Ibrahim, O. Recent Fuzzy Generalisations of Rough Sets Theory: A Systematic Review and Methodological Critique of the Literature. *Complexity* **2017**, *2017*, 33.
33. Liang, R.X.; Wang, J.Q.; Zhang, H.Y. A multi-criteria decision-making method based on single-valued trapezoidal neutrosophic preference relations with complete weight information. *Neural Comput. Appl.* **2017**, 1–16, doi:10.1007/s00521-017-2925-8.
34. Liang, R.; Wang, J.; Zhang, H. Evaluation of e-commerce websites: An integrated approach under a single-valued trapezoidal neutrosophic environment. *Knowl.-Based Syst.* **2017**, *135*, 44–59.
35. Peng, H.G.; Zhang, H.Y.; Wang, J.Q. Probability multi-valued neutrosophic sets and its application in multi-criteria group decision-making problems. *Neural Comput. Appl.* **2016**, 1–21, doi:10.1007/s00521-016-2702-0.
36. Wang, L.; Zhang, H.Y.; Wang, J.Q. Frank Choquet Bonferroni mean operators of bipolar neutrosophic sets and their application to multi-criteria decision-making problems. *Int. J. Fuzzy Syst.* **2018**, *20*, 13–28.
37. Zavadskas, E.K.; Bausys, R.; Kaklauskas, A.; Ubarte, I.; Kuzminske, A.; Gudiene, N. Sustainable market valuation of buildings by the single-valued neutrosophic MAMVA method. *Appl. Soft Comput.* **2017**, *57*, 74–87.
38. Li, Y.; Liu, P.; Chen, Y. Some single valued neutrosophic number heronian mean operators and their application in multiple attribute group decision making. *Informatica* **2016**, *27*, 85–110.

axioms

MDPI

Article

# Neutrosophic Soft Rough Graphs with Application

**Muhammad Akram [1,*], Hafsa M. Malik [1], Sundas Shahzadi [1] and Florentin Smarandache [2]**

[1] Department of Mathematics, University of the Punjab, New Campus, Lahore 54590, Pakistan; hafsa.masood.malik@gmail.com (H.M.M.); sundas1011@gmail.com (S.S.)

[2] Mathematics & Science Department, University of New Mexico, 705 Gurley Ave., Gallup, NM 87301, USA; fsmarandache@gmail.com

* Correspondence: m.akram@pucit.edu.pk; Tel.: +92-42-99231241

Received: 27 January 2018; Accepted: 23 February 2018; Published: 26 February 2018

**Abstract:** Neutrosophic sets (NSs) handle uncertain information while fuzzy sets (FSs) and intuitionistic fuzzy sets (IFs) fail to handle indeterminate information. Soft set theory, neutrosophic set theory, and rough set theory are different mathematical models for handling uncertainties and they are mutually related. The neutrosophic soft rough set (NSRS) model is a hybrid model by combining neutrosophic soft sets with rough sets. We apply neutrosophic soft rough sets to graphs. In this research paper, we introduce the idea of neutrosophic soft rough graphs (NSRGs) and describe different methods of their construction. We consider the application of NSRG in decision-making problems. In particular, we develop efficient algorithms to solve decision-making problems.

**Keywords:** neutrosophic soft rough sets; neutrosophic soft rough graphs; decision-making; algorithm

## 1. Introduction

Smarandache [1] initiated the concept of neutrosophic set (NS). Smarandache's NS is characterized by three parts: truth, indeterminacy, and falsity. Truth, indeterminacy and falsity membership values behave independently and deal with problems having uncertain, indeterminant and imprecise data. Wang et al. [2] gave a new concept of single valued neutrosophic sets (SVNSs) and defined the set theoretic operators on an instance of NS called SVNS. Peng et al. [3] discussed the operations of simplified neutrosophic numbers and introduced an outranking idea of simplified neutrosophic numbers.

Molodtsov [4] introduced the notion of soft set (SS) as a novel mathematical approach for handling uncertainties. Molodtsov's SSs gave us a new technique for dealing with uncertainty from the viewpoint of parameters. Maji et al. [5–7] introduced neutrosophic soft sets (NSSs), intuitionistic fuzzy soft sets and fuzzy soft sets (FSSs). In [8], Sahin and Kucuk presented NSS in the form of neutrosophic relations.

Theory of rough set (RS) was proposed by Pawlak [9] in 1982. Rough set theory is used to study the intelligence systems containing incomplete, uncertain or inexact information. The lower and upper approximation operators of RSs are used for managing hidden information in a system. Feng et al. [10] took a significant step to introduce parametrization tools in RSs. Meng et al. [11] provide further discussion of the combination of SSs, RSs and FSs. The existing results of RSs and other extended RSs such as rough fuzzy sets, generalized rough fuzzy sets, soft fuzzy rough sets and intuitionistic fuzzy rough sets based decision-making models have their advantages and limitations [12,13]. In a different way, rough set approximations have been constructed into the intuitionistic fuzzy environment and are known as intuitionistic fuzzy rough sets and rough intuitionistic fuzzy sets [14,15]. Zhang et al. [16,17] presented the notions of soft rough sets, soft rough intuitionistic fuzzy sets, intuitionistic fuzzy soft rough sets, its application in decision-making, and also introduced generalized intuitionistic fuzzy soft rough sets. Broumi et al. [18,19] developed a hybrid structure by combining

RSs and NSs, called RNSs, they also presented interval valued neutrosophic soft rough sets by combining interval valued neutrosophic soft sets and RSs. Yang et al. [20] proposed single valued neutrosophic rough sets (SVNRSs) by combining SVNSs and RSs and defined SVNRSs on two universes and established an algorithm for a decision-making problem based on SVNRSs on two universes. Akram and Nawaz [21] have introduced the concept of soft graphs and some operation on soft graphs. Certain concepts of fuzzy soft graphs and intuitionistic fuzzy soft graphs are discussed in [22–24]. Akram and Shahzadi [25] have introduced neutrosophic soft graphs. Zafar and Akram [26] introduced a rough fuzzy digraph and several basic notions concerning rough fuzzy digraphs. In this research paper, a neutrosophic soft rough set is a generalization of a neutrosophic rough set, and we introduce the idea of neutrosophic soft rough graphs (NSRGs) that are made by combining NSRSs with graphs and describe different methods of their construction. We consider the application of NSRG in decision-making problems and resolve the problem. In particular, we develop efficient algorithms to solve decision-making problems.

For other notations, terminologies and applications not mentioned in the paper, the readers are referred to [27–35].

## 2. Neutrosophic Soft Rough Information

In this section, we will introduce the notions of neutrosophic soft rough relation (NSRR), and NSRGs.

**Definition 1.** *Let $Y$ be an initial universal set, $\mathbb{P}$ a universal set of parameters and $\mathbb{M} \subseteq \mathbb{P}$. For an arbitrary neutrosophic soft relation $Q$ over $Y \times \mathbb{M}$, $(Y, \mathbb{M}, Q)$ is called neutrosophic soft approximation space (NSAS).*

*For any NS $A \in \mathcal{N}(\mathbb{M})$, we define the upper neutrosophic soft rough approximation (UNSRA) and the lower neutrosophic soft rough approximation (LNSRA) operators of $A$ with respect to $(Y, \mathbb{M}, Q)$ denoted by $\overline{Q}(A)$ and $\underline{Q}(A)$, respectively as follows:*

$$\begin{aligned}
\overline{Q}(A) &= \{(u, T_{\overline{Q}(A)}(u), I_{\overline{Q}(A)}(u), F_{\overline{Q}(A)}(u)) \mid u \in Y\},\\
\underline{Q}(A) &= \{(u, T_{\underline{Q}(A)}(u), I_{\underline{Q}(A)}(u), F_{\underline{Q}(A)}(u)) \mid u \in Y\},
\end{aligned}$$

*where*

$$\begin{aligned}
T_{\overline{Q}(A)}(u) &= \bigvee_{e\in\mathbb{M}} (T_{Q(A)}(u,e) \wedge T_A(e)), \quad I_{\overline{Q}(A)}(u) = \bigwedge_{e\in\mathbb{M}} (I_{Q(A)}(u,e) \vee I_A(e)),\\
F_{\overline{Q}(A)}(u) &= \bigwedge_{e\in\mathbb{M}} (F_{Q(A)}(u,e) \vee F_A(e)); \quad T_{\underline{Q}(A)}(u) = \bigwedge_{e\in\mathbb{M}} (F_{Q(A)}(u,e) \vee T_A(e)),\\
I_{\underline{Q}(A)}(u) &= \bigvee_{e\in\mathbb{M}} ((1 - I_{Q(A)}(u,e)) \wedge I_A(e)), \; F_{\underline{Q}(A)}(u) = \bigvee_{e\in\mathbb{M}} (T_{Q(A)}(u,e) \wedge F_A(e)).
\end{aligned}$$

*The pair $(\underline{Q}(A), \overline{Q}(A))$ is called NSRS of $A$ w.r.t $(Y, \mathbb{M}, Q)$, $\underline{Q}$ and $\overline{Q}$ are referred to as the LNSRA and the UNSRA operators, respectively.*

**Example 1.** *Suppose that $Y = \{w_1, w_2, w_3, w_4\}$ is the set of careers under consideration, and Mr. X wants to select the best suitable career. $\mathbb{M} = \{e_1, e_2, e_3\}$ is a set of decision parameters. Mr. X describes the "most suitable career" by defining a neutrosophic soft set $(Q, \mathbb{M})$ on $Y$ that is a neutrosophic relation from $Y$ to $\mathbb{M}$ as shown in Table 1.*

**Table 1.** Neutrosophic soft relation $Q$.

| $Q$ | $w_1$ | $w_2$ | $w_3$ | $w_4$ |
|---|---|---|---|---|
| $e_1$ | (0.3, 0.4, 0.5) | (0.4, 0.2, 0.3) | (0.1, 0.5, 0.4) | (0.2, 0.3, 0.4) |
| $e_2$ | (0.1, 0.5, 0.4) | (0.3, 0.4, 0.6) | (0.4, 0.4, 0.3) | (0.5, 0.3, 0.8) |
| $e_3$ | (0.3, 0.4, 0.4) | (0.4, 0.6, 0.7) | (0.3, 0.5, 0.4) | (0.5, 0.4, 0.6) |

*Now, Mr. X gives the most favorable decision object A, which is an NS on $\mathbb{M}$ defined as follows:* $A = \{(e_1, 0.5, 0.2, 0.4), (e_2, 0.2, 0.3, 0.1), (e_3, 0.2, 0.4, 0.6)\}$. *By Definition 1, we have*

$$T_{\overline{Q}(A)}(w_1) = 0.3, \quad I_{\overline{Q}(A)}(w_1) = 0.4, \quad F_{\overline{Q}(A)}(w_1) = 0.4,$$
$$T_{\overline{Q}(A)}(w_2) = 0.4, \quad I_{\overline{Q}(A)}(w_2) = 0.2, \quad F_{\overline{Q}(A)}(w_2) = 0.4,$$
$$T_{\overline{Q}(A)}(w_3) = 0.2, \quad I_{\overline{Q}(A)}(w_3) = 0.4, \quad F_{\overline{Q}(A)}(w_3) = 0.3,$$
$$T_{\overline{Q}(A)}(w_4) = 0.2, \quad I_{\overline{Q}(A)}(w_4) = 0.3, \quad F_{\overline{Q}(A)}(w_4) = 0.4.$$

*Similarly,*

$$T_{\underline{Q}(A)}(w_1) = 0.4, \quad I_{\underline{Q}(A)}(w_1) = 0.4, \quad F_{\underline{Q}(A)}(w_1) = 0.3,$$
$$T_{\underline{Q}(A)}(w_2) = 0.5, \quad I_{\underline{Q}(A)}(w_2) = 0.4, \quad F_{\underline{Q}(A)}(w_2) = 0.4,$$
$$T_{\underline{Q}(A)}(w_3) = 0.4, \quad I_{\underline{Q}(A)}(w_3) = 0.4, \quad F_{\underline{Q}(A)}(w_3) = 0.3,$$
$$T_{\underline{Q}(A)}(w_4) = 0.5, \quad I_{\underline{Q}(A)}(w_4) = 0.4, \quad F_{\underline{Q}(A)}(w_4) = 0.5.$$

*Thus, we obtain*

$$\begin{aligned}\overline{Q}(A) &= \{(w_1, 0.3, 0.4, 0.4), (w_2, 0.4, 0.2, 0.4), (w_3, 0.2, 0.4, 0.3), (w_4, 0.2, 0.3, 0.4)\}, \\ \underline{Q}(A) &= \{(w_1, 0.4, 0.4, 0.3), (w_2, 0.5, 0.4, 0.4), (w_3, 0.4, 0.4, 0.3), (w_4, 0.5, 0.4, 0.5)\}.\end{aligned}$$

*Hence,* $(\underline{Q}(A), \overline{Q}(A))$ *is an NSRS of A.*

The conventional neutrosophic soft set is a mapping from a parameter to the neutrosophic subset of the universe and letting $(Q, \mathbb{M})$ be neutrosophic soft set. Now, we present the constructive definition of neutrosophic soft rough relation by using a neutrosphic soft relation $S$ from $\mathbb{M} \times \mathbb{M} = \acute{\mathbb{M}}$ to $\mathcal{N}(Y \times Y = \acute{Y})$, where $Y$ is a universal set and $\mathbb{M}$ be a set of parameters.

**Definition 2.** *A neutrosophic soft rough relation* $(\underline{S}(B), \overline{S}(B))$ *on Y is an NSRS,* $S : \acute{\mathbb{M}} \to \mathcal{N}(\acute{Y})$ *is a neutrosophic soft relation on Y defined by*

$$\begin{aligned}S(e_ie_j) &= \{u_iu_j \mid \exists u_i \in Q(e_i), u_j \in Q(e_j)\},\ u_iu_j \in \acute{Y}, \ \textit{such that} \\ T_S(u_iu_j, e_ie_j) &\leqslant \min\{T_Q(u_i, e_i), T_Q(u_j, e_j)\} \\ I_S(u_iu_j, e_ie_j) &\leqslant \max\{I_Q(u_i, e_i), I_Q(u_j, e_j)\} \\ F_S(u_iu_j, e_ie_j) &\leqslant \max\{F_Q(u_i, e_i), F_Q(u_j, e_j)\}. \\ \textit{For any } B \in \mathcal{N}(\acute{\mathbb{M}}),\ B &= \{(e_ie_j, T_B(e_ie_j), I_B(e_ie_j), F_B(e_ie_j))\ u_iu_j \in \acute{\mathbb{M}}\}, \\ T_B(e_ie_j) &\leqslant \min\{T_A(e_i), T_A(e_j)\}, \\ I_B(e_ie_j) &\leqslant \max\{I_A(e_i), I_A(e_j)\}, \\ F_B(e_ie_j) &\leqslant \max\{F_A(e_i), F_A(e_j)\}.\end{aligned}$$

*The UNSA and the LNSA of B* w.r.t $(\acute{Y}, \acute{\mathbb{M}}, S)$ *are defined as follows:*

$$\overline{S}(B) = \{(u_iu_j, T_{\overline{S}(B)}(u_iu_j), I_{\overline{S}(B)}(u_iu_j), F_{\overline{S}(B)}(u_iu_j)) \mid u_iu_j \in \acute{Y}\},$$

$$\underline{S}(B) = \{(u_iu_j, T_{\underline{S}(B)}(u_iu_j), I_{\underline{S}(B)}(u_iu_j), F_{\underline{S}(B)}(u_iu_j)) \mid u_iu_j \in \acute{Y}\},$$

*where*

$$\begin{aligned}
T_{\overline{S}(B)}(u_iu_j) &= \bigvee_{e_ie_j \in \acute{\mathbb{M}}} (T_S(u_iu_j, e_ie_j) \wedge T_B(e_ie_j)), \\
I_{\overline{S}(B)}(u_iu_j) &= \bigwedge_{e_ie_j \in \tilde{\mathbb{M}}} (I_S(u_iu_j, e_ie_j) \vee I_B(e_ie_j)), \\
F_{\overline{S}(B)}(u_iu_j) &= \bigwedge_{e_ie_j \in \tilde{\mathbb{M}}} (F_S(u_iu_j, e_ie_j) \vee F_B(e_ie_j));
\end{aligned}$$

$$\begin{aligned}
T_{\underline{S}(B)}(u_iu_j) &= \bigwedge_{e_ie_j \in \acute{\mathbb{M}}} (F_S(u_iu_j, e_ie_j) \vee T_B(e_ie_j)), \\
I_{\underline{S}(B)}(u_iu_j) &= \bigvee_{e_ie_j \in \tilde{\mathbb{M}}} ((1 - I_S(u_iu_j, e_ie_j)) \wedge I_B(e_ie_j)), \\
F_{\underline{S}(B)}(u_iu_j) &= \bigvee_{e_ie_j \in \tilde{\mathbb{M}}} (T_S(u_iu_j, e_ie_j) \wedge F_B(e_ie_j)).
\end{aligned}$$

*The pair $(\underline{S}(B), \overline{S}(B))$ is called NSRR and $\underline{S}, \overline{S} : \mathcal{N}(\acute{\mathbb{M}}) \to \mathcal{N}(\acute{Y})$ are called the LNSRA and the UNSRA operators, respectively.*

**Remark 1.** *Consider an NS B on $\acute{\mathbb{M}}$ and an NS A on $\mathbb{M}$, according to the definition of NSRR, we get*

$$\begin{aligned}
T_{\overline{S}(B)}(u_iu_j) &\leqslant \min\{T_{\overline{S}(A)}(u_i), T_{\overline{S}(A)}(u_j)\}, \\
I_{\overline{S}(B)}(u_iu_j) &\leqslant \max\{I_{\overline{S}(A)}(u_i), I_{\overline{S}(A)}(u_j)\}, \\
F_{\overline{S}(B)}(u_iu_j) &\leqslant \max\{F_{\overline{S}(A)}(u_i).F_{\overline{S}(A)}(u_j)\}.
\end{aligned}$$

*Similarly, for LNSRA operator $\underline{S}(B)$,*

$$\begin{aligned}
T_{\underline{S}(B)}(u_iu_j) &\leqslant \min\{T_{\underline{S}(A)}(u_i), T_{\underline{S}(A)}(u_j)\}, \\
I_{\underline{S}(B)}(u_iu_j) &\leqslant \max\{I_{\underline{S}(A)}(u_i), I_{\underline{S}(A)}(u_j)\}, \\
F_{\underline{S}(B)}(u_iu_j) &\leqslant \max\{F_{\underline{S}(A)}(u_i).F_{\underline{S}(A)}(u_j)\}.
\end{aligned}$$

**Example 2.** *Let $Y = \{u_1, u_2, u_3\}$ be a universal set and $\mathbb{M} = \{e_1, e_2, e_3\}$ a set of parameters. A neutrosophic soft set $(Q, \mathbb{M})$ on Y can be defined in tabular form in Table 2 as follows:*

**Table 2.** Neutrosophic soft set $(Q, \mathbb{M})$.

| $Q$ | $u_1$ | $u_2$ | $u_3$ |
|---|---|---|---|
| $e_1$ | (0.4, 0.5, 0.6) | (0.7, 0.3, 0.2) | (0.6, 0.3, 0.4) |
| $e_2$ | (0.5, 0.3, 0.6) | (0.3, 0.4, 0.3) | (0.7, 0.2, 0.3) |
| $e_3$ | (0.7, 0.2, 0.3) | (0.6, 0.5, 0.4) | (0.7, 0.2, 0.4) |

*Let $E = \{u_1u_2, u_2u_3, u_2u_2, u_3u_2\} \subseteq \acute{Y}$ and $L = \{e_1e_3, e_2e_1, e_3e_2\} \subseteq \acute{\mathbb{M}}$.*
*Then, a soft relation S on E (from L to E) can be defined in Table 3 as follows:*

**Table 3.** Neutrosophic soft relation $S$.

| $S$ | $u_1u_2$ | $u_2u_3$ | $u_2u_2$ | $u_3u_2$ |
|---|---|---|---|---|
| $e_1e_3$ | $(0.4, 0.4, 0.5)$ | $(0.6, 0.3, 0.4)$ | $(0.5, 0.4, 0.2)$ | $(0.5, 0.4, 0.3)$ |
| $e_2e_1$ | $(0.3, 0.3, 0.4)$ | $(0.3, 0.2, 0.3)$ | $(0.2, 0.3, 0.3)$ | $(0.7, 0.2, 0.2)$ |
| $e_3e_2$ | $(0.3, 0.3, 0.2)$ | $(0.5, 0.3, 0.2)$ | $(0.2, 0.4, 0.4)$ | $(0.3, 0.4, 0.4)$ |

*Let* $A = \{(e_1, 0.2, 0.4, 0.6), (e_2, 0.4, 0.5, 0.2), (e_3, 0.1, 0.2, 0.4)\}$ *be an NS on* $\mathbb{M}$, *then*
$\overline{S}(A) = \{(u_1, 0.4, 0.2, 0.4), (u_2, 0.3, 0.4, 0.3), (u_3, 0.4, 0.2, 0.3)\}$,
$\underline{S}(A) = \{(u_1, 0.3, 0.5, 0.4), (u_2, 0.2, 0.5, 0.6), (u_3, 0.4, 0.5, 0.6)\}$.
*Let* $B = \{(e_1e_3, 0.1, 0.3, 0.5), (e_2e_1, 0.2, 0.4, 0.3), (e_3e_2, 0.1, 0.2, 0.3)\}$ *be an NS on L, then*
$\overline{S}(B) = \{(u_1u_2, 0.2, 0.3, 0.3), (u_2u_3, 0.2, 0.3, 0.3), (u_2u_2, 0.2, 0.4, 0.3), (u_3u_2, 0.2, 0.4, 0.3)\}$,
$\underline{S}(B) = \{(u_1u_2, 0.2, 0.4, 0.4), (u_2u_3, 0.2, 0.4, 0.5), (u_2u_2, 0.3, 0.4, 0.5), (u_3u_2, 0.2, 0.4, 0.5)\}$.
*Hence,* $S(B) = (\underline{S}(B), \overline{S}(B))$ *is NSRR.*

**Definition 3.** *A neutrosophic soft rough graph (NSRG) on a non-empty* $V$ *is an 4-ordered tuple* $(V, \mathbb{M}, Q(A), S(B))$ *such that*

(i) $\mathbb{M}$ *is a set of parameters,*
(ii) $Q$ *is an arbitrary neutrosophic soft relation over* $V \times \mathbb{M}$,
(iii) $S$ *is an arbitrary neutrosophic soft relation over* $\acute{V} \times \acute{\mathbb{M}}$,
(vi) $Q(A) = (\underline{Q}A, \overline{Q}A)$ *is an NSRS of* $A$,
(v) $S(B) = (\underline{S}B, \overline{S}B)$ *is an NSRR on* $\acute{V} \subset V \times V$,
(iv) $G = (Q(A), S(B))$ *is a neutrosophic soft rough graph, where* $\underline{G} = (\underline{Q}A, \underline{S}B)$ *and* $\overline{G} = (\overline{Q}A, \overline{S}B)$ *are lower neutrosophic approximate graph (LNAG) and upper neutrosophic approximate graph (UNAG), respectively of neutrosophic soft rough graph (NSRG)* $G = (Q(A), S(B))$.

**Example 3.** *Let* $V = \{v_1, v_2, v_3, v_4, v_5, v_6\}$ *be a vertex set and* $\mathbb{M} = \{e_1, e_2, e_3\}$ *a set of parameters. A neutrosophic soft relation over* $V \times \mathbb{M}$ *can be defined in tabular form in Table 4 as follows:*

**Table 4.** Neutrosophic soft relation $Q$.

| $Q$ | $v_1$ | $v_2$ | $v_3$ | $v_4$ | $v_5$ | $v_6$ |
|---|---|---|---|---|---|---|
| $e_1$ | $(0.4, 0.5, 0.6)$ | $(0.7, 0.3, 0.5)$ | $(0.6, 0.2, 0.3)$ | $(0.4, 0.4, 0.2)$ | $(0.5, 0.5, 0.6)$ | $(0.4, 0.5, 0.6)$ |
| $e_2$ | $(0.5, 0.4, 0.2)$ | $(0.6, 0.4, 0.5)$ | $(0.7, 0.3, 0.4)$ | $(0.5, 0.3, 0.2)$ | $(0.4, 0.5, 0.4)$ | $(0.6, 0.5, 0.4)$ |
| $e_3$ | $(0.5, 0.4, 0.1)$ | $(0.6, 0.3, 0.2)$ | $(0.5, 0.4, 0.3)$ | $(0.6, 0.2, 0.3)$ | $(0.5, 0.4, 0.4)$ | $(0.7, 0.3, 0.5)$ |

*Let* $A = \{(e_1, 0.5, 0.4, 0.6), (e_2, 0.7, 0.4, 0.5), (e_3, 0.6, 0.2, 0.5)\}$ *be an NS on* $\mathbb{M}$, *then*

$$\begin{aligned}\overline{S}(A) &= \{(v_1, 0.5, 0.4, 0.5), (v_2, 0.6, 0.3, 0.5), (v_3, 0.7, 0.4, 0.5), (v_4, 0.6, 0.2, 0.5), (v_5, 0.5, \\ &\quad 0.4, 0.5), (v_6, 0.6, 0.3, 0.5)\}, \\ \underline{S}(A) &= \{(v_1, 0.6, 0.4, 0.5), (v_2, 0.5, 0.4, 0.6), (v_3, 0.5, 0.4, 0.6), (v_4, 0.5, 0.4, 0.5), (v_5, 0.6, \\ &\quad 0.4, 0.5), (v_6, 0.6, 0.4, 0.5)\}.\end{aligned}$$

*Let* $E = \{v_1v_1, v_1v_2, v_2v_1, v_2v_3, v_4v_5, v_3v_4, v_5v_2, v_5v_6\} \subseteq \acute{V}$ *and* $L = \{e_1e_3, e_2e_1, e_3e_2\} \subseteq \acute{\mathbb{M}}$. *Then, a neutrosophic soft relation S on E (from L to E) can be defined in Tables 5 and 6 as follows:*

**Table 5.** Neutrosophic soft relation $S$.

| $S$ | $v_1v_1$ | $v_1v_2$ | $v_2v_1$ | $v_2v_3$ |
|---|---|---|---|---|
| $e_1e_2$ | $(0.4, 0.4, 0.2)$ | $(0.4, 0.4, 0.5)$ | $(0.4, 0.4, 0.5)$ | $(0.6, 0.3, 0.4)$ |
| $e_2e_3$ | $(0.5, 0.4, 0.1)$ | $(0.4, 0.3, 0.2)$ | $(0.4, 0.3, 0.2)$ | $(0.5, 0.3, 0.2)$ |
| $e_1e_3$ | $(0.4, 0.4, 0.1)$ | $(0.4, 0.2, 0.2)$ | $(0.4, 0.2, 0.2)$ | $(0.5, 0.3, 0.3)$ |

**Table 6.** Neutrosophic soft relation $S$.

| $S$ | $v_3v_4$ | $v_4v_5$ | $v_5v_2$ | $v_5v_6$ |
|---|---|---|---|---|
| $e_1e_2$ | $(0.4,0.2,0.2)$ | $(0.4,0.4,0.2)$ | $(0.4,0.3,0.4)$ | $(0.3,0.2,0.3)$ |
| $e_2e_3$ | $(0.6,0.2,0.4)$ | $(0.3,0.2,0.1)$ | $(0.4,0.3,0.2)$ | $(0.4,0.3,0.4)$ |
| $e_1e_3$ | $(0.4,0.2,0.3)$ | $(0.4,0.3,0.1)$ | $(0.5,0.3,0.2)$ | $(0.5,0.3,0.5)$ |

$$\begin{aligned}
\textit{Let } B &= \{(e_1e_2,0.4,0.4,0.5;),(e_2e_3,0.5,0.4,0.5),(e_1e_3,0.5,0.2,0.5)\} \textit{ be an NS on L, then}\\
\overline{S}B &= \{(v_1v_1,0.5,0.4,0.5),(v_1v_2,0.4,0.2,0.5),(v_2v_1,0.4,0.2,0.5),(v_2v_3,0.5,0.3,0.5),\\
&\quad (v_3v_4,0.5,0.2,0.5),(v_4v_5,0.4,0.3,0.5),(v_5v_2,0.5,0.3,0.5),(v_5v_6,0.5,0.3,0.5)\},\\
\underline{S}B &= \{(v_1v_1,0.4,0.4,0.5)(v_1v_2,0.5,0.4,0.4),(v_2v_1,0.5,0.4,0.4),(v_2v_3,0.4,0.4,0.5),\\
&\quad (v_3v_4,0.4,0.4,0.5),(v_4v_5,0.4,0.4,0.4),(v_5v_2,0.4,0.4,0.5),(v_5v_6,0.4,0.4,0.5)\}.\\
\textit{Hence, } S(B) &= (\underline{S}B,\overline{S}B) \textit{ is NSRR on } \acute{V}.
\end{aligned}$$

*Thus,* $\underline{G} = (\underline{Q}A,\underline{S}B)$ *and* $\overline{G} = (\overline{Q}A,\overline{S}B)$ *are LNAG and UNAG, respectively, are shown in Figure 1.*

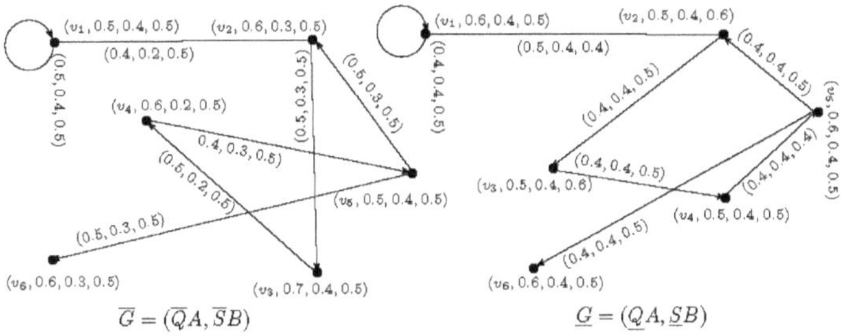


**Figure 1.** Neutrosophic soft rough graph $G = (\underline{G},\overline{G})$

*Hence,* $G = (\underline{G},\overline{G})$ *is NSRG.*

**Definition 4.** *Let* $G = (V,\mathbb{M},Q,S)$ *be a neutrosophic soft rough graph on a non-empty set* $V$*. The order of* $G$ *can be denoted by* $\mathbf{O}(G)$*, defined by*

$$\begin{aligned}
\mathbf{O}(G) &= \mathbf{O}(\overline{G}) + \mathbf{O}(\underline{G}), \text{ where}\\
\mathbf{O}(\overline{G}) &= \sum_{v\in V} \overline{Q}A(v), \mathbf{O}(\underline{G}) = \sum_{v\in V} \underline{Q}A(v).
\end{aligned}$$

*The size of neutrosophic soft rough graph* $G$*, denoted by* $\mathbf{S}(G)$*, defined by*

$$\begin{aligned}
\mathbf{S}(G) &= (\mathbf{S}\overline{G} + \mathbf{S}\underline{G}), \text{ where}\\
\mathbf{S}(\overline{G}) &= \sum_{uv\in E} \overline{S}B(uv), \mathbf{S}(\underline{G}) = \sum_{uv\in E} \underline{S}B(uv).
\end{aligned}$$

**Example 4.** *Let* $G$ *be a neutrosophic soft rough graph as shown in Figure 1. Then,*

$$\begin{aligned}
\mathbf{O}(\overline{G}) &= (3.5,2.0,3.0), \ \mathbf{O}(\underline{G}) = (3.3,2.4,3.2),\\
\mathbf{O}(G) &= \mathbf{O}(\overline{G}) + \mathbf{O}(\underline{G}) = (6.8,4.4,6.2), \text{ and}\\
\mathbf{S}(\overline{G}) &= (3.2,1.8,3.0) \ \mathbf{S}(\underline{G}) = (2.5,2.4,2.8)\\
\mathbf{S}(G) &= \mathbf{S}(\overline{G}) + \mathbf{S}(\underline{G}) = (5.7,4.2,5.8).
\end{aligned}$$

**Definition 5.** *Let $G_1 = (\underline{G}_1, \overline{G}_1)$ and $G_2 = (\underline{G}_2, \overline{G}_2)$ be two neutrosophic soft rough graphs on $V$. The union of $G_1$ and $G_2$ is a neutrosophic soft rough graph $G = G_1 \cup G_2 = (\underline{G}_1 \cup \underline{G}_2, \overline{G}_1 \cup \overline{G}_2)$, where $\underline{G}_1 \cup \underline{G}_2 = (\underline{Q}A_1 \cup \underline{Q}A_2, \underline{S}B_1 \cup \underline{S}B_2)$ and $\overline{G}_1 \cup \overline{G}_2 = (\overline{Q}A_1 \cup \overline{Q}A_2, \overline{S}B_1 \cup \overline{S}B_2)$ are neutrosophic graphs, such that*

*(i)* $\forall v \in QA_1$ *but* $v \notin QA_2$.

$$T_{\overline{Q}A_1 \cup \overline{Q}A_2}(v) = T_{\overline{Q}A_1}(v), T_{\underline{Q}A_1 \cup \underline{Q}A_2}(v) = T_{\underline{Q}A_1}(v),$$
$$I_{\overline{Q}A_1 \cup \overline{Q}A_2}(v) = I_{\overline{Q}A_1}(v), \; I_{\underline{Q}A_1 \cup \underline{Q}A_2}(v) = I_{\underline{Q}A_1}(v),$$
$$F_{\overline{Q}A_1 \cup \overline{Q}A_2}(v) = F_{\overline{Q}A_1}(v), F_{\underline{Q}A_1 \cup \underline{Q}A_2}(v) = F_{\underline{Q}A_1}(v).$$

*(ii)* $\forall v \notin QA_1$ *but* $v \in QA_2$.

$$T_{\overline{Q}A_1 \cup \overline{Q}A_2}(v) = T_{\overline{Q}A_2}(v), T_{\underline{Q}A_1 \cup \underline{Q}A_2}(v) = T_{\underline{Q}A_2}(v),$$
$$I_{\overline{Q}A_1 \cup \overline{Q}A_2}(v) = I_{\overline{Q}A_2}(v), \; I_{\underline{Q}A_1 \cup \underline{Q}A_2}(v) = I_{\underline{Q}A_2}(v),$$
$$F_{\overline{Q}A_1 \cup \overline{Q}A_2}(v) = F_{\overline{Q}A_2}(v), F_{\underline{Q}A_1 \cup \underline{Q}A_2}(v) = F_{\underline{Q}A_2}(v).$$

*(iii)* $\forall v \in QA_1 \cap \underline{Q}A_2$

$$T_{\overline{Q}A_1 \cup \overline{Q}A_2}(v) = \max\{T_{\overline{Q}A_1}(v), T_{\overline{Q}A_2}(v)\}, T_{\underline{Q}A_1 \cup \underline{Q}A_2}(v) = \max\{T_{\underline{Q}A_1}(v), T_{\underline{Q}A_2}(v)\},$$
$$I_{\overline{Q}A_1 \cup \overline{Q}A_2}(v) = \min\{I_{\overline{Q}A_1}(v), I_{\overline{Q}A_2}(v)\}, \; I_{\underline{Q}A_1 \cup \underline{Q}A_2}(v) = \min\{I_{\underline{Q}A_1}(v), I_{\underline{Q}A_2}(v)\},$$
$$F_{\overline{Q}A_1 \cup \overline{Q}A_2}(v) = \min\{F_{\overline{Q}A_1}(v), F_{\overline{Q}A_2}(v)\}, F_{\underline{Q}A_1 \cup \underline{Q}A_2}(v) = \min\{F_{\underline{Q}A_1}(v), F_{\underline{Q}A_2}(v)\}.$$

*(iv)* $\forall vu \in SB_1$ *but* $vu \notin SB_2$.

$$T_{\overline{S}B_1 \cup \overline{S}B_2}(vu) = T_{\overline{S}B_1}(vu), T_{\underline{S}B_1 \cup \underline{S}B_2}(vu) = T_{\underline{S}B_1}(vu),$$
$$I_{\overline{S}B_1 \cup \overline{S}B_2}(vu) = I_{\overline{S}B_1}(vu), \; I_{\underline{S}B_1 \cup \underline{S}B_2}(vu) = I_{\underline{S}B_1}(vu),$$
$$F_{\overline{S}B_1 \cup \overline{S}B_2}(vu) = F_{\overline{S}B_1}(vu), F_{\underline{S}B_1 \cup \underline{S}B_2}(vu) = F_{\underline{S}B_1}(vu).$$

*(v)* $\forall vu \notin SB_1$ *but* $vu \in SB_2$

$$T_{\overline{S}B_1 \cup \overline{S}B_2}(vu) = T_{\overline{S}B_2}(vu), T_{\underline{S}B_1 \cup \underline{S}B_2}(vu) = T_{\underline{S}B_2}(vu),$$
$$I_{\overline{S}B_1 \cup \overline{S}B_2}(vu) = I_{\overline{S}B_2}(vu), \; I_{\underline{S}B_1 \cup \underline{S}B_2}(vu) = I_{\underline{S}B_2}(vu),$$
$$F_{\overline{S}B_1 \cup \overline{S}B_2}(vu) = F_{\overline{S}B_2}(vu), F_{\underline{S}B_1 \cup \underline{S}B_2}(vu) = F_{\underline{S}B_2}(vu).$$

*(vi)* $\forall vu \in SB_1 \cap \underline{S}B_2$

$$T_{\overline{S}B_1 \cup \overline{S}B_2}(vu) = \max\{T_{\overline{S}B_1}(vu), T_{\overline{S}B_2}(vu)\}, T_{\underline{S}B_1 \cup \underline{S}B_2}(vu) = \max\{T_{\underline{S}}B_1(vu), T_{\underline{S}B_2}(vu)\},$$
$$I_{\overline{S}B_1 \cup \overline{S}B_2}(vu) = \min\{I_{\overline{S}B_1}(vu), I_{\overline{S}B_2}(vu)\}, \; I_{\underline{S}B_1 \cup \underline{S}B_2}(vu) = \min\{I_{\underline{S}B_1}(vu), I_{\underline{S}B_2}(vu)\},$$
$$F_{\overline{S}B_1 \cup \overline{S}B_2}(vu) = \min\{F_{\overline{S}B_1}(vu), F_{\overline{S}B_2}(vu)\}, F_{\underline{S}B_1 \cup \underline{S}B_2}(vu) = \min\{F_{\underline{S}B_1}(vu), F_{\underline{S}B_2}(vu)\}.$$

**Example 5.** *Let $V = \{v_1, v_2, v_3, v_4\}$ be a set of universes, and $\mathbb{M} = \{e_1, e_2, e_3\}$ a set of parameters. Then, a neutrosophic soft relation over $V \times \mathbb{M}$ can be written as in Table 7.*

**Table 7.** Neutrosophic soft relation $Q$.

| $Q$ | $v_1$ | $v_2$ | $v_3$ | $v_4$ |
|---|---|---|---|---|
| $e_1$ | (0.5, 0.4, 0.3) | (0.7, 0.6, 0.5) | (0.7, 0.6, 0.4) | (0.5, 0.7, 0.4) |
| $e_2$ | (0.3, 0.5, 0.6) | (0.4, 0.5, 0.1) | (0.3, 0.6, 0.5) | (0.4, 0.8, 0.2) |
| $e_3$ | (0.7, 0.5, 0.8) | (0.2, 0.3, 0.8) | (0.7, 0.3, 0.5) | (0.6, 0.4, 0.3) |

*Let* $A_1 = \{(e_1, 0.5, 0.7, 0.8), (e_2, 0.7, 0.5, 0.3), (e_3, 0.4, 0.5, 0.3)\}$, *and* $A_2 = \{(e_1, 0.6, 0.3, 0.5), (e_2, 0.5, 0.8, 0.2), (e_3, 0.5, 0.7, 0.2)\}$ *are two neutrosophic sets on* $\mathbb{M}$, *Then,* $Q(A_1) = (\underline{Q}(A_1), \overline{Q}(A_1))$ *and* $Q(A_2) = (\underline{Q}(A_2), \overline{Q}(A_2))$ *are NSRSs, where*

$$\begin{aligned}
\underline{Q}(A_1) &= \{(v_1, 0.5, 0.6, 0.5), (v_2, 0.5, 0.5, 0.7)(v_3, 0.5, 0.5, 0.7), (v_4 0.4, 0.5, 0.5)\},\\
\overline{Q}(A_1) &= \{(v_1, 0.5, 0.5, 0.6), (v_2, 0.5, 0.5, 0.3), (v_3, 0.5, 0.5, 0.5), (v_4 0.5, 0.5, 0.3)\},\\
\underline{Q}(A_2) &= \{(v_1, 0.6, 0.5, 0.5), (v_2, 0.5, 0.7, 0.5), (v_3, 0.5, 0.7, 0.5), (v_4, 0.5, 0.6, 0.5)\},\\
\overline{Q}(A_2) &= \{(v_1, 0.5, 0.4, 0.5), (v_2, 0.6, 0.6, 0.2), (v_3, 0.6, 0.6, 0.5), (v_4, 0.5, 0.7, 0.2)\}.
\end{aligned}$$

*Let* $E = \{v_1v_2, v_1v_4, v_2v_2, v_2v_3, v_3v_3, v_3v_4\} \subseteq V \times V$, *and* $L = \{e_1e_2, e_1e_3, e_2e_3\} \subset \acute{\mathbb{M}}$. *Then, a neutrosophic soft relation on E can be written as in Table 8.*

**Table 8.** Neutrosophic soft relation $S$.

| $S$ | $v_1v_2$ | $v_1v_4$ | $v_2v_2$ | $v_2v_3$ | $v_3v_3$ | $v_3v_4$ |
|---|---|---|---|---|---|---|
| $e_1e_2$ | (0.3, 0.4 ,0.1) | (0.4, 0.4, 0.2) | (0.4, 0.5, 0.1) | (0.3, 0.5, 0.4) | (0.3, 0.4, 0.4) | (0.4, 0.5, 0.2) |
| $e_1e_3$ | (0.2 ,0.3 ,0.3) | (0.4, 0.3, 0.2) | (0.2, 0.3, 0.5) | (0.4, 0.3, 0.3) | (0.5, 0.3, 0.3) | (0.5, 0.4, 0.3) |
| $e_2e_3$ | (0.2,0.3,0.5) | (0.3, 0.3, 0.3) | (0.2, 0.3, 0.1) | (0.4, 0.3, 0.1) | (0.3, 0.3, 0.5) | (0.3, 0.4, 0.3) |

*Let* $B_1 = \{(e_1e_2, 0.5, 0.4, 0.5), (e_1e_3, 0.3, 0.4, 0.5), (e_2e_3, 0.4, 0.4, 0.3)\}$, *and* $B_2 = \{(e_1e_2, 0.5, 0.3, 0.2), (e_1e_3, 0.4, 0.3, 0.3), (e_2e_3, 0.4, 0.6, 0.2)\}$ *are two neutrosophic sets on L, Then,* $S(B_1) = (\underline{S}(B_1), \overline{S}(B_1))$ *and* $S(B_2) = (\underline{S}(B_2), \overline{S}(B_2))$ *are NSRRs, where*

$$\begin{aligned}
\underline{S}(B_1) &= \{(v_1v_2, 0.3, 0.4, 0.3), (v_1v_4, 0.3, 0.4, 0.4), (v_2v_2, 0.4, 0.4, 0.4), (v_2v_3, 0.3, 0.4, 0.4),\\
&\quad (v_3v_3, 0.3, 0.4, 0.5), (v_3v_4, 0.3, 0.4, 0.5)\},\\
\overline{S}(B_1) &= \{(v_1v_2, 0.3, 0.4, 0.5), (v_1v_4, 0.4, 0.4, 0.3), (v_2v_2, 0.4, 0.4, 0.3), (v_2v_3, 0.4, 0.4, 0.3),\\
&\quad (v_3v_3, 0.3, 0.4, 0.5), (v_3v_4, 0.4, 0.4, 0.3)\};\\
\underline{S}(B_2) &= \{(v_1v_2, 0.4, 0.6, 0.2), (v_1v_4, 0.4, 0.6, 0.3), (v_2v_2, 0.4, 0.6, 0.2), (v_2v_3, 0.4, 0.6, 0.3),\\
&\quad (v_3v_3, 0.4, 0.6, 0.3), (v_3v_4, 0.4, 0.6, 0.3)\},\\
\overline{S}(B_2) &= \{(v_1v_2, 0.3, 0.3, 0.2), (v_1v_4, 0.4, 0.3, 0.2), (v_2v_2, 0.4, 0.3, 0.2), (v_2v_3, 0.4, 0.3, 0.2),\\
&\quad (v_3v_3, 0.4, 0.3, 0.3), (v_3v_4, 0.4, 0.4, 0.2)\}.
\end{aligned}$$

*Thus,* $G_1 = (\underline{G}_1, \overline{G}_1)$ *and* $G_2 = (\underline{G}_2, \overline{G}_2)$ *are NSRGs, where* $\underline{G}_1 = (\underline{Q}(A_1), \underline{S}(B_1))$, $\overline{G}_1 = (\overline{Q}(A_1), \overline{S}(B_1))$ *as shown in Figure 2.*

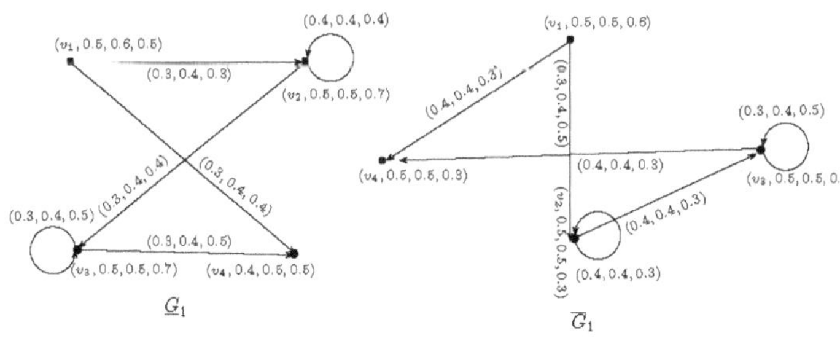


**Figure 2.** Neutrosophic soft rough graph $G_1 = (\underline{G}_1, \overline{G}_1)$

$\underline{G}_2 = (\underline{Q}(A_2), \underline{S}(B_2))$, $\overline{G}_2 = (\overline{Q}(A_2), \overline{S}(B_2))$ *as shown in Figure 3.*

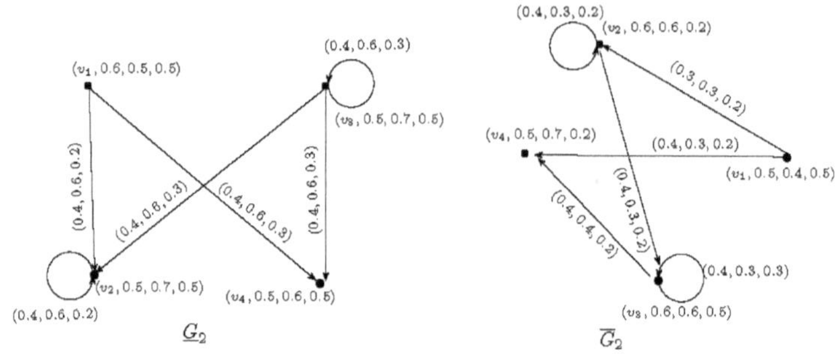


**Figure 3.** Neutrosophic soft rough graph $G_2 = (\underline{G}_2, \overline{G}_2)$

*The union of* $G_1 = (\underline{G}_1, \overline{G}_1)$ *and* $G_2 = (\underline{G}_2, \overline{G}_2)$ *is NSRG* $G = G_1 \cup G_2 = (\underline{G}_1 \cup \underline{G}_2, \overline{G}_1 \cup \overline{G}_2)$ *as shown in Figure 4.*

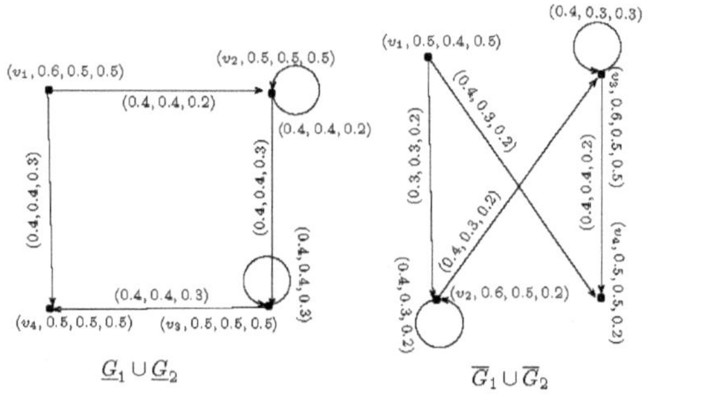


**Figure 4.** Neutrosophic soft rough graph $G_1 \cup G_2 = (\underline{G}_1 \cup \underline{G}_2, \overline{G}_1 \cup \overline{G}_2)$

**Definition 6.** *Let* $G_1 = (\underline{G}_1, \overline{G}_1)$ *and* $G_2 = (\underline{G}_2, \overline{G}_2)$ *be two NSRGs on* $V$. *The intersection of* $G_1$ *and* $G_2$ *is a neutrosophic soft rough graph* $G = G_1 \cap G_2 = (\underline{G}_1 \cap \underline{G}_2, \overline{G}_1 \cap \overline{G}_2)$, *where* $\underline{G}_1 \cap \underline{G}_2 = (\underline{Q}A_1 \cap \underline{Q}A_2, \underline{S}B_1 \cap \underline{S}B_2)$ *and* $\overline{G}_1 \cap \overline{G}_2 = (\overline{Q}A_1 \cap \overline{Q}A_2, \overline{S}B_1 \cap \overline{S}B_2)$ *are neutrosophic graphs, respectively, such that*

*(i)* $\forall v \in QA_1$ *but* $v \notin QA_2$.

$$T_{\overline{Q}A_1 \cap \overline{Q}A_2}(v) = T_{\overline{Q}A_1}(v), T_{\underline{Q}A_1 \cap \underline{Q}A_2}(v) = T_{\underline{Q}A_1}(v),$$
$$I_{\overline{Q}A_1 \cap \overline{Q}A_2}(v) = I_{\overline{Q}A_1}(v), I_{\underline{Q}A_1 \cap \underline{Q}A_2}(v) = I_{\underline{Q}A_1}(v),$$
$$F_{\overline{Q}A_1 \cap \overline{Q}A_2}(v) = F_{\overline{Q}A_1}(v), F_{\underline{Q}A_1 \cap \underline{Q}A_2}(v) = F_{\underline{Q}A_1}(v).$$

*(ii)* $\forall v \notin QA_1$ *but* $v \in QA_2$.

$$T_{\overline{Q}A_1 \cap \overline{Q}A_2}(v) = T_{\overline{Q}A_2}(v), T_{\underline{Q}A_1 \cap \underline{Q}A_2}(v) = T_{\underline{Q}A_2}(v),$$
$$I_{\overline{Q}A_1 \cap \overline{Q}A_2}(v) = I_{\overline{Q}A_2}(v), I_{\underline{Q}A_1 \cap \underline{Q}A_2}(v) = I_{\underline{Q}A_2}(v),$$
$$F_{\overline{Q}A_1 \cap \overline{Q}A_2}(v) = F_{\overline{Q}A_2}(v), F_{\underline{Q}A_1 \cap \underline{Q}A_2}(v) = F_{\underline{Q}A_2}(v).$$

(iii) $\forall v \in QA_1 \cap \underline{Q}A_2$

$$T_{\overline{Q}A_1 \cap \overline{Q}A_2}(v) = \min\{T_{\overline{Q}A_1}(v), T_{\overline{Q}A_2}(v)\}, \; T_{\underline{Q}A_1 \cap \underline{Q}A_2}(v) = \min\{T_{\underline{Q}A_1}(v), T_{\underline{Q}A_2}(v)\},$$
$$I_{\overline{Q}A_1 \cap \overline{Q}A_2}(v) = \max\{I_{\overline{Q}A_1}(v), I_{\overline{Q}A_2}(v)\}, \; I_{\underline{Q}A_1 \cap \underline{Q}A_2}(v) = \max\{I_{\underline{Q}A_1}(v), I_{\underline{Q}A_2}(v)\},$$
$$F_{\overline{Q}A_1 \cap \overline{Q}A_2}(v) = \max\{F_{\overline{Q}A_1}(v), F_{\overline{Q}A_2}(v)\}, F_{\underline{Q}A_1 \cap \underline{Q}A_2}(v) = \max\{F_{\underline{Q}A_1}(v), F_{\underline{Q}A_2}(v)\}.$$

(iv) $\forall vu \in SB_1$ *but* $vu \notin SB_2$.

$$T_{\overline{S}B_1 \cap \overline{S}B_2}(vu) = T_{\overline{S}B_1}(vu), T_{\underline{S}B_1 \cap \underline{S}B_2}(vu) = T_{\underline{S}B_1}(vu),$$
$$I_{\overline{S}B_1 \cap \overline{S}B_2}(vu) = I_{\overline{S}B_1}(vu), \; I_{\underline{S}B_1 \cap \underline{S}B_2}(vu) = I_{\underline{S}B_1}(vu),$$
$$F_{\overline{S}B_1 \cap \overline{S}B_2}(vu) = F_{\overline{S}B_1}(vu), F_{\underline{S}B_1 \cap \underline{S}B_2}(vu) = F_{\underline{S}B_1}(vu).$$

(v) $\forall vu \notin SB_1$ *but* $vu \in SB_2$

$$T_{\overline{S}B_1 \cap \overline{S}B_2}(vu) = T_{\overline{S}B_2}(vu), T_{\underline{S}B_1 \cap \underline{S}B_2}(vu) = T_{\underline{S}B_2}(vu),$$
$$I_{\overline{S}B_1 \cap \overline{S}B_2}(vu) = I_{\overline{S}B_2}(vu), \; I_{\underline{S}B_1 \cap \underline{S}B_2}(vu) = I_{\underline{S}B_2}(vu),$$
$$F_{\overline{S}B_1 \cap \overline{S}B_2}(vu) = F_{\overline{S}B_2}(vu), F_{\underline{S}B_1 \cap \underline{S}B_2}(vu) = F_{\underline{S}B_2}(vu).$$

(vi) $\forall vu \in SB_1 \cap \underline{S}B_2$

$$T_{\overline{S}B_1 \cap \overline{S}B_2}(vu) = \min\{T_{\overline{S}B_1}(vu), T_{\overline{S}B_2}(vu)\}, \; T_{\underline{S}B_1 \cap \underline{S}B_2}(vu) = \min\{T_{\underline{S}}B_1(vu), T_{\underline{S}B_2}(vu)\},$$
$$I_{\overline{S}B_1 \cap \overline{S}B_2}(vu) = \max\{I_{\overline{S}B_1}(vu), I_{\overline{S}B_2}(vu)\}, \; I_{\underline{S}B_1 \cap \underline{S}B_2}(vu) = \max\{I_{\underline{S}B_1}(vu), I_{\underline{S}B_2}(vu)\},$$
$$F_{\overline{S}B_1 \cap \overline{S}B_2}(vu) = \max\{F_{\overline{S}B_1}(vu), F_{\overline{S}B_2}(vu)\}, F_{\underline{S}B_1 \cap \underline{S}B_2}(vu) = \max\{F_{\underline{S}B_1}(vu), F_{\underline{S}B_2}(vu)\}.$$

**Definition 7.** *Let* $G_1 = (\underline{G}_1, \overline{G}_1)$ *and* $G_2 = (\underline{G}_2, \overline{G}_2)$ *be two neutrosophic soft rough graphs on* $V$. *The join of* $G_1$ *and* $G_2$ *is a neutrosophic soft rough graph* $G = G_1 + G_2 = (\underline{G}_1 + \underline{G}_2, \overline{G}_1 + \overline{G}_2)$, *where* $\underline{G}_1 + \underline{G}_2 = (\underline{Q}A_1 + \underline{Q}A_2, \underline{S}B_1 + \underline{S}B_2)$ *and* $\overline{G}_1 + \overline{G}_2 = (\overline{Q}A_1 + \overline{Q}A_2, \overline{S}B_1 + \overline{S}B_2)$ *are neutrosophic graph, respectively, such that*

(i) $\forall v \in QA_1$ *but* $v \notin QA_2$.

$$T_{\overline{Q}A_1 + \overline{Q}A_2}(v) = T_{\overline{Q}A_1}(v), T_{\underline{Q}A_1 + \underline{Q}A_2}(v) = T_{\underline{Q}A_1}(v),$$
$$I_{\overline{Q}A_1 + \overline{Q}A_2}(v) = I_{\overline{Q}A_1}(v), \; I_{\underline{Q}A_1 + \underline{Q}A_2}(v) = I_{\underline{Q}A_1}(v),$$
$$F_{\overline{Q}A_1 + \overline{Q}A_2}(v) = F_{\overline{Q}A_1}(v), F_{\underline{Q}A_1 + \underline{Q}A_2}(v) = F_{\underline{Q}A_1}(v).$$

(ii) $\forall v \notin QA_1$ *but* $v \in QA_2$.

$$T_{\overline{Q}A_1 + \overline{Q}A_2}(v) = T_{\overline{Q}A_2}(v), T_{\underline{Q}A_1 + \underline{Q}A_2}(v) = T_{\underline{Q}A_2}(v),$$
$$I_{\overline{Q}A_1 + \overline{Q}A_2}(v) = I_{\overline{Q}A_2}(v), \; I_{\underline{Q}A_1 + \underline{Q}A_2}(v) = I_{\underline{Q}A_2}(v),$$
$$F_{\overline{Q}A_1 + \overline{Q}A_2}(v) = F_{\overline{Q}A_2}(v), F_{\underline{Q}A_1 + \underline{Q}A_2}(v) = F_{\underline{Q}A_2}(v).$$

(iii) $\forall v \in QA_1 \cap \underline{Q}A_2$

$$T_{\overline{Q}A_1 + \overline{Q}A_2}(v) = \max\{T_{\overline{Q}A_1}(v), T_{\overline{Q}A_2}(v)\}, T_{\underline{Q}A_1 + \underline{Q}A_2}(v) = \max\{T_{\underline{Q}A_1}(v), T_{\underline{Q}A_2}(v)\},$$
$$I_{\overline{Q}A_1 + \overline{Q}A_2}(v) = \min\{I_{\overline{Q}A_1}(v), I_{\overline{Q}A_2}(v)\}, \; I_{\underline{Q}A_1 + \underline{Q}A_2}(v) = \min\{I_{\underline{Q}A_1}(v), I_{\underline{Q}A_2}(v)\},$$
$$F_{\overline{Q}A_1 + \overline{Q}A_2}(v) = \min\{F_{\overline{Q}A_1}(v), F_{\overline{Q}A_2}(v)\}, F_{\underline{Q}A_1 + \underline{Q}A_2}(v) = \min\{F_{\underline{Q}A_1}(v), F_{\underline{Q}A_2}(v)\}.$$

*(iv)* $\forall vu \in SB_1$ *but* $vu \notin SB_2$.

$$\begin{aligned}
T_{\overline{S}B_1+\overline{S}B_2}(vu) &= T_{\overline{S}B_1}(vu), T_{\underline{S}B_1+\underline{S}B_2}(vu) = T_{\underline{S}B_1}(vu),\\
I_{\overline{S}B_1+\overline{S}B_2}(vu) &= I_{\overline{S}B_1}(vu), \ I_{\underline{S}B_1+\underline{S}B_2}(vu) = I_{\underline{S}B_1}(vu),\\
F_{\overline{S}B_1+\overline{S}B_2}(vu) &= F_{\overline{S}B_1}(vu), F_{\underline{S}B_1+\underline{S}B_2}(vu) = F_{\underline{S}B_1}(vu).
\end{aligned}$$

*(v)* $\forall vu \notin SB_1$ *but* $vu \in SB_2$

$$\begin{aligned}
T_{\overline{S}B_1+\overline{S}B_2}(vu) &= T_{\overline{S}B_2}(vu), T_{\underline{S}B_1+\underline{S}B_2}(vu) = T_{\underline{S}B_2}(vu),\\
I_{\overline{S}B_1+\overline{S}B_2}(vu) &= I_{\overline{S}B_2}(vu), \ I_{\underline{S}B_1+\underline{S}B_2}(vu) = I_{\underline{S}B_2}(vu),\\
F_{\overline{S}B_1+\overline{S}B_2}(vu) &= F_{\overline{S}B_2}(vu), F_{\underline{S}B_1+\underline{S}B_2}(vu) = F_{\underline{S}B_2}(vu).
\end{aligned}$$

*(vi)* $\forall vu \in SB_1 \cap \underline{S}B_2$

$$\begin{aligned}
T_{\overline{S}B_1+\overline{S}B_2}(vu) &= \max\{T_{\overline{S}B_1}(vu), T_{\overline{S}B_2}(vu)\}, T_{\underline{S}B_1+\underline{S}B_2}(vu) = \max\{T_{\underline{S}}B_1(vu), T_{\underline{S}B_2}(vu)\},\\
I_{\overline{S}B_1+\overline{S}B_2}(vu) &= \min\{I_{\overline{S}B_1}(vu), I_{\overline{S}B_2}(vu)\}, \ I_{\underline{S}B_1+\underline{S}B_2}(vu) = \min\{I_{\underline{S}B_1}(vu), I_{\underline{S}B_2}(vu)\},\\
F_{\overline{S}B_1+\overline{S}B_2}(vu) &= \min\{F_{\overline{S}B_1}(vu), F_{\overline{S}B_2}(vu)\}, F_{\underline{S}B_1+\underline{S}B_2}(vu) = \min\{F_{\underline{S}B_1}(vu), F_{\underline{S}B_2}(vu)\}.
\end{aligned}$$

*(vii)* $\forall vu \in \tilde{E}$, *where* $\tilde{E}$ *is the set of edges joining vertices of* $QA_1$ *and* $QA_2$.

$$\begin{aligned}
T_{\overline{S}B_1+\overline{S}B_2}(vu) &= \min\{T_{\overline{Q}A_1}(v), T_{\overline{Q}A_2}(u)\}, \ T_{\underline{S}B_1+\underline{S}B_2}(vu) = \min\{T_{\underline{Q}A_1}(v), T_{\underline{Q}A_2}(u)\},\\
I_{\overline{S}B_1+\overline{S}B_2}(vu) &= \max\{I_{\overline{Q}A_1}(v), I_{\overline{Q}A_2}(u)\}, \ I_{\underline{S}B_1+\underline{S}B_2}(vu) = \max\{I_{\underline{Q}A_1}(v), I_{\underline{Q}A_2}(u)\},\\
F_{\overline{S}B_1+\overline{S}B_2}(vu) &= \max\{F_{\overline{Q}A_1}(v), F_{\overline{Q}A_2}(u)\}, F_{\underline{S}B_1+\underline{S}B_2}(vu) = \max\{F_{\underline{Q}A_1}(v), F_{\underline{Q}A_2}(u)\}.
\end{aligned}$$

**Definition 8.** *The Cartesian product of* $G_1$ *and* $G_2$ *is a* $G = G_1 \ltimes G_2 = (\underline{G}_1 \ltimes \underline{G}_2, \overline{G}_1 \ltimes \overline{G}_2)$, *where* $\underline{G}_1 \ltimes \underline{G}_2 = (\underline{Q}A_1 \ltimes \underline{Q}A_2, \underline{S}B_1 \ltimes \underline{S}B_2)$ *and* $\overline{G}_1 \ltimes \overline{G}_2 = (\overline{Q}A_1 \ltimes \overline{Q}A_2, \overline{S}B_1 \ltimes \overline{S}B_2)$ *are neutrosophic digraph, such that*

*(i)* $\forall\, (v_1, v_2) \in QA_1 \times QA_2$.

$$\begin{aligned}
T_{(\overline{Q}A_1 \ltimes \overline{Q}A_2)}(v_1, v_2) &= \min\{T_{\overline{Q}A_1}(v_1), T_{\overline{Q}A_2}(v_1)\}, \ T_{(\underline{Q}A_1 \ltimes \underline{Q}A_2)}(v_1, v_2) = \min\{T_{\underline{Q}A_1}(v_1), T_{\underline{Q}A_2}(v_1)\},\\
I_{(\overline{Q}A_1 \ltimes \overline{Q}A_2)}(v_1, v_2) &= \max\{I_{\overline{Q}A_1}(v_1), I_{\overline{Q}A_2}(v_1)\}, \ I_{(\underline{Q}A_1 \ltimes \underline{Q}A_2)}(v_1, v_2) = \max\{I_{\underline{Q}A_1}(v_1), I_{\underline{Q}A_2}(v_1)\},\\
F_{(\overline{Q}A_1 \ltimes \overline{Q}A_2)}(v_1, v_2) &= \max\{F_{\overline{Q}A_1}(v_1), F_{\overline{Q}A_2}(v_1)\}, F_{(\underline{Q}A_1 \ltimes \underline{Q}A_2)}(v_1, v_2) = \max\{F_{\underline{Q}A_1}(v_1), F_{\underline{Q}A_2}(v_1)\}.
\end{aligned}$$

*(ii)* $\forall v_1 v_2 \in SB_2, v \in QA_1$.

$$\begin{aligned}
T_{(\overline{S}B_1 \ltimes \overline{S}B_2)}\big((v, v_1)(v, v_2)\big) &= \min\{T_{\overline{Q}A_1}(v), T_{\overline{S}B_2}(v_1 v_2)\},\\
T_{(\underline{S}B_1 \ltimes \underline{S}B_2)}\big((v, v_1)(v, v_2)\big) &= \min\{T_{\underline{Q}A_1}(v), T_{\underline{S}B_2}(v_1 v_2)\},\\
I_{(\overline{S}B_1 \ltimes \overline{S}B_2)}\big((v, v_1)(v, v_2)\big) &= \max\{I_{\overline{Q}A_1}(v), I_{\overline{S}B_2}(v_1 v_2)\},\\
I_{(\underline{S}B_1 \ltimes \underline{S}B_2)}\big((v, v_1)(v, v_2)\big) &= \max\{I_{\underline{Q}A_1}(v), I_{\underline{S}B_2}(v_1 v_2)\},\\
F_{(\overline{S}B_1 \ltimes \overline{S}B_2)}\big((v, v_1)(v, v_2)\big) &= \max\{F_{\overline{Q}A_1}(v), F_{\overline{S}B_2}(v_1 v_2)\},\\
F_{(\underline{S}B_1 \ltimes \underline{S}B_2)}\big((v, v_1)(v, v_2)\big) &= \max\{F_{\underline{Q}A_1}(v), F_{\underline{S}B_2}(v_1 v_2)\}.
\end{aligned}$$

(iii) $\forall v_1 v_2 \in SB_1, v \in QA_2.$

$$\begin{aligned}
T_{(\underline{S}B_1 \ltimes \underline{S}B_2)}((v_1,v)(v_2,v)) &= \min\{T_{\underline{S}B_1}(v_1v_2), T_{\underline{Q}A_2}(v)\},\\
T_{(\overline{S}B_1 \ltimes \overline{S}B_2)}((v_1,v)(v_2,v)) &= \min\{T_{\overline{S}B_1}(v_1v_2), T_{\overline{Q}A_2}(v)\},\\
I_{(\overline{S}B_1 \ltimes \overline{S}B_2)}((v_1,v)(v_2,v)) &= \max\{I_{\overline{S}B_1}(v_1v_2), I_{\overline{Q}A_2}(v)\},\\
I_{(\underline{S}B_1 \ltimes \underline{S}B_2)}((v_1,v)(v_2,v)) &= \max\{I_{\underline{S}B_1}(v_1v_2), I_{\underline{Q}A_2}(v)\},\\
F_{(\overline{S}B_1 \ltimes \overline{S}B_2)}((v_1,v)(v_2,v)) &= \max\{F_{\overline{S}B_1}(v_1v_2), F_{\overline{Q}A_2}(v)\},\\
F_{(\underline{S}B_1 \ltimes \underline{S}B_2)}((v_1,v)(v_2,v)) &= \max\{F_{\underline{S}B_1}(v_1v_2), F_{\underline{Q}A_2}(v)\}.
\end{aligned}$$

**Definition 9.** *The cross product of $G_1$ and $G_2$ is a neutrosophic soft rough graph $G = G_1 \circledcirc G_2 = (\underline{G}_1 \circledcirc \underline{G}_2, \overline{G}_1 \circledcirc \overline{G}_2)$, where $\underline{G}_1 \circledcirc \underline{G}_2 = (\underline{Q}A_1 \circledcirc \underline{Q}A_2, \underline{S}B_1 \circledcirc \underline{S}B_2)$ and $\overline{G}_1 \circledcirc \overline{G}_2 = (\overline{Q}A_1 \circledcirc \overline{Q}A_2, \overline{S}B_1 \circledcirc \overline{S}B_2)$ are neutrosophic graphs, respectively, such that*

(i) $\forall\ (v_1, v_2) \in QA_1 \times QA_2.$

$$\begin{aligned}
T_{(\overline{Q}A_1 \circledcirc \overline{Q}A_2)}(v_1,v_2) &= \min\{T_{\overline{Q}A_1}(v_1), T_{\overline{Q}A_2}(v_1)\},\ T_{(\underline{Q}A_1 \circledcirc \underline{Q}A_2)}(v_1,v_2) = \min\{T_{\underline{Q}A_1}(v_1), T_{\underline{Q}A_2}(v_1)\},\\
I_{(\overline{Q}A_1 \circledcirc \overline{Q}A_2)}(v_1,v_2) &= \max\{I_{\overline{Q}A_1}(v_1), I_{\overline{Q}A_2}(v_1)\},\ I_{(\underline{Q}A_1 \circledcirc \underline{Q}A_2)}(v_1,v_2) = \max\{I_{\underline{Q}A_1}(v_1), I_{\underline{Q}A_2}(v_1)\},\\
F_{(\overline{Q}A_1 \circledcirc \overline{Q}A_2)}(v_1,v_2) &= \max\{F_{\overline{Q}A_1}(v_1), F_{\overline{Q}A_2}(v_1)\}, F_{(\underline{Q}A_1 \circledcirc \underline{Q}A_2)}(v_1,v_2) = \max\{F_{\underline{Q}A_1}(v_1), F_{\underline{Q}A_2}(v_1)\}.
\end{aligned}$$

(ii) $\forall v_1 u_1 \in SB_1, v_2 u_2 \in SB_2.$

$$\begin{aligned}
T_{(\overline{S}B_1 \circledcirc \overline{S}B_2)}((v_1,v_2)(u_1,u_2)) &= \min\{T_{\overline{S}B_1}(v_1u_1), T_{\overline{S}B_2}(v_1u_2)\},\\
T_{(\underline{S}B_1 \circledcirc \underline{S}B_2)}((v_1,v_2)(u_1,u_2)) &= \min\{T_{\underline{S}B_1}(v_1u_1), T_{\underline{S}B_2}(v_1u_2)\},\\
I_{(\overline{S}B_1 \circledcirc \overline{S}B_2}((v_1,v_2)(u_1,u_2)) &= \max\{I_{\overline{S}B_1}(v_1u_1), I_{\overline{S}B_2}(v_1u_2)\},\\
I_{(\underline{S}B_1 \circledcirc \underline{S}B_2)}((v_1,v_2)(u_1,u_2)) &= \max\{I_{\underline{S}B_1}(v_1u_1), I_{\underline{S}B_2}(v_1u_2)\},\\
F_{(\overline{S}B_1 \circledcirc \overline{S}B_2}((v_1,v_2)(u_1,u_2)) &= \max\{F_{\overline{S}B_1}(v_1u_1), F_{\overline{S}B_2}(v_1u_2)\},\\
F_{(\underline{S}B_1 \circledcirc \underline{S}B_2)}((v_1,v_2)(u_1,u_2)) &= \max\{F_{\underline{S}B_1}(v_1u_1), F_{\underline{S}B_2}(v_1u_2)\}.
\end{aligned}$$

**Definition 10.** *The rejection of $G_1$ and $G_2$ is a neutrosophic soft rough graph $G = G_1|G_2 = (\underline{G}_1|\underline{G}_2, \overline{G}_1|\overline{G}_2)$, where $\underline{G}_1|\underline{G}_2 = (\underline{S}A_1|\underline{S}A_2, \underline{S}B_1|\underline{S}B_2)$ and $\overline{G}_1|\overline{G}_2 = (\overline{S}A_1|\overline{S}A_2, \overline{S}B_1|\overline{S}B_2)$ are neutrosophic graphs such that*

(i) $\forall\ (v_1, v_2) \in QA_1 \times QA_2.$

$$\begin{aligned}
T_{(\overline{Q}A_1|\overline{Q}A_2)}(v_1,v_2) &= \min\{T_{\overline{Q}A_1}(v_1), T_{\overline{Q}A_2}(v_2)\}, T_{(\underline{Q}A_1|\underline{Q}A_2)}(v_1,v_2) = \min\{T_{\underline{Q}A_1}(v_1), T_{\underline{Q}A_2}(v_2)\},\\
I_{(\overline{Q}A_1|\overline{Q}A_2)}(v_1,v_2) &= \max\{I_{\overline{Q}A_1}(v_1), I_{\overline{Q}A_2}(v_2)\},\ I_{(\underline{Q}A_1|\underline{Q}A_2)}(v_1,v_2) = \max\{I_{\underline{Q}A_1}(v_1), I_{\underline{Q}A_2}(v_2)\},\\
F_{(\overline{Q}A_1|\overline{Q}A_2)}(v_1,v_2) &= \max\{F_{\overline{Q}A_1}(v_1), F_{\overline{Q}A_2}(v_2)\}, F_{(\underline{Q}A_1|\underline{Q}A_2)}(v_1,v_2) = \max\{F_{\underline{Q}A_1}(v_1), F_{\underline{Q}A_2}(v_2)\}.
\end{aligned}$$

(ii) $\forall v_2 u_2 \notin SB_2,\ v \in QA_1.$

$$\begin{aligned}
T_{(\overline{S}B_1|\overline{S}B_2)}((v,v_2)(v,u_2)) &= \min\{T_{\overline{Q}A_1}(v), T_{\overline{Q}A_2}(v_2), T_{\overline{Q}A_2}(u_2)\},\\
T_{(\underline{S}B_1|\underline{Q}B_2)}((v,v_2)(v,u_2)) &= \min\{T_{\underline{Q}A_1}(v), T_{\underline{Q}A_2}(v_2), T_{\underline{Q}A_2}(u_2)\},\\
(I_{\overline{S}B_1|\overline{S}B_2)}((v,v_2)(v,u_2)) &= \max\{I_{\overline{Q}A_1}(v), I_{\overline{Q}A_2}(v_2), I_{\overline{Q}A_2}(u_2)\},\\
(I_{\underline{S}B_1|\underline{S}B_2)}((v,v_2)(v,u_2)) &= \max\{I_{\underline{Q}A_1}(v), I_{\underline{Q}A_2}(v_2), I_{\underline{Q}A_2}(u_2)\},\\
(F_{\overline{S}B_1|\overline{S}B_2)}((v,v_2)(v,u_2)) &= \max\{F_{\overline{Q}A_1}(v), F_{\overline{Q}A_2}(v_2), F_{\overline{Q}A_2}(u_2)\},\\
(F_{\underline{S}B_1|\underline{S}B_2)}((v,v_2)(v,u_2)) &= \max\{F_{\underline{Q}A_1}(v), F_{\underline{Q}A_2}(v_2), F_{\underline{Q}A_2}(u_2)\}.
\end{aligned}$$

*(iii)* $\forall v_1u_1 \notin SB_1,\ v \in QA_2,$

$$\begin{aligned}
T_{(\underline{S}B_1|\underline{S}B_2)}((v_1,v)(u_1,v)) &= \min\{T_{\underline{Q}A_1}(v_1), T_{\underline{Q}A_1}(u_1), T_{\underline{Q}A_2}(v)\},\\
I_{(\underline{S}B_1|\underline{S}B_2)}((v_1,v)(u_1,v)) &= \max\{I_{\underline{Q}A_1}(v_1), I_{\underline{Q}A_1}(u_1), I_{\underline{Q}A_2}(v)\},\\
F_{(\underline{S}B_1|\underline{S}B_2)}((v_1,v)(u_1,v)) &= \max\{F_{\underline{Q}A_1}(v_1), F_{\underline{Q}A_1}(u_1), F_{\underline{Q}A_2}(v)\},\\
T_{(\overline{S}B_1|\overline{S}B_2)}((v_1,v)(u_1,v)) &= \min\{T_{\overline{Q}A_1}(v_1), T_{\overline{Q}A_1}(u_1), T_{\overline{Q}A_2}(v)\},\\
I_{(\overline{S}B_1|\overline{S}B_2)}((v_1,v)(u_1,v)) &= \max\{I_{\overline{Q}A_1}(v_1), I_{\overline{Q}A_1}(u_1), I_{\overline{Q}A_2}(v)\},\\
F_{(\overline{S}B_1|\overline{S}B_2)}((v_1,v)(u_1,v)) &= \max\{F_{\overline{Q}A_1}(v_1), F_{\overline{Q}A_1}(u_1), F_{\overline{Q}A_2}(v)\}.
\end{aligned}$$

*(iv)* $\forall v_1u_1 \notin \overline{S}B_1,\ v_2u_2 \notin \overline{S}B_2,\ v_1 = u_1.$

$$\begin{aligned}
T_{(\underline{S}B_1|\underline{S}B_2)}((v_1,v_2)(u_1,u_2)) &= \min\{T_{\underline{Q}A_1}(v_1), T_{\underline{Q}A_1}(u_1), T_{\underline{Q}A_2}(v_2), T_{\underline{Q}A_2}(u_2)\},\\
I_{(\underline{S}B_1|\underline{S}B_2)}((v_1,v_2)(u_1,u_2)) &= \max\{I_{\underline{Q}A_1}(v_1), I_{\underline{Q}A_1}(u_1), I_{\underline{Q}A_2}(v_2), I_{\underline{Q}A_2}(u_2)\},\\
F_{(\underline{S}B_1|\underline{S}B_2)}((v_1,v_2)(u_1,u_2)) &= \max\{F_{\underline{Q}A_1}(v_1), F_{\underline{Q}A_1}(u_1), F_{\underline{Q}A_2}(v_2), F_{\underline{Q}A_2}(u_2)\},\\
T_{(\overline{S}B_1|\overline{S}B_2)}((v_1,v_2)(u_1,u_2)) &= \min\{T_{\overline{Q}A_1}(v_1), T_{\overline{Q}A_1}(u_1), T_{\overline{Q}A_2}(v_2), T_{\overline{Q}A_2}(u_2)\},\\
I_{(\overline{S}B_1|\overline{S}B_2)}((v_1,v_2)(u_1,u_2)) &= \max\{I_{\overline{Q}A_1}(v_1), I_{\overline{Q}A_1}(u_1), I_{\overline{Q}A_2}(v_2), I_{\overline{Q}A_2}(u_2)\},\\
F_{(\overline{S}B_1|\overline{S}B_2)}((v_1,v_2)(u_1,u_2)) &= \max\{F_{\overline{Q}A_1}(v_1), F_{\overline{Q}A_1}(u_1), F_{\overline{Q}A_2}(v_2), F_{\overline{Q}A_2}(u_2)\},
\end{aligned}$$

**Example 6.** *Let* $G_1 = (\underline{G}_1, \overline{G}_1)$ *and* $G_2 = (\underline{G}_2, \overline{G}_2)$ *be two neutrosophic soft rough graphs on* $V$*, where* $\underline{G}_1 = (\underline{Q}A_1, \underline{S}B_1)$ *and* $\overline{G}_1 = (\overline{Q}A_1, \overline{S}B_1)$ *are neutrosophic graphs as shown in Figure 2 and* $\underline{G}_2 = (\underline{Q}A_2, \underline{S}B_2)$ *and* $\overline{G}_2 = (\overline{Q}A_2, \overline{S}B_2)$ *are neutrosophic graphs as shown in Figure 3. The Cartesian product of* $G_1 = (\underline{G}_1, \overline{G}_1)$ *and* $G_2 = (\underline{G}_2, \overline{G}_2)$ *is NSRG* $G = G_1 \times G_2 = (\underline{G}_1 \times \underline{G}_2, \overline{G}_1 \times \overline{G}_2)$ *as shown in Figure 5.*

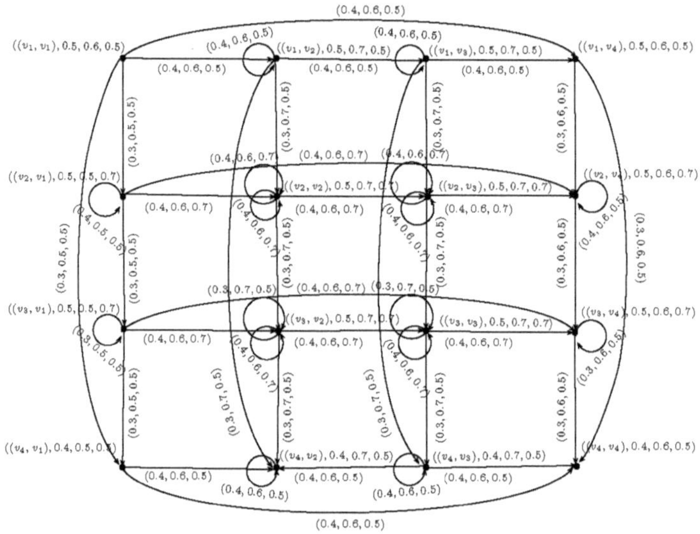


$\underline{G}_1 \times \underline{G}_2$

**(a)**

**Figure 5.** *Cont.*

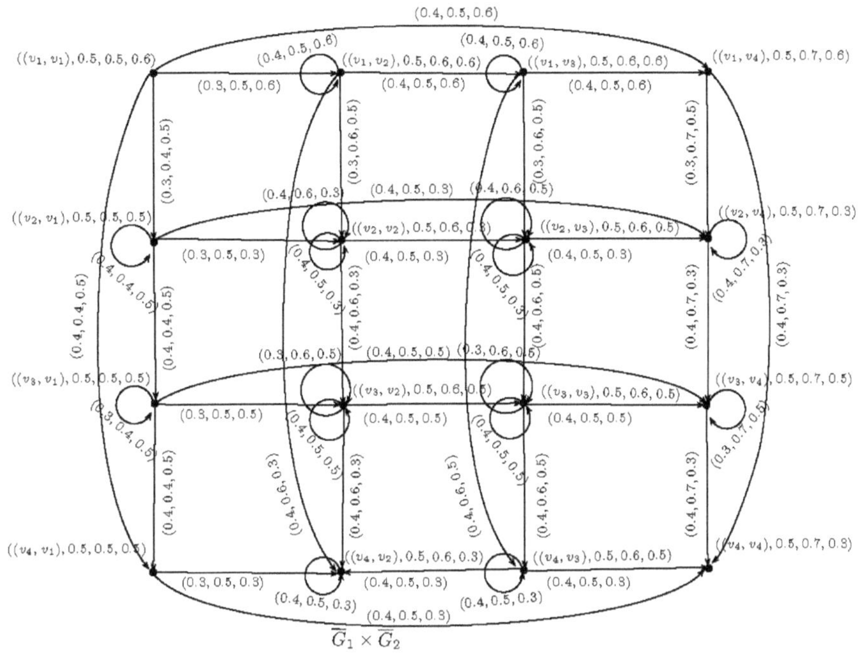


**(b)**

**Figure 5.** Cartesian product of two neutrosophic soft rough graphs $G_1 \times G_2$

**Definition 11.** *The symmetric difference of $G_1$ and $G_2$ is a neutrosophic soft rough graph $G = G_1 \oplus G_2 = (\underline{G}_1 \oplus \underline{G}_2, \overline{G}_1 \oplus \overline{G}_2)$, where $\underline{G}_1 \oplus \underline{G}_2 = (\underline{Q}A_1 \oplus \underline{Q}A_2, \underline{S}B_1 \oplus \underline{S}B_2)$ and $\overline{G}_1 \oplus \overline{G}_2 = (\overline{Q}A_1 \oplus \overline{Q}A_2, \overline{S}B_1 \oplus \overline{S}B_2)$ are neutrosophic graphs, respectively, such that*

*(i)* $\forall\ (v_1, v_2) \in QA_1 \times QA_2$.

$$T_{(\overline{Q}A_1 \oplus \overline{Q}A_2)}(v_1, v_2) = \min\{T_{\overline{Q}A_1}(v_1), T_{\overline{Q}A_2}(v_2)\},\ T_{(\underline{Q}A_1 \oplus \underline{Q}A_2)}(v_1, v_2) = \min\{T_{\underline{Q}A_1}(v_1), T_{\underline{Q}A_2}(v_2)\},$$
$$I_{(\overline{Q}A_1 \oplus \overline{Q}A_2)}(v_1, v_2) = \max\{I_{\overline{Q}A_1}(v_1), I_{\overline{Q}A_2}(v_2)\},\ I_{(\underline{Q}A_1 \oplus \underline{Q}A_2)}(v_1, v_2) = \max\{I_{\underline{Q}A_1}(v_1), I_{\underline{Q}A_2}(v_2)\},$$
$$F_{(\overline{Q}A_1 \oplus \overline{Q}A_2)}(v_1, v_2) = \max\{F_{\overline{Q}A_1}(v_1), F_{\overline{Q}A_2}(v_2)\}, F_{(\underline{Q}A_1 \oplus \underline{Q}A_2)}(v_1, v_2) = \max\{F_{\underline{Q}A_1}(v_1), F_{\underline{Q}A_2}(v_2)\}.$$

*(ii)* $\forall v_1 v_2 \in SB_2,\ v \in QA_1$.

$$T_{(\overline{S}B_1 \oplus \overline{S}B_2)}((v, v_1)(v, v_2)) = \min\{T_{\overline{Q}A_1}(v), T_{\overline{S}B_2}(v_1 v_2)\},$$
$$T_{(\underline{S}B_1 \oplus \underline{S}B_2)}((v, v_1)(v, v_2)) = \min\{T_{\underline{Q}A_1}(v), T_{\underline{S}B_2}(v_1 v_2)\},$$
$$I_{(\overline{S}B_1 \oplus \overline{S}B_2)}((v, v_1)(v, v_2)) = \max\{I_{\overline{Q}A_1}(v), I_{\overline{S}B_2}(v_1 v_2)\},$$
$$I_{(\underline{S}B_1 \oplus \underline{S}B_2)}((v, v_1)(v, v_2)) = \max\{I_{\underline{Q}A_1}(v), I_{\underline{S}B_2}(v_1 v_2)\},$$
$$F_{(\overline{S}B_1 \oplus \overline{S}B_2)}((v, v_1)(v, v_2)) = \max\{F_{\overline{Q}A_1}(v), F_{\overline{S}B_2}(v_1 v_2)\},$$
$$F_{(\underline{S}B_1 \oplus \underline{S}B_2)}((v, v_1)(v, v_2)) = \max\{F_{\underline{Q}A_1}(v), F_{\underline{S}B_2}(v_1 v_2)\}.$$

*(iii)* $\forall v_1v_2 \in SB_1, v \in QA_2$.

$$\begin{aligned}
T_{(\overline{S}B_1 \oplus \overline{S}B_2)}((v_1,v)(v_2,v)) &= \min\{T_{\overline{S}B_1}(v_1v_2), T_{\overline{Q}A_2}(v)\},\\
T_{(\underline{S}B_1 \oplus \underline{S}B_2)}((v_1,v)(v_2,v)) &= \min\{T_{\underline{S}B_1}(v_1v_2), T_{\underline{Q}A_2}(v)\},\\
I_{(\overline{S}B_1 \oplus \overline{S}B_2)}((v_1,v)(v_2,v)) &= \max\{I_{\overline{S}B_1}(v_1v_2), I_{\overline{Q}A_2}(v)\},\\
I_{(\underline{S}B_1 \oplus \underline{S}B_2)}((v_1,v)(v_2,v)) &= \max\{I_{\underline{S}B_1}(v_1v_2), I_{\underline{Q}A_2}(v)\},\\
F_{(\overline{S}B_1 \oplus \overline{S}B_2)}((v_1,v)(v_2,v)) &= \max\{F_{\overline{S}B_1}(v_1v_2), F_{\overline{Q}A_2}(v)\},\\
F_{(\underline{S}B_1 \oplus \underline{S}B_2)}((v_1,v)(v_2,v)) &= \max\{F_{\underline{S}B_1}(v_1v_2), F_{\underline{Q}A_2}(v)\}.
\end{aligned}$$

*(iv)* $\forall v_1u_1 \notin SB_1,\ v_2u_2 \in SB_2$.

$$\begin{aligned}
T_{(\overline{S}B_1 \oplus \overline{S}B_2)}((v_1,v_2)(u_1,u_2)) &= \min\{T_{\overline{S}B_1}(v_1u_1), T_{\overline{Q}A_2}(v_2), T_{\overline{Q}A_2}(u_2)\},\\
T_{(\underline{S}B_1 \oplus \underline{S}B_2)}((v_1,v_2)(u_1,u_2)) &= \min\{T_{\underline{S}B_1}(v_1u_1), T_{\underline{Q}A_2}(v_2), T_{\underline{Q}A_2}(u_2)\},\\
I_{(\overline{S}B_1 \oplus \overline{S}B_2)}((v_1,v_2)(u_1,u_2)) &= \max\{I_{\overline{S}B_1}(v_1u_1), I_{\overline{Q}A_2}(v_2), I_{\overline{Q}A_2}(u_2)\},\\
I_{(\underline{S}B_1 \oplus \underline{S}B_2)}((v_1,v_2)(u_1,u_2)) &= \max\{I_{\underline{S}B_1}(v_1u_1), I_{\underline{Q}A_2}(v_2), I_{\underline{Q}A_2}(u_2)\},\\
F_{(\overline{S}B_1 \oplus \overline{S}B_2)}((v_1,v_2)(u_1,u_2)) &= \max\{F_{\overline{S}B_1}(v_1u_1), F_{\overline{Q}A_2}(v_2), F_{\overline{Q}A_2}(u_2)\},\\
F_{(\underline{S}B_1 \oplus \underline{S}B_2)}((v_1,v_2)(u_1,u_2)) &= \max\{F_{\underline{S}B_1}(v_1u_1), F_{\underline{Q}A_2}(v_2), F_{\underline{Q}A_2}(u_2)\}.
\end{aligned}$$

*(v)* $\forall v_1u_1 \notin SB_1,\ v_2u_2 \in SB_2$.

$$\begin{aligned}
T_{(\overline{S}B_1 \oplus \overline{S}B_2)}((v_1,v_2)(u_1,u_2)) &= \min\{T_{\overline{Q}A_1}(v_1), T_{\overline{Q}A_1}(u_1), T_{\overline{S}B_2}(v_2u_2)\},\\
T_{(\underline{S}B_1 \oplus \underline{S}B_2)}((v_1,v_2)(u_1,u_2)) &= \min\{T_{\underline{Q}A_1}(v_1), T_{\underline{Q}A_1}(u_1), T_{\underline{S}B_2}(v_2u_2)\},\\
I_{(\overline{S}B_1 \oplus \overline{S}B_2)}((v_1,v_2)(u_1,u_2)) &= \max\{I_{\overline{Q}A_1}(v_1), I_{\overline{Q}A_1}(u_1), I_{\overline{S}B_2}(v_2u_2)\},\\
I_{(\underline{S}B_1 \oplus \underline{S}B_2)}((v_1,v_2)(u_1,u_2)) &= \max\{I_{\underline{Q}A_1}(v_1), I_{\underline{Q}A_1}(u_1), I_{\underline{S}B_2}(v_2u_2)\},\\
F_{(\overline{S}B_1 \oplus \overline{S}B_2)}((v_1,v_2)(u_1,u_2)) &= \max\{F_{\overline{Q}A_1}(v_1), F_{\overline{Q}A_1}(u_1), F_{\overline{S}B_2}(v_2u_2)\},\\
F_{(\underline{S}B_1 \oplus \underline{S}B_2)}((v_1,v_2)(u_1,u_2)) &= \max\{F_{\underline{Q}A_1}(v_1), F_{\underline{Q}A_1}(u_1), F_{\underline{S}B_2}(v_2u_2)\}.
\end{aligned}$$

**Example 7.** *Let $G_1 = (\underline{G}_1, \overline{G}_1)$ and $G_2 = (\underline{G}_2, \overline{G}_2)$ be two neutrosophic soft rough graphs on $V$, where $\underline{G}_1 = (\underline{Q}A_1, \underline{S}B_1)$ and $\overline{G}_1 = (\overline{Q}A_1, \overline{S}B_1)$ are neutrosophic graphs as shown in Figure 6 and $\underline{G}_2 = (\underline{Q}A_2, \underline{S}B_2)$ and $\overline{G}_2 = (\overline{Q}A_2, \overline{S}B_2)$ are neutrosophic graphs as shown in Figure 7.*

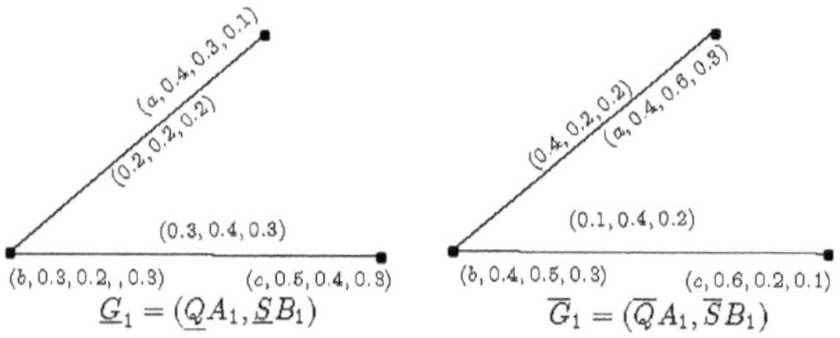


**Figure 6.** Neutrosophic soft rough graph $G_1 = (\underline{G}_1, \overline{G}_1)$

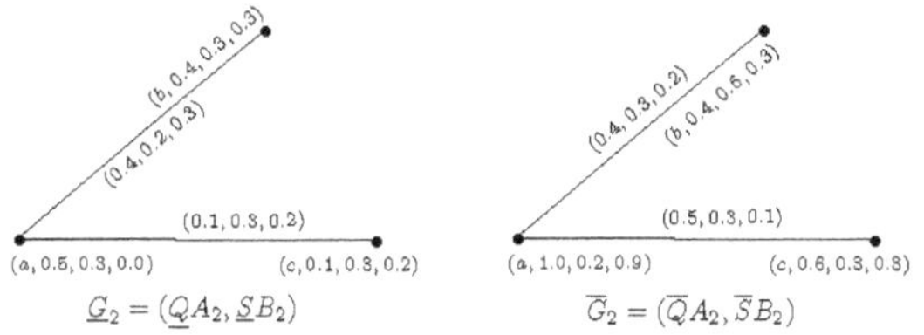


**Figure 7.** Neutrosophic soft rough graph $G_2 = (\underline{G}_2, \overline{G}_2)$

*The symmetric difference of* $G_1$ *and* $G_2$ *is* $G = G_1 \oplus G_2 = (\underline{G}_1 \oplus \underline{G}_2, \overline{G}_1 \oplus \overline{G}_2)$, *where* $\underline{G}_1 \oplus \underline{G}_2 = (\underline{Q}A_1 \oplus \underline{Q}A_2, \underline{S}B_1 \oplus \underline{S}B_2)$ *and* $\overline{G}_1 \oplus \overline{G}_2 = (\overline{Q}A_1 \oplus \overline{Q}A_2, \overline{S}B_1 \oplus \overline{S}B_2)$ *are neutrosophic graphs as shown in Figure 8.*

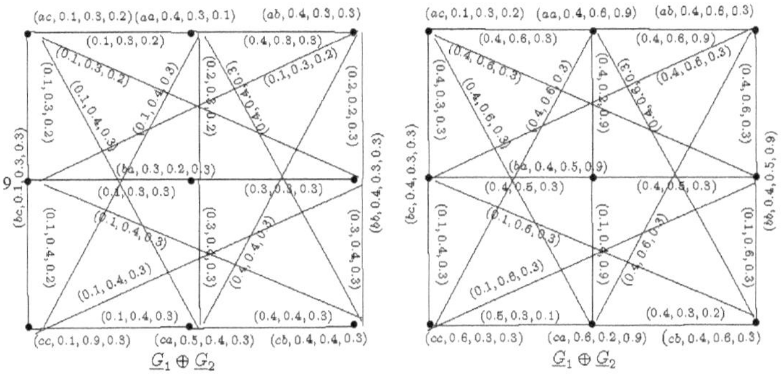


**Figure 8.** Neutrosophic soft rough graph $G_1 \oplus G_2 = (\underline{G}_1 \oplus \underline{G}_2, \overline{G}_1 \oplus \overline{G}_2)$

**Definition 12.** *The lexicographic product of* $G_1$ *and* $G_2$ *is a neutrosophic soft rough graph* $G = G_1 \odot G_2 = (G_{1*} \odot G_{2*}, G_1^* \odot G_2^*)$, *where* $G_{1*} \odot G_{2*} = (\underline{Q}A_1 \odot \underline{Q}A_2, \underline{S}B_1 \odot \underline{S}B_2)$ *and* $G_1^* \odot G_2^* = (\overline{Q}A_1 \odot \overline{Q}A_2, \overline{S}B_1 \odot \overline{S}B_2)$ *are neutrosophic graphs, respectively, such that*

*(i)* $\forall (v_1, v_2) \in QA_1 \times QA_2$.

$$T_{(\overline{Q}A_1 \odot \overline{Q}A_2)}(v_1, v_2) = \min\{T_{\overline{Q}A_1}(v_1), T_{\overline{Q}A_2}(v_2)\}, T_{(\underline{Q}A_1 \odot \underline{Q}A_2)}(v_1, v_2) = \min\{T_{\underline{Q}A_1}(v_1), T_{\underline{Q}A_2}(v_2)\},$$
$$I_{(\overline{Q}A_1 \odot \overline{Q}A_2)}(v_1, v_2) = \max\{I_{\overline{Q}A_1}(v_1), I_{\overline{Q}A_2}(v_2)\}, I_{(\underline{Q}A_1 \odot \underline{Q}A_2)}(v_1, v_2) = \max\{I_{\underline{Q}A_1}(v_1), I_{\underline{Q}A_2}(v_2)\},$$
$$F_{(\overline{Q}A_1 \odot \overline{Q}A_2)}(v_1, v_2) = \max\{F_{\overline{Q}A_1}(v_1), F_{\overline{Q}A_2}(v_2)\}, F_{(\underline{Q}A_1 \odot \underline{Q}A_2)}(v_1, v_2) = \max\{F_{\underline{Q}A_1}(v_1), F_{\underline{Q}A_2}(v_2)\}.$$

*(ii)* $\forall v_1 v_2 \in SB_2,\ v \in QA_1$.

$$T_{(\overline{S}B_1 \odot \overline{S}B_2)}((v, v_1)(v, v_2)) = \min\{T_{\overline{Q}A_1}(v), T_{\overline{S}B_2}(v_1 v_2)\},$$
$$T_{(\underline{S}B_1 \odot \underline{S}B_2)}((v, v_1)(v, v_2)) = \min\{T_{\underline{Q}A_1}(v), T_{\underline{S}B_2}(v_1 v_2)\},$$
$$I_{(\overline{S}B_1 \odot \overline{S}B_2)}((v, v_1)(v, v_2)) = \max\{I_{\overline{Q}A_1}(v), I_{\overline{S}B_2}(v_1 v_2)\},$$
$$I_{(\underline{S}B_1 \odot \underline{S}B_2)}((v, v_1)(v, v_2)) = \max\{I_{\underline{Q}A_1}(v), I_{\underline{S}B_2}(v_1 v_2)\},$$
$$F_{(\overline{S}B_1 \odot \overline{S}B_2)}((v, v_1)(v, v_2)) = \max\{F_{\overline{Q}A_1}(v), F_{\overline{S}B_2}(v_1 v_2)\},$$
$$F_{(\underline{S}B_1 \odot \underline{S}B_2)}((v, v_1)(v, v_2)) = \max\{F_{\underline{Q}A_1}(v), F_{\underline{S}B_2}(v_1 v_2)\}.$$

*(iii)* $\forall v_1u_1 \in SB_1, v_1u_2 \in SB_2$.

$$
\begin{aligned}
T_{(\overline{S}B_1 \odot \overline{S}B_2)}((v_1,v_1)(u_1,u_2)) &= \min\{T_{\overline{S}B_1}(v_1u_1), T_{\overline{S}B_2}(v_1u_2)\},\\
T_{(\underline{S}B_1 \odot \underline{S}B_2)}((v_1,v_1)(u_1,u_2)) &= \min\{T_{\underline{S}B_1}(v_1u_1), T_{\underline{S}B_2}(v_1u_2)\},\\
I_{(\overline{S}B_1 \odot \overline{S}B_2}((v_1,v_1)(u_1,u_2)) &= \max\{I_{\overline{S}B_1}(v_1u_1), I_{\overline{S}B_2}(v_1u_2)\},\\
I_{(\underline{S}B_1 \odot \underline{S}B_2)}((v_1,v_1)(u_1,u_2)) &= \max\{I_{\underline{S}B_1}(v_1u_1), I_{\underline{S}B_2}(v_1u_2)\},\\
F_{(\overline{S}B_1 \odot \overline{S}B_2}((v_1,v_1)(u_1,u_2)) &= \max\{F_{\overline{S}B_1}(v_1u_1), F_{\overline{S}B_2}(v_1u_2)\},\\
F_{(\underline{S}B_1 \odot \underline{S}B_2)}((v_1,v_1)(u_1,u_2)) &= \max\{F_{\underline{S}B_1}(v_1u_1), F_{\underline{S}B_2}(v_1u_2)\}.
\end{aligned}
$$

**Definition 13.** *The strong product of $G_1$ and $G_2$ is a neutrosophic soft rough graph $G = G_1 \otimes G_2 = (G_{1*} \otimes G_{2*}, G_1^* \otimes G_2^*)$, where $G_{1*} \otimes G_{2*} = (\underline{Q}A_1 \otimes \underline{Q}A_2, \underline{S}B_1 \otimes \underline{S}B_2)$ and $G_1^* \otimes G_2^* = (\overline{Q}A_1 \otimes \overline{Q}A_2, \overline{S}B_1 \otimes \overline{S}B_2)$ are neutrosophic graphs, respectively, such that*

*(i)* $\forall\,(v_1,v_2) \in QA_1 \times QA_2$.

$$
\begin{aligned}
T_{(\overline{Q}A_1 \otimes \overline{Q}A_2)}(v_1,v_2) &= \min\{T_{\overline{Q}A_1}(v_1), T_{\overline{Q}A_2}(v_2)\}, \; T_{(\underline{Q}A_1 \otimes \underline{Q}A_2)}(v_1,v_2) = \min\{T_{\underline{Q}A_1}(v_1), T_{\underline{Q}A_2}(v_2)\},\\
I_{(\overline{Q}A_1 \otimes \overline{Q}A_2)}(v_1,v_2) &= \max\{I_{\overline{Q}A_1}(v_1), I_{\overline{Q}A_2}(v_2)\}, \; I_{(\underline{Q}A_1 \otimes \underline{Q}A_2)}(v_1,v_2) = \max\{I_{\underline{Q}A_1}(v_1), I_{\underline{Q}A_2}(v_2)\},\\
F_{(\overline{Q}A_1 \otimes \overline{Q}A_2)}(v_1,v_2) &= \max\{F_{\overline{Q}A_1}(v_1), F_{\overline{Q}A_2}(v_2)\}, F_{(\underline{Q}A_1 \otimes \underline{Q}A_2)}(v_1,v_2) = \max\{F_{\underline{Q}A_1}(v_1), F_{\underline{Q}A_2}(v_2)\}.
\end{aligned}
$$

*(ii)* $\forall v_1v_2 \in SB_2,\; v \in QA_1$.

$$
\begin{aligned}
T_{(\overline{S}B_1 \otimes \overline{S}B_2)}((v,v_1)(v,v_2)) &= \min\{T_{\overline{Q}A_1}(v), T_{\overline{S}B_2}(v_1v_2)\},\\
T_{(\underline{S}B_1 \otimes \underline{S}B_2)}((v,v_1)(v,v_2)) &= \min\{T_{\underline{Q}A_1}(v), T_{\underline{S}B_2}(v_1v_2)\},\\
I_{(\overline{S}B_1 \otimes \overline{S}B_2)}((v,v_1)(v,v_2)) &= \max\{I_{\overline{Q}A_1}(v), I_{\overline{S}B_2}(v_1v_2)\},\\
I_{(\underline{S}B_1 \otimes \underline{S}B_2)}((v,v_1)(v,v_2)) &= \max\{I_{\underline{Q}A_1}(v), I_{\underline{S}B_2}(v_1v_2)\},\\
F_{(\overline{S}B_1 \otimes \overline{S}B_2)}((v,v_1)(v,v_2)) &= \max\{F_{\overline{Q}A_1}(v), F_{\overline{S}B_2}(v_1v_2)\},\\
F_{(\underline{S}B_1 \otimes \underline{S}B_2)}((v,v_1)(v,v_2)) &= \max\{F_{\underline{Q}A_1}(v), F_{\underline{S}B_2}(v_1v_2)\}.
\end{aligned}
$$

*(iii)* $\forall v_1v_2 \in SB_1,\; v \in QA_2$.

$$
\begin{aligned}
T_{(\overline{S}B_1 \otimes \overline{S}B_2)}((v_1,v)(v_2,v)) &= \min\{T_{\overline{S}B_1}(v_1v_2), T_{\overline{Q}A_2}(v)\},\\
T_{(\underline{S}B_1 \otimes \underline{S}B_2)}((v_1,v)(v_2,v)) &= \min\{T_{\underline{S}B_1}(v_1v_2), T_{\underline{Q}A_2}(v)\},\\
I_{(\overline{S}B_1 \otimes \overline{S}B_2)}((v_1,v)(v_2,v)) &= \max\{I_{\overline{S}B_1}(v_1v_2), I_{\overline{Q}A_2}(v)\},\\
I_{(\underline{S}B_1 \otimes \underline{S}B_2)}((v_1,v)(v_2,v)) &= \max\{I_{\underline{S}B_1}(v_1v_2), I_{\underline{Q}A_2}(v)\},\\
F_{(\overline{S}B_1 \otimes \overline{S}B_2)}((v_1,v)(v_2,v)) &= \max\{F_{\overline{S}B_1}(v_1v_2), F_{\overline{Q}A_2}(v)\},\\
F_{(\underline{S}B_1 \otimes \underline{S}B_2)}((v_1,v)(v_2,v)) &= \max\{F_{\underline{S}B_1}(v_1v_2), F_{\underline{Q}A_2}(v)\}.
\end{aligned}
$$

*(iv)* $\forall v_1u_1 \in SB_1, v_1u_2 \in SB_2$.

$$\begin{aligned}
T_{(\overline{SB}_1 \otimes \overline{SB}_2)}((v_1,v_1)(u_1,u_2)) &= \min\{T_{\overline{SB}_1}(v_1u_1), T_{\overline{SB}_2}(v_1u_2)\},\\
T_{(\underline{SB}_1 \otimes \underline{SB}_2)}((v_1,v_1)(u_1,u_2)) &= \min\{T_{\underline{SB}_1}(v_1u_1), T_{\underline{SB}_2}(v_1u_2)\},\\
I_{(\overline{SB}_1 \otimes \overline{SB}_2}((v_1,v_1)(u_1,u_2)) &= \max\{I_{\overline{SB}_1}(v_1u_1), I_{\overline{SB}_2}(v_1u_2)\},\\
I_{(\underline{SB}_1 \otimes \underline{SB}_2)}((v_1,v_1)(u_1,u_2)) &= \max\{I_{\underline{SB}_1}(v_1u_1), I_{\underline{SB}_2}(v_1u_2)\},\\
F_{(\overline{SB}_1 \otimes \overline{SB}_2}((v_1,v_1)(u_1,u_2)) &= \max\{F_{\overline{SB}_1}(v_1u_1), F_{\overline{SB}_2}(v_1u_2)\},\\
F_{(\underline{SB}_1 \otimes \underline{SB}_2)}((v_1,v_1)(u_1,u_2)) &= \max\{F_{\underline{SB}_1}(v_1u_1), F_{\underline{SB}_2}(v_1u_2)\}.
\end{aligned}$$

**Definition 14.** *The composition of $G_1$ and $G_2$ is a neutrosophic soft rough graph $G = G_1[G_2] = (G_{1*}[G_{2*}], G_1^*[G_2^*])$, where $G_{1*}[G_{2*}] = (\underline{QA}_1[\underline{QA}_2], \underline{SB}_1[\underline{SB}_2])]$ and $G_1^*[G_2^*] = (\overline{QA}_1[\overline{QA}_2], \overline{SB}_1[\overline{SB}_2])$ are neutrosophic graphs, respectively, such that*

*(i)* $\forall (v_1, v_2) \in QA_1 \times QA_2$.

$$\begin{aligned}
T_{(\overline{QA}_1 \times \overline{QA}_2)}(v_1,v_2) &= \min\{T_{\overline{QA}_1}(v_1), T_{\overline{QA}_2}(v_2)\},\ T_{(\underline{QA}_1 \times \underline{QA}_2)}(v_1,v_2) = \min\{T_{\underline{QA}_1}(v_1), T_{\underline{QA}_2}(v_2)\},\\
I_{(\overline{QA}_1 \times \overline{QA}_2)}(v_1,v_2) &= \max\{I_{\overline{QA}_1}(v_1), I_{\overline{QA}_2}(v_2)\},\ I_{(\underline{QA}_1 \times \underline{QA}_2)}(v_1,v_2) = \max\{I_{\underline{QA}_1}(v_1), I_{\underline{QA}_2}(v_2)\},\\
F_{(\overline{QA}_1 \times \overline{QA}_2)}(v_1,v_2) &= \max\{F_{\overline{QA}_1}(v_1), F_{\overline{QA}_2}(v_2)\}, F_{(\underline{QA}_1 \times \underline{QA}_2)}(v_1,v_2) = \max\{F_{\underline{QA}_1}(v_1), F_{\underline{QA}_2}(v_2)\}.
\end{aligned}$$

*(ii)* $\forall v_1v_2 \in SB_2,\ v \in QA_1$.

$$\begin{aligned}
T_{(\overline{SB}_1 \times \overline{SB}_2)}((v,v_1)(v,v_2)) &= \min\{T_{\overline{QA}_1}(v), T_{\overline{SB}_2}(v_1v_2)\},\\
T_{(\underline{SB}_1 \times \underline{SB}_2)}((v,v_1)(v,v_2)) &= \min\{T_{\underline{QA}_1}(v), T_{\underline{SB}_2}(v_1v_2)\},\\
I_{(\overline{SB}_1 \times \overline{SB}_2)}((v,v_1)(v,v_2)) &= \max\{I_{\overline{QA}_1}(v), I_{\overline{SB}_2}(v_1v_2)\},\\
I_{(\underline{SB}_1 \times \underline{SB}_2)}((v,v_1)(v,v_2)) &= \max\{I_{\underline{QA}_1}(v), I_{\underline{SB}_2}(v_1v_2)\},\\
F_{(\overline{SB}_1 \times \overline{SB}_2)}((v,v_1)(v,v_2)) &= \max\{F_{\overline{QA}_1}(v), F_{\overline{SB}_2}(v_1v_2)\},\\
F_{(\underline{SB}_1 \times \underline{SB}_2)}((v,v_1)(v,v_2)) &= \max\{F_{\underline{QA}_1}(v), F_{\underline{SB}_2}(v_1v_2)\}.
\end{aligned}$$

*(iii)* $\forall v_1v_2 \in SB_1,\ v \in QA_2$.

$$\begin{aligned}
T_{(\overline{SB}_1 \times \overline{SB}_2)}((v_1,v)(v_2,v)) &= \min\{T_{\overline{SB}_1}(v_1v_2), T_{\overline{QA}_2}(v)\},\\
T_{(\underline{SB}_1 \times \underline{SB}_2)}((v_1,v)(v_2,v)) &= \min\{T_{\underline{SB}_1}(v_1v_2), T_{\underline{QA}_2}(v)\},\\
I_{(\overline{SB}_1 \times \overline{SB}_2)}((v_1,v)(v_2,v)) &= \max\{I_{\overline{SB}_1}(v_1v_2), I_{\overline{QA}_2}(v)\},\\
I_{(\underline{SB}_1 \times \underline{SB}_2)}((v_1,v)(v_2,v)) &= \max\{I_{\underline{SB}_1}(v_1v_2), I_{\underline{QA}_2}(v)\},\\
F_{(\overline{SB}_1 \times \overline{SB}_2)}((v_1,v)(v_2,v)) &= \max\{F_{\overline{SB}_1}(v_1v_2), F_{\overline{QA}_2}(v)\},\\
F_{(\underline{SB}_1 \times \underline{SB}_2)}((v_1,v)(v_2,v)) &= \max\{F_{\underline{SB}_1}(v_1v_2), F_{\underline{QA}_2}(v)\}.
\end{aligned}$$

*(iv)* $\forall v_1u_1 \in SB_1,\ v_1 \neq u_2 \in QA_2.$

$$
\begin{aligned}
T_{(\overline{S}B_1\times\overline{S}B_2)}\big((v_1,v_1)(u_1,u_2)\big) &= \min\{T_{\overline{S}B_1}(v_1u_1), T_{\overline{S}B_2}(v_1u_2)\},\\
T_{(\underline{S}B_1\times\underline{S}B_2)}\big((v_1,v_1)(u_1,u_2)\big) &= \min\{T_{\underline{S}B_1}(v_1u_1), T_{\underline{S}B_2}(v_1u_2)\},\\
I_{(\overline{S}B_1\times\overline{S}B_2}\big((v_1,v_1)(u_1,u_2)\big) &= \max\{I_{\overline{S}B_1}(v_1u_1), I_{\overline{S}B_2}(v_1u_2)\},\\
I_{(\underline{S}B_1\times\underline{S}B_2)}\big((v_1,v_1)(u_1,u_2)\big) &= \max\{I_{\underline{S}B_1}(v_1u_1), I_{\underline{S}B_2}(v_1u_2)\},\\
F_{(\overline{S}B_1\times\overline{S}B_2}\big((v_1,v_1)(u_1,u_2)\big) &= \max\{F_{\overline{S}B_1}(v_1u_1), F_{\overline{S}B_2}(v_1u_2)\},\\
F_{(\underline{S}B_1\times\underline{S}B_2)}\big((v_1,v_1)(u_1,u_2)\big) &= \max\{F_{\underline{S}B_1}(v_1u_1), F_{\underline{S}B_2}(v_1u_2)\}.
\end{aligned}
$$

**Definition 15.** *Let $G = (\underline{G}, \overline{G})$ be a neutrosophic soft rough graph. The complement of $G$, denoted by $\underline{\acute{G}} = (\underline{\acute{G}}, \overline{\acute{G}})$ is a neutrosophic soft rough graph, where $\underline{\acute{G}} = (\underline{Q}\acute{A}, \underline{S}\acute{B})$ and $\overline{\acute{G}} = (\overline{Q}\acute{A}, \overline{S}\acute{B})$ are neutrosophic graphs such that*

*(i)* $\forall v \in QA.$

$$
\begin{aligned}
&T'_{\overline{Q}A}(v) = T_{\overline{Q}A(v)},\ I'_{\overline{Q}A}(v) = I_{\overline{Q}A(v)},\ F'_{\overline{Q}A}(v) = F_{\overline{Q}A(v)},\\
&T'_{\underline{Q}A}(v) = T_{\underline{Q}A(v)},\ I'_{\underline{Q}A}(v) = I_{\underline{Q}A(v)},\ F'_{\underline{Q}A}(v) = F_{\underline{Q}A(v)}.
\end{aligned}
$$

*(ii)* $\forall\, v,u \in QA.$

$$
\begin{aligned}
T'_{\overline{S}B}(vu) &= \min\{T_{\overline{Q}A}(v), T_{\overline{Q}A}(u)\} - T_{\overline{S}B}(vu),\\
I'_{\overline{S}B}(vu) &= \max\{I_{\overline{Q}A}(v), I_{\overline{Q}A}(u)\} - I_{\overline{S}B}(vu),\\
F'_{\overline{S}B}(vu) &= \max\{F_{\overline{Q}A}(v), F_{\overline{Q}A}(u)\} - F_{\overline{S}B}(vu),\\
T'_{\underline{S}B}(vu) &= \min\{T_{\underline{Q}A}(v), T_{\underline{Q}A}(u)\} - T_{\underline{S}B}(vu),\\
I'_{\underline{S}B}(vu) &= \max\{I_{\underline{Q}A}(v), I_{\underline{Q}A}(u)\} - I_{\overline{S}B}(vu),\\
F'_{\underline{S}B}(vu) &= \max\{F_{\underline{Q}A}(v), F_{\underline{Q}A}(u)\} - F_{\underline{S}B}(vu).
\end{aligned}
$$

**Example 8.** *Consider an NSRGs $G$ as shown in Figure 9.*

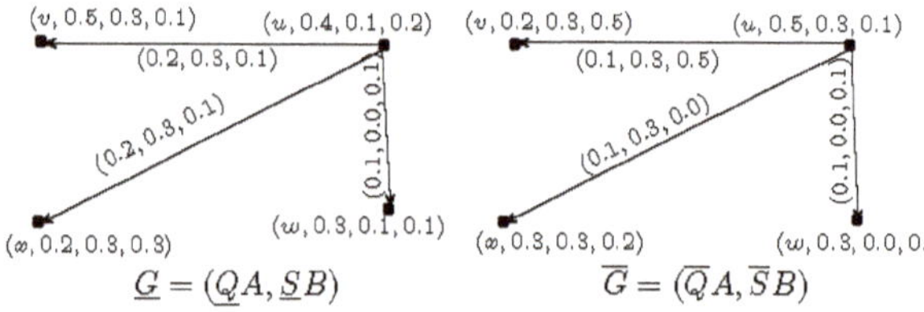


**Figure 9.** Neutrosophic soft rough graph $G = (\underline{G}, \overline{G})$

*The complement of $G$ is $\acute{G} = (\underline{\acute{G}}, \overline{\acute{G}})$ is obtained by using the Definition 15, where $\underline{\acute{G}} = (\underline{Q}\acute{A}, \underline{S}\acute{B})$ and $\overline{\acute{G}} = (\overline{Q}\acute{A}, \overline{S}\acute{B})$ are neutrosophic graphs as shown in Figure 10.*

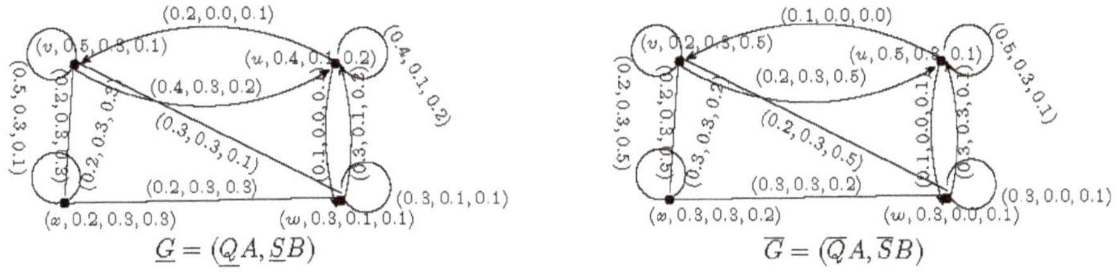


**Figure 10.** Neutrosophic soft rough graph $\acute{G} = (\underline{\acute{G}}, \overline{\acute{G}})$

**Definition 16.** *A graph G is called self complement, if* $G = \acute{G}$, *i.e.,*

*(i)* $\forall v \in QA$.

$$T'_{\overline{QA}}(v) = T_{\overline{QA}(v)},\ I'_{\overline{QA}}(v) = I_{\overline{QA}(v)}, F'_{\overline{QA}}(v) = F_{\overline{QA}(v)},$$
$$T'_{\underline{QA}}(v) = T_{\underline{QA}(v)},\ I'_{\underline{QA}}(v) = I_{\underline{QA}(v)},\ F'_{\underline{QA}}(v) = F_{\underline{QA}(v)}.$$

*(ii)* $\forall\ v, u \in QA$.

$$T'_{\overline{SB}}(vu) = T_{\overline{SB}}(vu),\ I'_{\overline{SB}}(vu) = I_{\overline{SB}}(vu),\ F'_{\overline{SB}}(vu) = F_{\overline{SB}}(vu),$$
$$T'_{\underline{SB}}(vu) = T_{\underline{SB}}(vu),\ I'_{\underline{SB}}(vu) = I_{\underline{SB}}(vu),\ F'_{\underline{SB}}(vu) = F_{\underline{SB}}(vu).$$

**Definition 17.** *A neutrosophic soft rough graph G is called strong neutrosophic soft rough graph if* $\forall uv \in SB$,

$$T_{\overline{SB}}(vu) = \min\{T_{\overline{QA}}(v), T_{\overline{QA}}(u)\},\ I_{\overline{SB}}(vu) = \max\{I_{\overline{QA}}(v), I_{\overline{QA}}(u)\}),\ F_{\overline{SB}}(vu) = \max\{F_{\overline{QA}}(v), F_{\overline{QA}}(u)\},$$
$$T_{\underline{SB}}(vu) = \min\{T_{\underline{QA}}(v), T_{\underline{QA}}(u)\},\ I_{\underline{SB}}(vu) = \max\{I_{\underline{QA}}(v), I_{\underline{QA}}(u)\},\ F_{\underline{SB}}(vu) = \max\{F_{\underline{QA}}(v), F_{\underline{QA}}(u)\}.$$

**Example 9.** *Consider a graph G such that* $V = \{u, v, w\}$ *and* $E = \{uv, vw, wu\}$, *as shown in Figure 11. Let QA be a neutrosophic soft rough set of V and let SB be a neutrosophic soft rough set of E defined in the Tables 9 and 10, respectively.*

**Table 9.** Neutrosophic soft rough set on $V$.

| $V$ | $\overline{QA}$ | $\underline{QA}$ |
|---|---|---|
| $u$ | (0.8, 0.5, 0.2) | (0.7, 0.5, 0.2) |
| $v$ | (0.9, 0.5, 0.1) | (0.7, 0.5, 0.2) |
| $w$ | (0.7, 0.5, 0.1) | (0.7, 0.5, 0.2) |

**Table 10.** Neutrosophic soft rough set on $E$.

| $E$ | $\overline{SB}$ | $\underline{SB}$ |
|---|---|---|
| $uv$ | (0.8, 0.5, 0.2) | (0.7, 0.5, 0.2) |
| $vw$ | (0.7, 0.5, 0.1) | (0.7, 0.5, 0.2) |
| $wu$ | (0.7, 0.5, 0.2) | (0.7, 0.5, 0.2) |

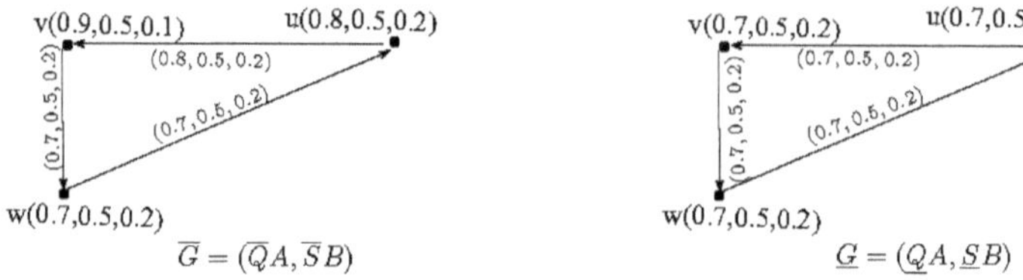


**Figure 11.** Strong neutrosophic soft rough graph $G = (QA, SB)$

*Hence, $G = (QA, SB)$ is a strong neutrosophic soft rough graph.*

**Definition 18.** *A neutrosophic soft rough graph G is called a complete neutrosophic soft rough graph if $\forall\ vu \in QA$,*

$$T_{\overline{S}B}(vu) = \min\{T_{\overline{Q}A}(v), T_{\overline{Q}A}(u)\}, I_{\overline{S}B}(vu) = \max\{I_{\overline{Q}A}(v), I_{\overline{Q}A}(u)\}, F_{\overline{S}B}(vu) = \max\{F_{\overline{Q}A}(v), F_{\overline{Q}A}(u)\},$$
$$T_{\underline{S}B}(vu) = \min\{T_{\underline{Q}A}(v), T_{\underline{Q}A}(u)\}, I_{\underline{S}B}(vu) = \max\{I_{\underline{Q}A}(v), I_{\underline{Q}A}(u)\}, F_{\underline{S}B}(vu) = \max\{F_{\underline{Q}A}(v), F_{\underline{Q}A}(u)\}.$$

**Remark 2.** *Every complete neutrosophic soft rough graph is a strong neutrosophic soft rough graph. However, the converse is not true.*

**Definition 19.** *A neutrosophic soft rough graph G is isolated, if $\forall x, y \in QA$.*

$$T_{\underline{S}B}(vu) = 0,\ I_{\underline{S}B}(vu) = 0,\ F_{\underline{S}B}(vu) = 0,\ T_{\overline{S}B}(vu) = 0,\ I_{\overline{S}B}(vu) = 0,\ F_{\overline{S}B}(vu) = 0.$$

**Theorem 1.** *The rejection of two neutrosophic soft rough graphs is a neutrosophic soft rough graph.*

**Proof.** Let $G_1 = (\underline{G_1}, \overline{G_1})$ and $G_2 = (\underline{G_2}, \overline{G_2})$ be two NSRGs. Let $G = G_1|G_2 = (\underline{G_1}|\underline{G_2}, \overline{G_1}|\overline{G_2})$ be the rejection of $G_1$ and $G_2$, where $\underline{G_1}|\underline{G_2} = (\underline{Q}A_1|\underline{Q}A_2, \underline{S}B_1|\underline{S}B_2)$ and $\overline{G_1}|\overline{G_2} = (\overline{Q}A_1|\overline{Q}A_2, \overline{S}B_1|\underline{S}B_2)$. We claim that $G = G_1|G_2$ is a neutrosophic soft rough graph. It is enough to show that $\underline{S}B_1|\underline{S}B_2$ and $\overline{S}B_1|\overline{S}B_2$ are neutrosophic relations on $\underline{Q}A_1|\underline{Q}A_2$ and $\overline{Q}A_1|\overline{Q}A_2$, respectively. First, we show that $\underline{S}B_1|\underline{S}B_2$ is a neutrosophic relation on $\underline{Q}A_1|\underline{Q}A_2$.
If $v \in \underline{Q}A_1$, $v_1v_2 \notin \underline{S}B_2$, then

$$\begin{aligned}
T_{(\underline{S}B_1|\underline{S}B_2)}((v,v_1)(v,v_2)) &= (T_{\underline{Q}A_1}(v) \wedge (T_{\underline{Q}A_2}(v_2) \wedge T_{\underline{Q}A_2}(v_2)))\\
&= (T_{\underline{Q}A_1}(v) \wedge T_{\underline{Q}A_2}(v_2)) \wedge (T_{\underline{Q}A_1}(v) \wedge T_{\underline{Q}A_2}(v_2))\\
&= T_{(\underline{Q}A_1|\underline{Q}A_2)}(v,v_1) \wedge T_{(\underline{Q}A_1|\underline{Q}A_2)}(v,v_2)\\
T_{(\underline{S}B_1|\underline{S}B_2)}((v,v_1)(v,v_2)) &= T_{(\underline{Q}A_1|\underline{Q}A_2)}(v,v_1) \wedge T_{(\underline{Q}A_1|\underline{Q}A_2)}(v,v_2)\\
\text{Similarly, } I_{(\underline{S}B_1|\underline{S}B_2)}((v,v_1)(v,v_2)) &= I_{(\underline{Q}A_1|\underline{Q}A_2)}(v,v_1) \vee I_{(\underline{Q}A_1|\underline{Q}A_2)}(v,v_2)\\
F_{(\underline{S}B_1|\underline{S}B_2)}((v,v_1)(v,v_2)) &= F_{(\underline{Q}A_1|\underline{Q}A_2)}(v,v_1) \vee F_{(\underline{Q}A_1|\underline{Q}A_2)}(v,v_2).
\end{aligned}$$

If $v_1v_2 \notin \underline{S}B_1$, $v \in \underline{Q}A_2$, then

$$\begin{aligned}
T_{(\underline{S}B_1|\underline{S}B_2)}((v_1,v)(v_2,v)) &= ((T_{\underline{Q}A_1}(v_1) \wedge T_{\underline{Q}A_1}(v_2)) \wedge T_{\underline{Q}A_2}(v))\\
&= ((T_{\underline{Q}A_1}(v_1) \wedge T_{\underline{Q}A_2}(v)) \wedge (T_{\underline{Q}A_1}(v_2) \wedge T_{\underline{Q}A_2}(v)))\\
&= T_{(\underline{Q}A_1|\underline{Q}A_2)}(v_1,v) \wedge T_{(\underline{Q}A_1|\underline{Q}A_2)}(v_2,v)\\
T_{(\underline{S}B_1|\underline{S}B_2)}((v_1,v)(v_2,v)) &= T_{(\underline{Q}A_1|\underline{Q}A_2)}(v_1,v) \wedge T_{(\underline{Q}A_1|\underline{Q}A_2)}(v_2,v)\\
\text{Similarly, } I_{(\underline{S}B_1|\underline{S}B_2)}((v_1,v)(v_2,v)) &= I_{(\underline{Q}A_1|\underline{Q}A_2)}(v_1,v) \vee I_{(\underline{Q}A_1|\underline{Q}A_2)}(v_2,v)\\
F_{(\underline{S}B_1|\underline{S}B_2)}((v_1,v)(v_2,v)) &= F_{(\underline{Q}A_1|\underline{Q}A_2)}(v_1,v) \vee F_{(\underline{Q}A_1|\underline{Q}A_2)}(v_2,v).
\end{aligned}$$

If $v_1v_2 \notin \underline{S}B_1$, $u_1, u_2 \notin \underline{S}B_2$, then

$$\begin{aligned}
T_{(\underline{S}B_1|\underline{S}B_2)}((v_1,u_1)(v_2,u_2)) &= ((T_{\underline{Q}A_1}(v_1) \wedge T_{\underline{Q}A_1}(v_2)) \wedge (T_{\underline{Q}A_2}(u_1) \wedge T_{\underline{Q}A_2}(u_2))) \\
&= (T_{\underline{Q}A_1}(v_1) \wedge T_{\underline{Q}A_2}(u_1)) \wedge (T_{\underline{Q}A_1}(v_2) \wedge T_{\underline{Q}A_2}(u_2)) \\
&= T_{(\underline{Q}A_1|\underline{Q}A_2)}(v_1,u_1) \wedge T_{(\underline{Q}A_1|\underline{Q}A_2)}(v_2,u_2) \\
T_{(\underline{S}B_1|\underline{S}B_2)}((v_1,u_1)(v_2,u_2)) &= T_{(\underline{Q}A_1|\underline{Q}A_2)}(v_1,u_1) \wedge T_{(\underline{Q}A_1|\underline{Q}A_2)}(u_1,u_2) \\
\text{Similarly, } I_{(\underline{S}B_1|\underline{S}B_2)}((v_1,u_1)(v_2,u_2)) &= I_{(\underline{Q}A_1|\underline{Q}A_2)}(v_1,u_1) \vee I_{(\underline{Q}A_1|\underline{Q}A_2)}(u_1,u_2) \\
F_{(\underline{S}B_1|\underline{S}B_2)}((v_1,u_1)(v_2,u_2)) &= F_{(\underline{Q}A_1|\underline{Q}A_2)}(v_1,u_1) \vee F_{(\underline{Q}A_1|\underline{Q}A_2)}(u_1,u_2).
\end{aligned}$$

Thus, $\underline{S}B_1|\underline{S}B_2$ is a neutrosophic relation on $\underline{Q}A_1|\underline{Q}A_2$. Similarly, we can show that $\overline{S}B_1|\overline{S}B_2$ is a neutrosophic relation on $\overline{Q}A_1|\overline{Q}A_2$. Hence, $G$ is a neutrosophic soft rough graph. □

**Theorem 2.** *The Cartesian product of two NSRGs is a neutrosophic soft rough graph.*

**Proof.** Let $G_1 = (\underline{G}_1, \overline{G}_1)$ and $G_2 = (\underline{G}_2, \overline{G}_2)$ be two NSRGs. Let $G = G_1 \ltimes G_2 = (\underline{G}_1 \ltimes \underline{G}_2, \overline{G}_1 \ltimes \overline{G}_2)$ be the Cartesian product of $G_1$ and $G_2$, where $\underline{G}_1 \ltimes \underline{G}_2 = (\underline{Q}A_1 \ltimes \underline{Q}A_2, \underline{S}B_1 \ltimes \underline{S}B_2)$ and $\overline{G}_1 \ltimes \overline{G}_2 = (\overline{Q}A_1 \ltimes \overline{Q}A_2, \overline{S}B_1 \ltimes \underline{S}B_2)$. We claim that $G = G_1 \ltimes G_2$ is a neutrosophic soft rough graph. It is enough to show that $\underline{S}B_1 \ltimes \underline{S}B_2$ and $\overline{S}B_1 \ltimes \overline{S}B_2$ are neutrosophic relations on $\underline{Q}A_1 \ltimes \underline{Q}A_2$ and $\overline{Q}A_1 \ltimes \overline{Q}A_2$, respectively. We have to show that $\underline{S}B_1 \ltimes \underline{S}B_2$ is a neutrosophic relation on $\underline{Q}A_1 \ltimes \underline{Q}A_2$.

If $v \in \underline{Q}A_1$, $v_1u_1 \in \underline{S}B_2$, then

$$\begin{aligned}
T_{(\underline{S}B_1 \ltimes \underline{S}B_2)}((v,v_1)(v,u_1)) &= T_{(\underline{Q}A_1)}(v) \wedge T_{(\underline{S}B_2)}(v_1u_1) \\
&\leqslant T_{(\underline{Q}A_1)}(v) \wedge (T_{(\underline{Q}A_2)}(v_1) \wedge T(\underline{Q}A_2)(u_1)) \\
&= (T_{(\underline{Q}A_1)}(v) \wedge T_{(\underline{Q}A_2)}(v_1)) \wedge (T_{(\underline{Q}A_1)}(v) \wedge T_{(\underline{Q}A_2)}(u_1)) \\
&= T_{(\underline{Q}A_1 \ltimes \underline{Q}A_2)}(v,v_1) \wedge T_{(\underline{Q}A_1 \ltimes \underline{Q}A_2)}(v,u_1) \\
T_{(\underline{S}B_1 \ltimes \underline{S}B_2)}((v,v_1)(v,u_1)) &\leqslant T_{(\underline{Q}A_1 \ltimes \underline{Q}A_2)}(v,v_1) \wedge T_{(\underline{Q}A_1 \ltimes \underline{Q}A_2)}(v,u_1) \\
\text{Similarly, } I_{(\underline{S}B_1 \ltimes \underline{S}B_2)}((v,v_1)(v,u_1)) &\leqslant I_{(\underline{Q}A_1 \ltimes \underline{Q}A_2)}(v,v_1) \vee I_{(\underline{Q}A_1 \ltimes \underline{Q}A_2)}(v,u_1) \\
F_{(\underline{S}B_1 \ltimes \underline{S}B_2)}((v,v_1)(v,u_1)) &\leqslant F_{(\underline{Q}A_1 \ltimes \underline{Q}A_2)}(v,v_1) \vee F_{(\underline{Q}A_1 \ltimes \underline{Q}A_2)}(v,u_1).
\end{aligned}$$

If $v_1u_1 \in \underline{S}B_1$, $z \in \underline{Q}A_2$, then

$$\begin{aligned}
T_{(\underline{S}B_1 \ltimes \underline{S}B_2)}((v_1,z)(u_1,z)) &= T_{(\underline{S}B_1)}(v_1u_1) \wedge T_{(\underline{Q}A_2)}(z) \\
&\leqslant (T_{(\underline{Q}A_1)(v_1) \wedge (\underline{Q}A_1)}(u_1)) \wedge T_{(\underline{Q}A_2)}(z) \\
&= T_{(\underline{Q}A_1 \ltimes \underline{Q}A_2)}(v_1,z) \wedge T_{(\underline{Q}A_1 \ltimes \underline{Q}A_2)}(u_1,z) \\
T_{(\underline{S}B_1 \ltimes \underline{S}B_2)}((v_1,z)(u_1,z)) &\leqslant T_{(\underline{Q}A_1 \ltimes \underline{Q}A_2)}(v_1,z) \wedge T_{(\underline{Q}A_1 \ltimes \underline{Q}A_2)}(u_1,z) \\
\text{Similarly, } I_{(\underline{S}B_1 \ltimes \underline{S}B_2)}((v_1,z)(u_1,z)) &\leqslant I_{(\underline{Q}A_1 \ltimes \underline{Q}A_2)}(v_1,z) \vee I_{(\underline{Q}A_1 \ltimes \underline{Q}A_2)}(u_1,z) \\
F_{(\underline{S}B_1 \ltimes \underline{S}B_2)}((v_1,z)(u_1,z)) &\leqslant F_{(\underline{Q}A_1 \ltimes \underline{Q}A_2)}(v_1,z) \vee F_{(\underline{Q}A_1 \ltimes \underline{Q}A_2)}(u_1,z).
\end{aligned}$$

Therefore, $\underline{S}B_1 \ltimes \underline{S}B_2$ is a neutrosophic relation on $\underline{Q}A_1 \ltimes \underline{Q}A_2$. Similarly, $\overline{S}B_1 \ltimes \overline{S}B_2$ is a neutrosophic relation on $\overline{Q}A_1 \ltimes \overline{Q}A_2$. Hence, $G$ is a neutrosophic rough graph. □

**Theorem 3.** *The cross product of two neutrosophic soft rough graphs is a neutrosophic soft rough graph.*

**Proof.** Let $G_1 = (\underline{G}_1, \overline{G}_1)$ and $G_2 = (\underline{G}_2, \overline{G}_2)$ be two NSRGs. Let $G = G_1 \circledcirc G_2 = (\underline{G}_1 \circledcirc \underline{G}_2, \overline{G}_1 \circledcirc \overline{G}_2)$ be the cross product of $G_1$ and $G_2$, where $\underline{G}_1 \circledcirc \underline{G}_2 = (\underline{Q}A_1 \circledcirc \underline{Q}A_2, \underline{S}B_1 \circledcirc \underline{S}B_2)$ and $\overline{G}_1 \circledcirc \overline{G}_2 = (\overline{Q}A_1 \circledcirc \overline{Q}A_2, \overline{S}B_1 \circledcirc \underline{S}B_2)$. We claim that $G = G_1 \circledcirc G_2$ is a neutrosophic soft rough graph. It is enough to show that $\underline{S}B_1 \circledcirc \underline{S}B_2$ and $\overline{S}B_1 \circledcirc \overline{S}B_2$ are neutrosophic relations on $\underline{Q}A_1 \circledcirc \underline{Q}A_2$ and $\overline{Q}A_1 \circledcirc \overline{Q}A_2$, respectively. First, we show that $\underline{S}B_1 \circledcirc \underline{S}B_2$ is a neutrosophic relation on $\underline{Q}A_1 \circledcirc \underline{Q}A_2$.

If $v_1u_1 \in \underline{S}B_1, v_1u_2 \in \underline{S}B_2$, then

$$\begin{aligned}
T_{(\underline{S}B_1 \circledcirc \underline{S}B_2)}((v_1,v_1)(u_1,u_2)) &= T_{(\underline{S}B_1)}(v_1u_1) \wedge T_{(\underline{S}B_2)}(v_1u_2) \\
&\leqslant (T_{(\underline{Q}A_1)}(v_1) \wedge T_{(\underline{Q}A_1)}(u_1) \wedge (T_{(\underline{Q}A_2)}(v_1) \wedge T_{(\underline{Q}A_2)}(u_2)) \\
&= (T_{(\underline{Q}A_1)}(v_1) \wedge T_{(\underline{Q}A_2)}(v_1)) \wedge (T_{(\underline{Q}A_1)}(u_1) \wedge T_{(\underline{Q}A_2)}(u_2)) \\
&= T_{(\underline{Q}A_1 \circledcirc \underline{Q}A_2)}(v_1,v_1) \wedge T_{(\underline{Q}A_1 \circledcirc \underline{Q}A_2)}(u_1,u_2) \\
T_{(\underline{S}B_1 \circledcirc \underline{S}B_2)}((v_1,v_1)(u_1,u_2)) &\leqslant T_{(\underline{Q}A_1 \circledcirc \underline{Q}A_2)}(v_1,v_1) \wedge T_{(\underline{Q}A_1 \circledcirc \underline{Q}A_2)}(v,u_2) \\
\text{Similarly, } I_{(\underline{S}B_1 \circledcirc \underline{S}B_2)}((v_1,v_1)(u_1,u_2)) &\leqslant I_{(\underline{Q}A_1 \circledcirc \underline{Q}A_2)}(v_1,v_1) \vee I_{(\underline{Q}A_1 \circledcirc \underline{Q}A_2)}(v,u_2) \\
F_{(\underline{S}B_1 \circledcirc \underline{S}B_2)}((v_1,v_1)(u_1,u_2)) &\leqslant F_{(\underline{Q}A_1 \circledcirc \underline{Q}A_2)}(v_1,v_1) \vee F_{(\underline{Q}A_1 \circledcirc \underline{Q}A_2)}(v,u_2).
\end{aligned}$$

Thus, $\underline{S}B_1 \circledcirc \underline{S}B_2$ is a neutrosophic relation on $\underline{Q}A_1 \circledcirc \underline{Q}A_2$. Similarly, we can show that $\overline{S}B_1 \circledcirc \overline{S}B_2$ is a neutrosophic relation on $\overline{Q}A_1 \circledcirc \overline{Q}A_2$. Hence, $G$ is a neutrosophic soft rough graph. □

## 3. Application

In this section, we apply the concept of NSRSs to a decision-making problem. In recent times, the object recognition problem has gained considerable importance. The object recognition problem can be considered as a decision-making problem, in which final identification of objects is founded on a given set of information. A detailed description of the algorithm for the selection of most suitable objects based on an available set of alternatives is given, and purposed decision-making method can be used to calculate lower and upper approximation operators to progress deep concerns of the problem. The presented algorithms can be applied to avoid lengthy calculations when dealing with large number of objects. This method can be applied in various domains for multi-criteria selection of objects.

*Selection of Most Suitable Generic Version of Brand Name Medicine*

In the pharmaceutical industry, different pharmaceutical companies develop, produce and discover pharmaceutical medicine (drugs) for use as medication. These pharmaceutical companies deals with "brand name medicine" and "generic medicine". Brand name medicine and generic medicine are bioequivalent, and have a generic medicine rate and element of absorption. Brand name medicine and generic medicine have the same active ingredients, but the inactive ingredients may differ. The most important difference is cost. Generic medicine is less expensive as compared to brand name comparators. Usually, generic drug manufacturers face competition to produce cost less products. The product may possibly be slightly dissimilar in color, shape, or markings. The major difference is cost. We consider a brand name drug "$u$ = Loratadine" used for seasonal allergies medication. Consider

$$\begin{aligned}
V = &\{u_1 = \text{Triamcinolone}, u_2 = \text{Cetirizine/Pseudoephedrine}, \\
&u_3 = \text{Pseudoephedrine}, u_4 = \text{loratadine/pseudoephedrine}, \\
&u_5 = \text{Fluticasone}\}
\end{aligned}$$

is a set of generic versions of "Loratadine". We want to select the most suitable generic version of Loratadine on the basis of parameters $e_1$ = Highly soluble, $e_2$ = Highly permeable, $e_3$ = Rapidly dissolving. $\mathbb{M} = \{e_1, e_2, e_3\}$ be a set of paraments. Let $Q$ be a neutrosophic soft relation from $V$ to parameter set $M$, and describe truth-membership, indeterminacy-membership and false-membership degrees of generic version medicine corresponding to the parameters as shown in Table 11.

**Table 11.** Neutrosophic soft set $(Q, M)$.

| $Q$ | $u_1$ | $u_2$ | $u_3$ | $u_4$ | $u_5$ |
|---|---|---|---|---|---|
| $e_1$ | $(0.4, 0.5, 0.6)$ | $(0.5, 0.3, 0.6)$ | $(0.7, 0.2, 0.3)$ | $(0.5, 0.7, 0.5)$ | $(0.6, 0.5, 0.4)$ |
| $e_2$ | $(0.7, 0.3, 0.2)$ | $(0.3, 0.4, 0.3)$ | $(0.6, 0.5, 0.4)$ | $(0.8, 0.4, 0.6)$ | $(0.7, 0.8, 0.5)$ |
| $e_3$ | $(0.6, 0.3, 0.4)$ | $(0.7, 0.2, 0.3)$ | $(0.7, 0.2, 0.4)$ | $(0.8, 0.7, 0.6)$ | $(0.7, 0.3, 0.5)$ |

Suppose $A = \{(e_1, 0.2, 0.4, 0.5), (e_2, 0.5, 0.6, 0.4), (e_3, 0.7, 0.5, 0.4)\}$ is the most favorable object that is an NS on the parameter set $M$ under consideration. Then, $(\underline{Q}(A), \overline{Q}(A))$ is an NSRS in NSAS $(V, M, Q)$, where

$$\begin{aligned}
\overline{Q}(A) &= \{(u_1, 0.6, 0.5, 0.4), (u_2, 0.7, 0.4, 0.4), (u_3, 0.7, 0.4, 0.4), (u_4, 0.7, 0.6, 0.5), (u_5, 0.7, 0.5, 0.5)\}, \\
\underline{Q}(A) &= \{(u_1, 0.5, 0.6, 0.4), (u_2, 0.5, 0.6, 0.5), (u_3, 0.3, 0.3, 0.5), (u_4, 0.5, 0.6, 0.5), (u_5, 0.4, 0.5, 0.5)\}.
\end{aligned}$$

Let $E = \{u_1v_2, u_1u_3, u_4u_1, u_2u_3, u_5u_3, u_2u_4, u_2u_5\} \subseteq \acute{V}$ and $L = \{e_1e_3, e_2e_1, e_3e_2\} \subseteq \acute{\mathbb{M}}$.
Then, a neutrosophic soft relation $S$ on $E$ (from $L$ to $E$) can be defined as follows in Table 12:

**Table 12.** Neutrosophic soft relation $S$.

| $S$ | $u_1u_2$ | $u_1u_3$ | $u_4u_1$ | $u_2u_3$ | $u_5u_3$ | $u_2u_4$ | $u_2u_5$ |
|---|---|---|---|---|---|---|---|
| $e_1e_2$ | (0.3, 0.4 ,0.2) | $(0.4, 0.4, 0.5)$ | $(0.4, 0.4, 0.5)$ | $(0.6, 0.3, 0.4)$ | $(0.4, 0.2, 0.2)$ | $(0.4, 0.4, 0.2)$ | $(0.4, 0.3, 0.4)$ |
| $e_2e_3$ | (0.5 ,0.4 ,0.1) | $(0.4, 0.3, 0.2)$ | $(0.4, 0.3, 0.2)$ | $(0.3, 0.3, 0.2)$ | $(0.6, 0.2, 0.4)$ | $(0.3, 0.2, 0.1)$ | $(0.3, 0.3, 0.2)$ |
| $e_1e_3$ | (0.4,0.4,0.1) | $(0.4, 0.2, 0.2)$ | $(0.4, 0.2, 0.2)$ | $(0.5, 0.3, 0.3)$ | $(0.4, 0.2, 0.3)$ | $(0.4, 0.3, 0.1)$ | $(0.5, 0.3, 0.2)$ |

Let $B = \{(e_1e_2, 0.2, 0.4, 0.5), (e_2e_3, 0.5, 0.4, 0.4), (e_1e_3, 0.5, 0.2, 0.5)\}$ be an NS on $L$ that describes some relationship between the parameters under consideration; then, $SB = (\underline{S}B, \overline{S}B)$ is an NSRR, where

$$\begin{aligned}
\overline{S}B &= \{(u_1u_2, 0.5, 0.4, 0.4), (u_1u_3, 0.4, 0.2, 0.4), (u_4u_1, 0.4, 0.2, 0.4), (u_2u_3, 0.5, 0.3, 0.4), \\
&\quad (u_5u_3, 0.5, 0.2, 0.4), (u_2u_4, 0.4, 0.3, 0.4), (u_2u_5, 0.5, 0.3, 0.4)\}, \\
\underline{S}B &= \{(u_1u_2, 0.2, 0.4, 0.4)(u_1u_3, 0.5, 0.4, 0.4), (u_4u_1, 0.5, 0.4, 0.4), (u_2u_3, 0.4, 0.4, 0.5), \\
&\quad (u_5u_3, 0.2, 0.4, 0.4), (u_2u_4, 0.2, 0.4, 0.4), (u_2u_5, 0.4, 0.4, 0.5)\}.
\end{aligned}$$

Thus, $G = (\underline{G}, \overline{G})$ is an NSRG as shown in Figure 12.

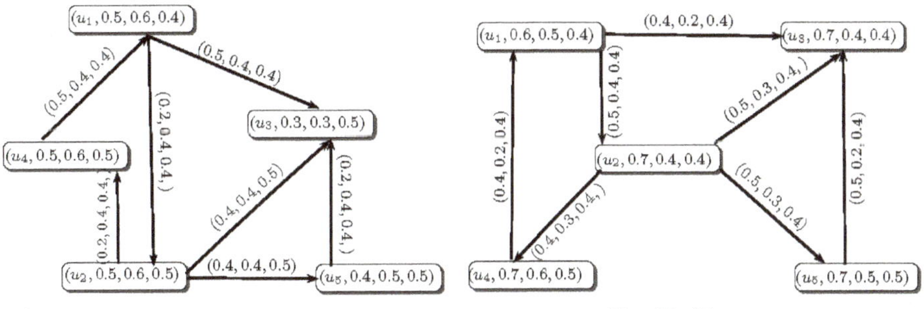


**Figure 12.** Neutrosophic soft rough graph $G = (\underline{G}, \overline{G})$

In [3], the sum of two neutrosophic numbers is defined.

**Definition 20.** [3] *Let C and D be two single valued neutrosophic numbers, and the sum of two single valued neutrosophic number is defined as follows:*

$$C \oplus D = < T_C + T_D - T_C \times T_D, I_C \times I_D, F_C \times F_D > . \tag{1}$$

**Algorithm 1:** Algorithm for selection of most suitable objects

1. Input the number of elements in vertex set $V = \{u_1, u_2, \ldots, u_n\}$.
2. Input the number of elements in parameter set $\mathbb{M} = \{e_1, e_2, \ldots, e_m\}$.
3. Input a neutrosophic soft relation $Q$ from $V$ to $\mathbb{M}$.
4. Input a neutrosophic set $A$ on $M$.
5. Compute neutrosophic soft rough vertex set $QA = (\underline{Q}A, \overline{Q}(A))$.
6. Input the number of elements in edge set $E = \{u_1u_1, u_1u_2, \ldots, u_ku_1\}$.
7. Input the number of elements in parameter set $\acute{\mathbb{M}} = \{e_1e_1, e_1e_2, \ldots, e_le_1\}$.
8. Input a neutrosophic soft relation $S$ from $\acute{V}$ to $\acute{\mathbb{M}}$.
9. Input a neutrosophic set $B$ on $\acute{\mathbb{M}}$.
10. Compute neutrosophic soft rough edge set $SB = (\underline{S}B, \overline{S}(B))$.
11. Compute neutrosophic set $\alpha = (T_\alpha(u_i), I_\alpha(u_i), F_\alpha(u_i))$, where

$$\begin{aligned} T_\alpha(u_i) &= T_{\overline{Q}(A)}(u_i) + T_{\underline{Q}(A)}(u_i) - T_{\overline{Q}(A)}(u_i) \times T_{\underline{Q}(A)}(u_i), \\ I_\alpha(u_i) &= T_{\overline{Q}(A)}(u_i) \times T_{\underline{Q}(A)}(u_i), \\ F_\alpha(u_i) &= F_{\overline{Q}(A)}(u_i) \times F_{\underline{Q}(A)}(u_i). \end{aligned}$$

12. Compute neutrosophic set $\beta = (T_\beta(u_iu_i), I_\beta(u_iu_j), F_\beta(u_iu_j))$, where

$$\begin{aligned} T_\beta(u_iu_j) &= T_{\overline{S}(B)}(u_iu_j) + T_{\underline{S}(B)}(u_iu_j) - T_{\overline{S}(B)}(u_iu_j) \times T_{\underline{S}(B)}(u_iu_j), \\ I_\beta(u_iu_j) &= T_{\overline{S}(B)}(u_iu_j) \times T_{\underline{S}(B)}(u_iu_j), \\ F_\beta(u_iu_j) &= F_{\overline{S}(B)}(u_iu_j) \times F_{\underline{S}(B)}(u_iu_j). \end{aligned}$$

13. Calculate the score values of each object $u_i$, and the score function is defined as follows:

$$\tilde{S}(u_i) = \sum_{u_iu_j \in E} \frac{T_\alpha(u_j) + I_\alpha(u_j) - F_\alpha(u_j)}{3 - (T_\beta(u_iu_j) + I_\beta(u_iu_j) - F_\beta(u_iu_j))}.$$

14. The decision is $S_i$ if $S_i = \max_{i=1}^{n} \tilde{S}_i$.
15. If $i$ has more than one value, then any one of $S_i$ may be chosen.

The sum of UNSRS $\overline{Q}A$ and the LNSRS $\underline{Q}A$ and sum of LNSRR $\underline{S}B$ and the UNSRR $\overline{S}B$ are NSs $\overline{Q}A \oplus \underline{Q}A$ and $\overline{S}B \oplus \underline{S}B$, respectively defined by

$$\begin{aligned} \alpha = \overline{Q}A \oplus \underline{Q}A &= \{(u_1, 0.8, 0.3, 0.16), (u_2, 0.85, 0.24, 0.2), (u_3, 0.79, 0.2, 0.2), (u_4, 0.85, 0.36, 0.25), \\ &\quad (u_5, 0.82, 0.25, 0.25)\}, \\ \beta = \overline{S}B \oplus \underline{S}B &= \{(u_1u_2, 0.6, 0.16, 0.16), (u_1u_3, 0.7, 0.8, 0.16), (u_4u_1, 0.7, 0.8, 0.16), (u_2u_3, 0.7, \\ &\quad 0.12, 0.2), (u_5u_3, 0.6, 0.08, 0.16), (u_2u_4, 0.52, 0.12, 0.16), (u_2u_5, 0.7, 0.12, 0.2)\}. \end{aligned}$$

The score function $\tilde{S}(u_k)$ defines for each generic version medicine $u_i \in V$,

$$\tilde{S}(u_i) = \sum_{u_iu_j \in E} \frac{T_\alpha(u_j) + I_\alpha(u_j) - F_\alpha(u_j)}{3 - (T_\beta(u_iu_j) + I_\beta(u_iu_j) - F_\beta(u_iu_j))} \tag{2}$$

and $u_k$ with the larger score value $u_k = \max_i S(u_i)$ is the most suitable generic version medicine. By calculations, we have

$$\tilde{S}(u_1) = 0.88, \tilde{S}(u_2) = 0.69, \ \tilde{S}(u_3) = 0.26 \ \tilde{S}(u_4) = 0.57, \ \text{and} \ \tilde{S}(u_5) = 0.33. \tag{3}$$

Here, $u_1$ is the optimal decision, and the most suitable generic version of "Loratadine" is "Triamcinolone". We have used software MATLAB (version 7, MathWorks, Natick, MA, USA) for calculating the required results in the application. The algorithm is given in Algorithm 1. The algorithm of the program is general for any number of objects with respect to certain parameters.

## 4. Conclusions

Rough set theory can be considered as an extension of classical set theory. Rough set theory is a very useful mathematical model to handle vagueness. NS theory, RS theory and SS theory are three useful distinguished approaches to deal with vagueness. NS and RS models are used to handle uncertainty, and combining these two models with another remarkable model of SSs gives more precise results for decision-making problems. In this paper, we have presented the notion of NSRGs and investigated some properties of NSRGs in detail. The notion of NSRGs can be utilized as a mathematical tool to deal with imprecise and unspecified information. In addition, a decision-making method based on NSRGs is proposed. This research work can be extended to (1) Rough bipolar neutrosophic soft sets; (2) Bipolar neutrosophic soft rough sets, (3) Interval-valued bipolar neutrosophic rough sets, and (4) Soft rough neutrosophic graphs.

**Author Contributions:** Muhammad Akram and Sundas Shahzadi conceived and designed the experiments; Hafsa M. Malik performed the experiments; Florentin Smarandache analyzed the data; Sundas Shahzadi and Hafsa M. Malik wrote the paper.

**Conflicts of Interest:** The authors declare no conflict of interest.

## References

1. Smarandache, F. Neutrosophic set, a generalisation of the Intuitionistic Fuzzy Sets. *Int. J. Pure Appl. Math.* **2010**, *24*, 289–297.
2. Wang, H.; Smarandache, F.; Zhang, Y.; Sunderraman, R. Single-valued neutrosophic sets. *Multispace Multistructure* **2010**, *4*, 410–413.
3. Peng, J.J.; Wang, J.Q.; Zhang, H.Y.; Chen, X.H. An outranking approach for multi-criteria decision-making problems with simplified neutrosophic sets. *Appl. Soft Comput.* **2014**, *25*, 336–346.
4. Molodtsov, D.A. Soft set theory-first results. *Comput. Math. Appl.* **1999**, *37*, 19–31.
5. Maji, P.K.; Biswas, R.; Roy, A.R. Fuzzy soft sets. *J. Fuzzy Math.* **2001**, *9*, 589–602.
6. Maji, P.K.; Biswas, R.; Roy, A.R. Intuitionistic fuzzy soft sets. *J. Fuzzy Math.* **2001**, *9*, 677–692.
7. Maji, P.K. Neutrosophic soft set. *Ann. Fuzzy Math. Inform.* **2013**, *5*, 157–168.
8. Sahin, R.; Kucuk, A. On similarity and entropy of neutrosophic soft sets. *J. Intell. Fuzzy Syst. Appl. Eng. Technol.* **2014**, *27*, 2417–2430.
9. Pawlak, Z. Rough sets. *Int. J. Comput. Inf. Sci.* **1982**, *11*, 341–356.
10. Feng, F.; Liu, X.; Leoreanu-Fotea, B.; Jun, Y.B. Soft sets and soft rough sets. *Inf. Sci.* **2011**, *181*, 1125–1137.
11. Meng, D.; Zhang, X.; Qin, K. Soft rough fuzzy sets and soft fuzzy rough sets. *Comput. Math. Appl.* **2011**, *62*, 4635–4645.
12. Sun, B.Z.; Ma, W. Soft fuzzy rough sets and its application in decision making. *Artif. Intell. Rev.* **2014**, *41*, 67–80.
13. Sun, B.Z.; Ma, W.; Liu, Q. An approach to decision making based on intuitionistic fuzzy rough sets over two universes. *J. Oper. Res. Soc.* **2013**, *64*, 1079–1089.
14. Zhang, X.; Zhou, B.; Li, Q. A general frame for intuitionistic fuzzy rough sets. *Inf. Sci.* **2012**, *216*, 34–49.
15. Zhang, X.; Dai, J.; Yu, Y. On the union and intersection operations of rough sets based on various approximation spaces. *Inf. Sci.* **2015**, *292*, 214–229.

16. Zhang, H.; Shu, L.; Liao, S. Intuitionistic fuzzy soft rough set and its application in decision making. *Abstr. Appl. Anal.* **2014**, *2014*, 287314.
17. Zhang, H.; Xiong, L.; Ma, W. Generalized intuitionistic fuzzy soft rough set and its application in decision making. *J. Comput. Anal. Appl.* **2016**, *20*, 750–766.
18. Broumi, S.; Smarandache, F. Interval-valued neutrosophic soft rough sets. *Int. J. Comput. Math.* **2015**, *2015*, 232919.
19. Broumi, S.; Smarandache, F.; Dhar, M. Rough neutrosophic sets. *Neutrosophic Sets Syst.* **2014**, *3*, 62–67.
20. Yang, H.L.; Zhang, C.L.; Guo, Z.L.; Liu, Y.L.; Liao, X. A hybrid model of single valued neutrosophic sets and rough sets: Single valued neutrosophic rough set model. *Soft Comput.* **2016**, doi:10.1007/s00500-016-2356-y .
21. Akram, M.; Nawaz, S. Operations on soft graphs. *Fuzzy Inf. Eng.* **2015**, *7*, 423–449.
22. Akram, M.; Nawaz, S. On fuzzy soft graphs. *Ital. J. Pure Appl. Math.* **2015**, *34*, 497–514.
23. Akram, M.; Shahzadi, S. Novel intuitionistic fuzzy soft multiple-attribute decision-making methods. *Neural Comput. Appl.* **2016**, doi:10.1007/s00521-016-2543-x 1–13.
24. Shahzadi, S.; Akram, M. Intuitionistic fuzzy soft graphs with applications. *J. Appl. Math. Comput.* **2016**, *55*, 369–392.
25. Akram, M.; Shahzadi, S. Neutrosophic soft graphs with application. *J. Intell. Fuzzy Syst.* **2017**, *2*, 841–858.
26. Zafar, F.; Akram, M. A novel decision-making method based on rough fuzzy information. *Int. J. Fuzzy Syst.* **2017**, 1–15, doi:10.1007/s40815-017-0368-0.
27. Peng, T.Z.; Qiang, W.J.; Yu, Z.H. Hybrid single-valued neutrosophic MCGDM with QFD for market segment evaluation and selection. *J. Intell. Fuzzy Syst.* **2018**, *34*, 177–187.
28. Qiang, W.J.; Xu, Z.; Yu, Z.H. Hotel recommendation approach based on the online consumer reviews using interval neutrosophic linguistic numbers. *J. Intell. Fuzzy Syst.* **2018**, *34*, 381–394.
29. Luo, S.Z.; Cheng, P.F.; Wang, J.Q.; Huang, Y.J. Selecting Project Delivery Systems Based on Simplified Neutrosophic Linguistic Preference Relations. *Symmetry* **2017**, *9*, 151, doi:10.3390/sym9070151.
30. Nie, R.X.; Wang, J.Q.; Zhang, H.Y. Solving solar-wind power station location problem using an extended WASPAS technique with Interval neutrosophic sets. *Symmetry* **2017**, *9*, 106, doi:10.3390/sym9070106.
31. Wu, X.; Wang, J.; Peng, J.; Qian, J. A novel group decision-making method with probability hesitant interval neutrosphic set and its application in middle level manager's selection. *Int. J. Uncertain. Quant.* **2018**, doi:10.1615/Int.J.UncertaintyQuantification.2017020671.
32. Medina, J.; Ojeda-Aciego, M. Multi-adjoint t-concept lattices. *Inf. Sci.* **2010**, *180*, 712–725.
33. Pozna, C.; Minculete, N.; Precup, R.E.; Kóczy, L.T.; Ballagi, Á. Signatures: Definitions, operators and applications to fuzzy modeling. *Fuzzy Sets Syst.* **2012**, *201*, 86–104.
34. Nowaková, J.; Prílepok, M.; Snášel, V. Medical image retrieval using vector quantization and fuzzy S-tree. *J. Med. Syst.* **2017**, *41*, 1–16.
35. Kumar, A.; Kumar, D.; Jarial, S.K. A hybrid clustering method based on improved artificial bee colony and fuzzy C-Means algorithm. *Int. J. Artif. Intell.* **2017**, *15*, 40–60.

axioms

MDPI

Article

# Neutrosophic Number Nonlinear Programming Problems and Their General Solution Methods under Neutrosophic Number Environments

**Jun Ye *, Wenhua Cui and Zhikang Lu**

Department of Electrical and Information Engineering, Shaoxing University, 508 Huancheng West Road, Shaoxing 312000, China; wenhuacui@usx.edu.cn (W.C.); luzhikang@usx.edu.cn (Z.L.)
* Correspondence: yehjun@aliyun.com or yejun@usx.edu.cn; Tel.: +86-575-8832-7323

Received: 22 January 2018; Accepted: 22 February 2018; Published: 24 February 2018

**Abstract:** In practical situations, we often have to handle programming problems involving indeterminate information. Building on the concepts of indeterminacy $I$ and neutrosophic number (NN) ($z = p + qI$ for $p, q \in \mathbb{R}$), this paper introduces some basic operations of NNs and concepts of NN nonlinear functions and inequalities. These functions and/or inequalities contain indeterminacy $I$ and naturally lead to a formulation of NN nonlinear programming (NN-NP). These techniques include NN nonlinear optimization models for unconstrained and constrained problems and their general solution methods. Additionally, numerical examples are provided to show the effectiveness of the proposed NN-NP methods. It is obvious that the NN-NP problems usually yield NN optimal solutions, but not always. The possible optimal ranges of the decision variables and NN objective function are indicated when the indeterminacy $I$ is considered for possible interval ranges in real situations.

**Keywords:** neutrosophic number; neutrosophic number function; neutrosophic number nonlinear programming; neutrosophic number optimal solution

## 1. Introduction

Traditional mathematical programming usually handles optimization problems involving deterministic objective functions and/or constrained functions. However, uncertainty also exists in real problems. Hence, many researchers have proposed uncertain optimization methods, such as approaches using fuzzy and stochastic logics, interval numbers, or uncertain variables [1–6]. Uncertain programming has been widely applied in engineering, management, and design problems. In existing uncertain programming methods, however, the objective functions or constrained functions are usually transformed into a deterministic or crisp programming problem to yield the optimal feasible crisp solution of the decision variables and the optimal crisp value of the objective function. Hence, existing uncertain linear or nonlinear programming methods are not really meaningful indeterminate methods because they only obtain optimal crisp solutions rather than indeterminate solutions necessary for real situations. However, indeterminate programming problems may also yield an indeterminate optimal solution for the decision variables and the indeterminate optimal value of the objective function suitable for real problems with indeterminate environments. Hence, it is necessary to understand how to handle indeterminate programming problems with indeterminate solutions.

Since there exists indeterminacy in the real world, Smarandache [7–9] first introduced a concept of indeterminacy—which is denoted by $I$, the imaginary value—and then he presented a neutrosophic number (NN) $z = p + qI$ for $p, q \in \mathbb{R}$ ($\mathbb{R}$ is all real numbers) by combining the determinate part $p$ with the indeterminate part $qI$. It is obvious that this is a useful mathematical concept for describing incomplete and indeterminate information. After their introduction, NNs were applied to decision-making [10,11] and fault diagnosis [12,13] under indeterminate environments.

In 2015, Smarandache [14] introduced a neutrosophic function (i.e., interval function or thick function), neutrosophic precalculus, and neutrosophic calculus to handle more indeterminate problems. He defined a neutrosophic thick function $g$: $\mathbb{R} \to G(\mathbb{R})$ ($G(\mathbb{R})$ is the set of all interval functions) as the form of an interval function $g(x) = [g_1(x), g_2(x)]$. After that, Ye et al. [15] introduced the neutrosophic functions in expressions for the joint roughness coefficient and the shear strength in the mechanics of rocks. Further, Ye [16] and Chen et al. [17,18] presented expressions and analyses of the joint roughness coefficient using NNs. Ye [19] proposed the use of neutrosophic linear equations and their solution methods in traffic flow problems with NN information.

Recently, NNs have been extended to linguistic expressions. For instance, Ye [20] proposed neutrosophic linguistic numbers and their aggregation operators for multiple attribute group decision-making. Further, Ye [21] presented hesitant neutrosophic linguistic numbers—based on both the neutrosophic linguistic numbers and the concept of hesitant fuzzy logic—calculated their expected value and similarity measure, and applied them to multiple attribute decision-making. Additionally, Fang and Ye [22] introduced linguistic NNs based on both the neutrosophic linguistic number and the neutrosophic set concept, and some aggregation operators of linguistic NNs for multiple attribute group decision-making.

In practical problems, the information obtained by decision makers or experts may be imprecise, uncertain, and indeterminate because of a lack of data, time pressures, measurement errors, or the decision makers' limited attention and knowledge. In these cases, we often have to solve programming problems involving indeterminate information (indeterminacy $I$). However, the neutrosophic functions introduced in [14,15] do not contain information about the indeterminacy $I$ and also cannot express functions involving indeterminacy $I$. Thus, it is important to define NN functions containing indeterminacy $I$ based on the concept of NNs, in order to handle programming problems under indeterminate environments. Jiang and Ye [23] and Ye [24] proposed NN linear and nonlinear programming models and their preliminary solution methods, but they only handled some simple/specified NN optimization problems and did not propose effective solution methods for complex NN optimization problems. To overcome this insufficiency, this paper first introduces some operations of NNs and concepts of NN linear and nonlinear functions and inequalities, which contain indeterminacy $I$. Then, various NN nonlinear programming (NN-NP) models and their general solution methods are proposed in order to obtain NN/indeterminate optimal solutions.

The rest of this paper is structured as follows. On the basis of some basic concept of NNs, Section 2 introduces some basic operations of NNs and concepts of NN linear and nonlinear functions and inequalities with indeterminacy $I$. Section 3 presents NN-NP problems, including NN nonlinear optimization models with unconstrained and constrained problems. In Section 4, general solution methods are introduced for various NN-NP problems, and then numerical examples are provided to illustrate the effectiveness of the proposed NN-NP methods. Section 5 contains some conclusions and future research.

## 2. Neutrosophic Numbers and Neutrosophic Number Functions

Smarandache [7–9] first introduced an NN, denoted by $z = p + qI$ for $p, q \in \mathbb{R}$, consisting of a determinate part $p$ and an indeterminate part $qI$, where $I$ is the indeterminacy. Clearly, it can express determinate information and indeterminate information as in real world situations. For example, consider the NN $z = 5 + 3I$ for $I \in [0, 0.3]$, which is equivalent to $z \in [5, 5.9]$. This indicates that the determinate part of $z$ is 5, the indeterminate part is $3I$, and the interval of possible values for the number $z$ is [5, 5.9]. If $I \in [0.1, 0.2]$ is considered as a possible interval range of indeterminacy $I$, then the possible value of $z$ is within the interval [5.3, 5.6]. For another example, the fraction 7/15 is within the interval [0.46, 0.47], which is represented as the neutrosophic number $z = 0.46 + 0.01I$ for $I \in [0, 1]$.

The NN $z$ indicates that the determinate value is 0.46, the indeterminate value is $0.01I$, and the possible value is within the interval [0.46, 0.47].

It is obvious that an NN $z = p + qI$ may be considered as the possible interval range (changeable interval number) $z = [p + q\cdot\inf\{I\}, p + q\cdot\sup\{I\}]$ for $p, q \in \mathbb{R}$ and $I \in [\inf\{I\}, \sup\{I\}]$. For convenience, $z$ is denoted by $z = [p + qI^L, p + qI^U]$ for $z \in Z$ ($Z$ is the set of all NNs) and $I \in [I^L, I^U]$ for short. In special cases, $z$ can be expressed as the determinate part $z = p$ if $qI = 0$ for the best case, and, also, $z$ can be expressed as the indeterminate part $z = qI$ if $p = 0$ for the worst case.

Let two NNs be $z_1 = p_1 + q_1I$ and $z_2 = p_2 + q_2I$ for $z_1, z_2 \in Z$, then their basic operational laws for $I \in [I^L, I^U]$ are defined as follows [23,24]:

(1) $z_1 + z_2 = p_1 + p_2 + (q_1 + q_2)I = [p_1 + p_2 + q_1I^L + q_2I^L, p_1 + p_2 + q_1I^U + q_2I^U]$;

(2) $z_1 - z_2 = p_1 - p_2 + (q_1 - q_2)I = [p_1 - p_2 + q_1I^L - q_2I^L, p_1 - p_2 + q_1I^U - q_2I^U]$;

(3)
$$\begin{aligned} z_1 \times z_2 &= p_1p_2 + (p_1q_2 + p_2q_1)I + q_1q_2I^2 \\ &= \begin{bmatrix} \min\begin{pmatrix} (p_1 + q_1I^L)(p_2 + q_2I^L), (p_1 + q_1I^L)(p_2 + q_2I^U), \\ (p_1 + q_1I^U)(p_2 + q_2I^L), (p_1 + q_1I^U)(p_2 + q_2I^U) \end{pmatrix}, \\ \max\begin{pmatrix} (p_1 + q_1I^L)(p_2 + q_2I^L), (p_1 + q_1I^L)(p_2 + q_2I^U), \\ (p_1 + q_1I^U)(p_2 + q_2I^L), (p_1 + q_1I^U)(p_2 + q_2I^U) \end{pmatrix} \end{bmatrix}; \end{aligned}$$

(4)
$$\begin{aligned} \frac{z_1}{z_2} &= \frac{p_1+q_1I}{p_2+q_2I} = \frac{[p_1+q_1I^L, p_1+q_1I^U]}{[p_2+q_2I^L, p_2+q_2I^U]} \\ &= \begin{bmatrix} \min\left(\frac{p_1+q_1I^L}{p_2+q_2I^U}, \frac{p_1+q_1I^L}{p_2+q_2I^L}, \frac{p_1+q_1I^U}{p_2+q_2I^U}, \frac{p_1+q_1I^U}{p_2+q_2I^L}\right), \\ \max\left(\frac{p_1+q_1I^L}{p_2+q_2I^U}, \frac{p_1+q_1I^L}{p_2+q_2I^L}, \frac{p_1+q_1I^U}{p_2+q_2I^U}, \frac{p_1+q_1I^U}{p_2+q_2I^L}\right) \end{bmatrix}. \end{aligned}$$

For a function containing indeterminacy $I$, we can define an NN function (indeterminate function) in $n$ variables (unknowns) as $F(x, I)$: $Z^n \to Z$ for $x = [x_1, x_2, \ldots, x_n]^T \in Z^n$ and $I \in [I^L, I^U]$, which is either an NN linear or an NN nonlinear function. For example, $F_1(x, I) = x_1 - Ix_2 + 1 + 2I$ for $x = [x_1, x_2]^T \in Z^2$ and $I \in [I^L, I^U]$ is an NN linear function, and $F_2(x) = x_1^2 + x_2^2 - 2Ix_1 - Ix_2 + 3I$ for $x = [x_1, x_2]^T \in Z^2$ and $I \in [I^L, I^U]$ is an NN nonlinear function.

For an NN function in $n$ variables (unknowns) $g(x, I)$: $Z^n \to Z$, we can define an NN inequality $g(x, I) \leq (\geq)\ 0$ for $x = [x_1, x_2, \ldots, x_n]^T \in Z^n$ and $I \in [I^L, I^U]$, where $g(x, I)$ is either an NN linear function or an NN nonlinear function. For example, $g_1(x, I) = 2x_1 - Ix_2 + 4 + 3I \leq 0$ and $g_2(x, I) = 2x_1^2 - x_2^2 + 2 + 5I \leq 0$ for $x = [x_1, x_2]^T \in Z^2$ and $I \in [I^L, I^U]$ are NN linear and NN nonlinear inequalities in two variables, respectively.

Generally, the values of $x$, $F(x, I)$, and $g(x, I)$ are NNs (usually but not always). In this study, we mainly research on NN-NP problems and their general solution methods.

## 3. Neutrosophic Number Nonlinear Programming Problems

An NN-NP problem is similar to a traditional nonlinear programming problem, which is composed of an objective function, general constraints, and decision variables. The difference is that an NN-NP problem includes at least one NN nonlinear function, which could be the objective function, or some or all of the constraints. In the real world, many real problems are inherently nonlinear and indeterminate. Hence, various NN optimization models need to be established to handle different NN-NP problems.

In general, NN-NP problems in $n$ decision variables can be expressed by the following NN mathematical models:

(1) Unconstrained NN optimization model:

$$\min F(x, I), x \in Z^n, \tag{1}$$

where $x = [x_1, x_2, \ldots, x_n]^T \in Z^n$, $F(x, I)$: $Z^n \to Z$, and $I \in [I^L, I^U]$.

(2) Constrained NN optimization model:

$$\begin{array}{c} \min F(x, I) \\ \text{s.t. } g_i(x, I) \leq 0, I = 1, 2, \ldots, m \\ h_j(x, I) = 0, j = 1, 2, \ldots, l \\ x \in Z^n, \end{array} \tag{2}$$

where $g_1(x, I), g_2(x, I), \ldots, g_m(x, I), h_1(x, I), h_2(x, I), \ldots, h_l(x, I)$: $Z^n \to Z$, and $I \in [I^L, I^U]$.

In special cases, if the NN-NP problem only contains the restrictions $h_j(x, I) = 0$ without inequality constraints, $g_i(x, I) \leq 0$, then the NN-NP problem is called the NN-NP problem with equality constraints. If the NN-NP problem only contains the restrictions $g_i(x, I) \leq 0$, without constraints $h_j(x, I) = 0$, then the NN-NP problem is called the NN-NP problem with inequality constraints. Finally, if the NN-NP problem does not contain either restrictions, $h_j(x, I) = 0$ or $g_i(x, I) \leq 0$, then the constrained NN-NP problem is reduced to the unconstrained NN-NP problem.

The NN optimal solution for the decision variables is feasible in an NN-NP problem if it satisfies all of the constraints. Usually, the optimal solution for the decision variables and the value of the NN objective function are NNs, but not always). When the indeterminacy $I$ is considered as a possible interval range (possible interval number), the optimal solution of all feasible intervals forms the feasible region or feasible set for $x$ and $I \in [I^L, I^U]$. In this case, the value of the NN objective function is an optimal possible interval (NN) for $F(x, I)$.

In the following section, we shall introduce general solution methods for NN-NP problems, including unconstrained NN and constrained NN nonlinear optimizations, based on methods of traditional nonlinear programming problems.

## 4. General Solution Methods for NN-NP Problems

### 4.1. One-Dimension Unconstrained NN Nonlinear Optimization

The simplest NN nonlinear optimization only has a nonlinear NN objective function with one variable and no constraints. Let us consider a single variable NN nonlinear objective function $F(x, I)$ for $x \in Z$ and $I \in [I^L, I^U]$. Then, for a differentiable NN nonlinear objective function $F(x, I)$, a local optimal solution $x^*$ satisfies the following two conditions:

(1) Necessary condition: The derivative is $dF(x^*, I)/dx = 0$ for $I \in [I^L, I^U]$;
(2) Sufficient condition: If the second derivative is $d^2F(x^*, I)/dx^2 < 0$ for $I \in [I^L, I^U]$, then $x^*$ is an optimal solution for the maximum $F(x^*, I)$; if the second derivative is $d^2F(x^*, I)/dx^2 > 0$, then $x^*$ is an optimal solution for the minimum $F(x^*, I)$.

**Example 1.** *An NN nonlinear objective function with one variable is $F(x, I) = 2Ix^2 + 5I$ for $x \in Z$ and $I \in [I^L, I^U]$. Based on the optimal conditions, we can obtain:*

$$\frac{dF(x, I)}{dx} = 4Ix = 0 \Rightarrow x^* = 0,$$

$$\frac{d^2F(x, I)}{dx^2}|_{x^*=0} = 4I.$$

*Assume that we consider a specific possible range of $I \in [I^L, I^U]$ according to real situations or actual requirements, then we can discuss its optimal possible value. If $I \in [1, 2]$ is considered as a possible interval range, then $d^2F(x^*, I)/dx^2 > 0$, and $x^* = 0$ is the optimal solution for the minimum $F(x^*, I)$. Thus, the minimum value of the NN objective function is $F(x^*, I) = [5, 10]$, which, in this case, is a possible interval range, but not always. Specifically if $I = 1$ (crisp value), then $F(x^*, I) = 5$.*

*4.2. Multi-Dimension Unconstrained NN Nonlinear Optimization*

Assume that a multiple variable NN function $F(\boldsymbol{x}, I)$ for $\boldsymbol{x} = [x_1, x_2, \ldots, x^n]^T \in Z^n$ and $I \in [I^L, I^U]$ is considered as an unconstrained differentiable NN nonlinear objective function in $n$ variables. Then, we can obtain the partial derivatives:

$$\nabla F(\boldsymbol{x}, I) = \left[\frac{\partial F(\boldsymbol{x}, I)}{\partial x_1}, \frac{\partial F(\boldsymbol{x}, I)}{\partial x_2}, \ldots, \frac{\partial F(\boldsymbol{x}, I)}{\partial x_n}\right]^T = 0 \Rightarrow \boldsymbol{x} = \boldsymbol{x}^*.$$

Further, the partial second derivatives, structured as the Hessian matrix $H(\boldsymbol{x}, I)$, are:

$$H(\boldsymbol{x}, I) = \begin{bmatrix} \frac{\partial^2 F(\boldsymbol{x},I)}{\partial x_1^2}, & \frac{\partial^2 F(\boldsymbol{x},I)}{\partial x_1 \partial x_2}, & \ldots, & \frac{\partial^2 F(\boldsymbol{x},I)}{\partial x_1 \partial x_n} \\ \frac{\partial^2 F(\boldsymbol{x},I)}{\partial x_2 \partial x_1}, & \frac{\partial^2 F(\boldsymbol{x},I)}{\partial x_2^2}, & \ldots, & \frac{\partial^2 F(\boldsymbol{x},I)}{\partial x_2 \partial x_n} \\ \vdots & \vdots & \vdots & \vdots \\ \frac{\partial^2 F(\boldsymbol{x},I)}{\partial x_n \partial x_1}, & \frac{\partial^2 F(\boldsymbol{x},I)}{\partial x_n \partial x_2}, & \ldots, & \frac{\partial^2 F(\boldsymbol{x},I)}{\partial x_n^2} \end{bmatrix}_{x=x^*}.$$

Then, the Hessian matrix $H(\boldsymbol{x}, I)$ is structured as its subsets $H_i(\boldsymbol{x}, I)$ ($i = 1, 2, \ldots, n$), where $H_i(\boldsymbol{x}, I)$ indicate the subset created by taking the first $i$ rows and columns of $H(\boldsymbol{x}, I)$. You calculate the determinant of each of the $n$ subsets at $\boldsymbol{x}^*$:

$$H_1(\boldsymbol{x}^*, I) = \left|\frac{\partial^2 F(\boldsymbol{x}^*, I)}{\partial x_1^2}\right|, H_2(\boldsymbol{x}^*, I) = \begin{vmatrix} \frac{\partial^2 F(\boldsymbol{x}^*,I)}{\partial x_1^2} & \frac{\partial^2 F(\boldsymbol{x}^*,I)}{\partial x_1 \partial x_2} \\ \frac{\partial^2 F(\boldsymbol{x}^*,I)}{\partial x_2 \partial x_1} & \frac{\partial^2 F(\boldsymbol{x}^*,I)}{\partial x_2^2} \end{vmatrix}, \ldots$$

from the sign patterns of the determinates of $H_i(\boldsymbol{x}^*, I)$ ($i = 1, 2, \ldots, n$) for $I \in [I^L, I^U]$, as follows:

(1) If $H_i(\boldsymbol{x}^*, I) > 0$, then $H(\boldsymbol{x}^*, I)$ is positive definite at $\boldsymbol{x}^*$;
(2) If $H_i(\boldsymbol{x}^*, I) < 0$ and the remaining $H_i(\boldsymbol{x}^*, I)$ alternate in sign, then $H(\boldsymbol{x}^*, I)$ is negative definite at $\boldsymbol{x}^*$;
(3) If some of the values which are supposed to be nonzero turn out to be zero, then $H(\boldsymbol{x}^*, I)$ can be positive semi-definite or negative semi-definite.

A local optimal value of $\boldsymbol{x}^*$ in neutrosophic nonlinear objective function $F(\boldsymbol{x}^*, I)$ for $I \in [I^L, I^U]$ can be determined by the following categories:

(1) $\boldsymbol{x}^*$ is a local maximum if $\nabla F(\boldsymbol{x}^*, I) = 0$ and $H(\boldsymbol{x}^*, I)$ is negative definite;
(2) $\boldsymbol{x}^*$ is a local minimum if $\nabla F(\boldsymbol{x}^*, I) = 0$ and $H(\boldsymbol{x}^*, I)$ is positive definite;
(3) $\boldsymbol{x}^*$ is a saddle point if $\nabla F(\boldsymbol{x}^*, I) = 0$ and $H(\boldsymbol{x}^*, I)$ is neither positive semi-definite nor negative semi-definite.

**Example 2.** *Consider an unconstrained NN nonlinear objective function with two variables $x_1$ and $x_2$ is $F(\boldsymbol{x}, I) = x_1^2 + x_2^2 - 4Ix_1 - 2Ix_2 + 5$ for $\boldsymbol{x} \in Z^2$ and $I \in [I^L, I^U]$. According to optimal conditions, we first obtain the following derivative and the optimal solution:*

$$\nabla F(\boldsymbol{x}, I) = \begin{bmatrix} \frac{\partial F(\boldsymbol{x},I)}{\partial x_1} \\ \frac{\partial F(\boldsymbol{x},I)}{\partial x_2} \end{bmatrix} = \begin{bmatrix} 2x_1 - 4I \\ 2x_2 - 2I \end{bmatrix} = 0 \Rightarrow \boldsymbol{x}^* = \begin{bmatrix} x_1^* \\ x_2^* \end{bmatrix} = \begin{bmatrix} 2I \\ I \end{bmatrix}.$$

*Then, the NN Hessian matrix is given as follows:*

$$H(\boldsymbol{x}^*, I) = \begin{bmatrix} \frac{\partial^2 F(\boldsymbol{x}^*,I)}{\partial x_1^2} & \frac{\partial^2 F(\boldsymbol{x}^*,I)}{\partial x_1 \partial x_2} \\ \frac{\partial^2 F(\boldsymbol{x}^*,I)}{\partial x_2 \partial x_1} & \frac{\partial^2 F(\boldsymbol{x}^*,I)}{\partial x_2^2} \end{bmatrix} = \begin{bmatrix} 2 & 0 \\ 0 & 2 \end{bmatrix}.$$

*Thus,* $|H_1(\mathbf{x}^*, I)| = 2 > 0$ *and* $|H(\mathbf{x}^*, I)| = \begin{vmatrix} 2 & 0 \\ 0 & 2 \end{vmatrix} = 4 > 0$. *Hence, the NN optimal solution is* $\mathbf{x}^* = [2I, I]^T$ *and the minimum value of the NN objective function is* $F(\mathbf{x}^*, I) = 5(1 - I^2)$ *in this optimization problem.*

*If the indeterminacy* $I \in [0, 1]$ *is considered as a possible interval range, then the optimal solution of* $\mathbf{x}$ *is* $x_1^* = [0, 2]$ *and* $x_2^* = [0, 1]$ *and the minimum value of the NN objective function is* $F(\mathbf{x}^*, I) = [0, 5]$. *Specifically, when* $I = 1$ *is a determinate value, then* $x_1^* = 2$, $x_2^* = 1$, *and* $F(\mathbf{x}^*, I) = 0$. *In this case, the NN nonlinear optimization is reduced to the traditional nonlinear optimization, which is a special case of the NN nonlinear optimization.*

### 4.3. NN-NP Problem Having Equality Constraints

Consider an NN-NP problem having NN equality constraints:

$$\begin{gathered} \min F(\mathbf{x}, I) \\ \text{s.t. } h_j(\mathbf{x}, I) = 0, j = 1, 2, \ldots, l \\ \mathbf{x} \in Z^n \end{gathered} \tag{3}$$

where $h_1(\mathbf{x}, I), h_2(\mathbf{x}, I), \ldots, h_l(\mathbf{x}, I)$: $Z^n \to Z$ and $I \in [I^L, I^U]$.

Here we consider Lagrange multipliers for the NN-NP problem. The Lagrangian function that we minimize is then given by:

$$L(\mathbf{x}, I, \boldsymbol{\lambda}) = F(\mathbf{x}, I) + \sum_{j=1}^{l} \lambda_j h_j(\mathbf{x}, I), \lambda \in Z^l, \mathbf{x} \in Z^n, \tag{4}$$

where $\lambda_j$ ($j = 1, 2, \ldots, l$) is a Lagrange multiplier and $I \in [I^L, I^U]$. It is obvious that this method transforms the constrained optimization into unconstrained optimization. Then, the necessary condition for this case to have a minimum is that:

$$\frac{\partial L(\mathbf{x}, I, \boldsymbol{\lambda})}{\partial x_i} = 0, i = 1, 2, \ldots, n,$$

$$\frac{\partial L(\mathbf{x}, I, \boldsymbol{\lambda})}{\partial \lambda_j} = 0, j = 1, 2, \ldots, l.$$

By solving $n + l$ equations above, we can obtain the optimum solution $\mathbf{x}^* = [x_1^*, x_2^*, \ldots, x_n^*]^T$ and the optimum multiplier values $\lambda_j^*$ ($j = 1, 2, \ldots, l$).

**Example 3.** *Let us consider an NN-NP problem having an NN equality constraint:*

$$\min F(\mathbf{x}, I) = 4Ix_1 + 5x_2$$

$$\text{s.t. } h(\mathbf{x}, I) = 2x_1 + 3x_2 - 6I = 0, \ \mathbf{x} \in Z^2.$$

*Then, we can construct the Lagrangian function:*

$$L(\mathbf{x}, I, \lambda) = 4Ix_1 + 5x_2 + \lambda(2x_1 + 3x_2 - 6I), \ \lambda \in Z, \mathbf{x} \in Z^2.$$

*The necessary condition for the optimal solution yields the following:*

$$\frac{\partial L(\mathbf{x}, I, \lambda)}{\partial x_1} = 8Ix_1 + 2\lambda = 0, \quad \frac{\partial L(\mathbf{x}, I, \lambda)}{\partial x_2} = 10x_2 + 3\lambda = 0, \text{ and } \frac{\partial L(\mathbf{x}, I, \lambda)}{\partial \lambda} = 2x_1 + 3x_2 - 6I = 0.$$

*By solving these equations, we obtain the results $x_1 = -\lambda/(4I)$, $x_2 = -3\lambda/10$, and $\lambda = -12I^2/(1 + 1.8I)$. Hence, the NN optimal solution is obtained by the results of $x_1^* = 3I/(1 + 1.8I)$ and $x_2^* = 18I^2/(5 + 9I)$. If the indeterminacy $I \in [1, 2]$ is considered as a possible interval range, then the optimal solution is $x_1^* = [0.6522, 4.2857]$ and $x_2^* = [0.7826, 5.1429]$. Specifically, if I = 1 (crisp value), then the optimal solution is $x_1^* = 1.0714$ and $x_2^* = 1.2857$, which are reduced to the crisp optimal solution in classical optimization problems.*

### 4.4. General Constrained NN-NP Problems

Now, we consider a general constrained NN-NP problem:

$$\begin{array}{c} \min F(\boldsymbol{x}, I) \\ \text{s.t. } g_k(\boldsymbol{x}, I) \leq 0, k = 1, 2, \ldots, m \\ h_j(\boldsymbol{x}, I) = 0, j = 1, 2, \ldots, l \\ \boldsymbol{x} \in Z^n \end{array} \quad (5)$$

where $g_1(\boldsymbol{x}, I), g_2(\boldsymbol{x}, I), \ldots, g_m(\boldsymbol{x}, I), h_1(\boldsymbol{x}, I), h_2(\boldsymbol{x}, I), \ldots, h_l(\boldsymbol{x}, I)$: $Z^n \to Z$ for $I \in [I^L, I^U]$. Then, we can consider the NN Lagrangian function for the NN-NP problem:

$$L(\boldsymbol{x}, I, \boldsymbol{\mu}, \boldsymbol{\lambda}) = F(\boldsymbol{x}, I) + \sum_{k=1}^{m} \mu_k g_k(\boldsymbol{x}, I) + \sum_{j=1}^{l} \lambda_j h_j(\boldsymbol{x}, I),\ \mu \in Z^m, \lambda \in Z^l, x \in Z^n. \quad (6)$$

The usual NN Karush–Kuhn–Tucker (KKT) necessary conditions yield:

$$\nabla F(\boldsymbol{x}, I) + \sum_{k=1}^{m} \{\mu_k \nabla g_k(\boldsymbol{x}, I)\} + \sum_{j=1}^{l} \{\lambda_j \nabla h_j(\boldsymbol{x}, I)\} = 0 \quad (7)$$

combined with the original constraints, complementary slackness for the inequality constraints, and $\mu_k \geq 0$ for $k = 1, 2, \ldots, m$.

**Example 4.** *Let us consider an NN-NP problem with one NN inequality constraint:*

$$\min F(\boldsymbol{x}, I) = Ix_1^2 + 2x_2^2$$

$$\text{s.t. } g(\boldsymbol{x}, I) = I - x_1 - x_2 \leq 0,\ x \in Z^2.$$

*Then, the NN Lagrangian function is constructed as:*

$$L(\boldsymbol{x}, I, \mu) = Ix_1^2 + 2x_2^2 + \mu(I - x_1 - x_2),\ \mu \in Z, x \in Z^2.$$

*The usual NN KKT necessary conditions yield:*

$$\frac{\partial L(\boldsymbol{x}, I, \mu)}{\partial x_1} = 2Ix_1 - \mu = 0,\ \frac{\partial L(\boldsymbol{x}, I, \mu)}{\partial x_2} = 4x_2 - \mu = 0, \text{ and } \mu(I - x_1 - x_2) = 0.$$

*By solving these equations, we can obtain the results of $x_1 = \mu/(2I)$, $x_2 = \mu/4$, and $\mu = 4I^2/(2 + I)$ ($\mu = 0$ yields an infeasible solution for $I > 0$). Hence, the NN optimal solution is obtained by the results of $x_1^* = 2I/(2 + I)$ and $x_2^* = I^2/(2 + I)$.*

*If the indeterminacy $I \in [1, 2]$ is considered as a possible interval range corresponding to some specific actual requirement, then the optimal solution is $x_1^* = [0.5, 1.3333]$ and $x_2^* = [0.25, 1.3333]$. As another case, if the indeterminacy $I \in [2, 3]$ is considered as a possible interval range corresponding to some specific actual requirement, then the optimal solution is $x_1^* = [0.8, 1.5]$ and $x_2^* = [0.8, 2.25]$. Specifically, if I = 2 (a crisp value), then the optimal solution is $x_1^* = 1$ and $x_2^* = 1$, which is reduced to the crisp optimal solution of the crisp/classical optimization problem.*

Compared with existing uncertain optimization methods [1–6], the proposed NN-NP methods can obtain ranges of optimal solutions (usually NN solutions but not always) rather than the crisp optimal solutions of previous uncertain optimization methods [1–6], which are not really meaningful in indeterminate programming of indeterminate solutions in real situations [23,24]. The existing uncertain optimization solutions are the special cases of the proposed NN-NP optimization solutions. Furthermore, the existing uncertain optimization methods in [1–6] cannot express and solve the NN-NP problems from this study. Obviously, the optimal solutions in the NN-NP problems are intervals corresponding to different specific ranges of the indeterminacy $I \in [I^L, I^U]$ and show the flexibility and rationality under indeterminate/NN environments, which is the main advantage of the proposed NN-NP methods.

## 5. Conclusions

On the basis of the concepts of indeterminacy $I$ and NNs, this paper introduced some basic operations of NNs and concepts of both NN linear and nonlinear functions and inequalities, which involve indeterminacy $I$. Then, we proposed NN-NP problems with unconstrained and constrained NN nonlinear optimizations and their general solution methods for various optimization models. Numerical examples were provided to illustrate the effectiveness of the proposed NN-NP methods. The main advantages are that: (1) some existing optimization methods like the Lagrange multiplier method and the KKT condition can be employed for NN-NP problems, (2) the indeterminate (NN) programming problems can show indeterminate (NN) optimal solutions which can indicate possible optimal ranges of the decision variables and NN objective function when indeterminacy $I \in [I^L, I^U]$ is considered as a possible interval range for real situations and actual requirements, and (3) NN-NP is the generalization of traditional nonlinear programming problems and is more flexible and more suitable than the existing unconcerned nonlinear programming methods under indeterminate environments. The proposed NN-NP methods provide a new effective way for avoiding crisp solutions of existing unconcerned programming methods under indeterminate environments.

It is obvious that the NN-NP methods proposed in this paper not only are the generalization of existing certain or uncertain nonlinear programming methods but also can deal with determinate and/or indeterminate mathematical programming problems. In the future, we shall apply these NN-NP methods to engineering fields, such as engineering design and engineering management.

**Acknowledgments:** This paper was supported by the National Natural Science Foundation of China (Nos. 71471172, 61703280).

**Author Contributions:** Jun Ye proposed the neutrosophic number nonlinear programming methods and Wenhua Cui and Zhikang Lu gave examples, calculations, and comparative analysis. All the authors wrote the paper.

**Conflicts of Interest:** The authors declare no conflict of interest.

## References

1. Jiang, C.; Long, X.Y.; Han, X.; Tao, Y.R.; Liu, J. Probability-interval hybrid reliability analysis for cracked structures existing epistemic uncertainty. *Eng. Fract. Mech.* **2013**, *112–113*, 148–164. [CrossRef]
2. Zhang, B.; Peng, J. Uncertain programming model for uncertain optimal assignment problem. *Appl. Math. Model.* **2013**, *37*, 6458–6468. [CrossRef]
3. Jiang, C.; Zhang, Z.G.; Zhang, Q.F.; Han, X.; Xie, H.C.; Liu, J. A new nonlinear interval programming method for uncertain problems with dependent interval variables. *Eur. J. Oper. Res.* **2014**, *238*, 245–253. [CrossRef]
4. Liu, B.D.; Chen, X.W. Uncertain multiobjective programming and uncertain goal programming. *J. Uncertain. Anal. Appl.* **2015**, *3*, 10. [CrossRef]
5. Veresnikov, G.S.; Pankova, L.A.; Pronina, V.A. Uncertain programming in preliminary design of technical systems with uncertain parameters. *Procedia Comput. Sci.* **2017**, *103*, 36–43. [CrossRef]
6. Chen, L.; Peng, J.; Zhang, B. Uncertain goal programming models for bicriteria solid transportation problem. *Appl. Soft Comput.* **2017**, *51*, 49–59. [CrossRef]

7. Smarandache, F. *Neutrosophy: Neutrosophic Probability, Set, and Logic*; American Research Press: Rehoboth, MA, USA, 1998.
8. Smarandache, F. *Introduction to Neutrosophic Measure, Neutrosophic Integral, and Neutrosophic Probability*; Sitech & Education Publisher: Columbus, OH, USA, 2013.
9. Smarandache, F. *Introduction to Neutrosophic Statistics*; Sitech & Education Publishing: Columbus, OH, USA, 2014.
10. Ye, J. Multiple-attribute group decision-making method under a neutrosophic number environment. *J. Intell. Syst.* **2016**, *25*, 377–386. [CrossRef]
11. Ye, J. Bidirectional projection method for multiple attribute group decision making with neutrosophic numbers. *Neural Comput. Appl.* **2017**, *28*, 1021–1029. [CrossRef]
12. Kong, L.W.; Wu, Y.F.; Ye, J. Misfire fault diagnosis method of gasoline engines using the cosine similarity measure of neutrosophic numbers. *Neutrosophic Sets Syst.* **2015**, *8*, 43–46.
13. Ye, J. Fault diagnoses of steam turbine using the exponential similarity measure of neutrosophic numbers. *J. Intell. Fuzzy Syst.* **2016**, *30*, 1927–1934. [CrossRef]
14. Smarandache, F. *Neutrosophic Precalculus and Neutrosophic Calculus*; EuropaNova: Brussels, Belgium, 2015.
15. Ye, J.; Yong, R.; Liang, Q.F.; Huang, M.; Du, S.G. Neutrosophic functions of the joint roughness coefficient (JRC) and the shear strength: A case study from the pyroclastic rock mass in Shaoxing City, China. *Math. Prob. Eng.* **2016**, *2016*, 4825709. [CrossRef]
16. Ye, J.; Chen, J.Q.; Yong, R.; Du, S.G. Expression and analysis of joint roughness coefficient using neutrosophic number functions. *Information* **2017**, *8*, 69. [CrossRef]
17. Chen, J.Q.; Ye, J.; Du, S.G.; Yong, R. Expressions of rock joint roughness coefficient using neutrosophic interval statistical numbers. *Symmetry* **2017**, *9*, 123. [CrossRef]
18. Chen, J.Q.; Ye, J.; Du, S.G. Scale effect and anisotropy analyzed for neutrosophic numbers of rock joint roughness coefficient based on neutrosophic statistics. *Symmetry* **2017**, *9*, 208. [CrossRef]
19. Ye, J. Neutrosophic linear equations and application in traffic flow problems. *Algorithms* **2017**, *10*, 133. [CrossRef]
20. Ye, J. Aggregation operators of neutrosophic linguistic numbers for multiple attribute group decision making. *SpringerPlus* **2016**, *5*, 1691. [CrossRef] [PubMed]
21. Ye, J. Multiple attribute decision-making methods based on expected value and similarity measure of hesitant neutrosophic linguistic numbers. *Cogn. Comput.* **2017**. [CrossRef]
22. Fang, Z.B.; Ye, J. Multiple attribute group decision-making method based on linguistic neutrosophic numbers. *Symmetry* **2017**, *9*, 111. [CrossRef]
23. Jiang, W.Z.; Ye, J. Optimal design of truss structures using a neutrosophic number optimization model under an indeterminate environment. *Neutrosophic Sets Syst.* **2016**, *14*, 93–97.
24. Ye, J. Neutrosophic number linear programming method and its application under neutrosophic number environments. *Soft Comput.* **2017**. [CrossRef]

axiomss

MDPI

Article

# NN-Harmonic Mean Aggregation Operators-Based MCGDM Strategy in a Neutrosophic Number Environment

**Kalyan Mondal [1], Surapati Pramanik [2,*], Bibhas C. Giri [1] and Florentin Smarandache [3]**

[1] Department of Mathematics, Jadavpur University, Kolkata-700032 West Bengal, India; kalyanmathematic@gmail.com (K.M.); bibhasc.giri@jadavpuruniversity.in (B.C.G.)
[2] Department of Mathematics, Nandalal Ghosh B.T. College, Panpur, PO-Narayanpur, and District: North 24 Parganas, Pin-743126 West Bengal, India
[3] Mathematics & Science Department, University of New Mexico, 705 Gurley Ave., Gallup, NM 87301, USA; smarand@unm.edu
* Correspondence: sura_pati@yahoo.co.in; Tel.: +91-94-7703-5544 or +91-33-2560-1826; Fax: +91-33-2560-1826

Received: 18 November 2017; Accepted: 11 February 2018; Published: 23 February 2018

**Abstract:** A neutrosophic number $(a + bI)$ is a significant mathematical tool to deal with indeterminate and incomplete information which exists generally in real-world problems, where $a$ and $bI$ denote the determinate component and indeterminate component, respectively. We define score functions and accuracy functions for ranking neutrosophic numbers. We then define a cosine function to determine the unknown weight of the criteria. We define the neutrosophic number harmonic mean operators and prove their basic properties. Then, we develop two novel multi-criteria group decision-making (MCGDM) strategies using the proposed aggregation operators. We solve a numerical example to demonstrate the feasibility, applicability, and effectiveness of the two proposed strategies. Sensitivity analysis with the variation of "$I$" on neutrosophic numbers is performed to demonstrate how the preference ranking order of alternatives is sensitive to the change of "$I$". The efficiency of the developed strategies is ascertained by comparing the results obtained from the proposed strategies with the results obtained from the existing strategies in the literature.

**Keywords:** neutrosophic number; neutrosophic number harmonic mean operator (NNHMO); neutrosophic number weighted harmonic mean operator (NNWHMO); cosine function; score function; multi-criteria group decision-making

## 1. Introduction

Multi-criteria decision-making (MCDM), and multi-criteria group decision-making (MCGDM) are significant branches of decision theories which have been commonly applied in many scientific fields. They have been developed in many directions, such as crisp environments [1,2], and uncertain environments, namely fuzzy environments [3–13], intuitionistic fuzzy environments [14–24], and neutrosophic set environments [25–45]. Smarandache [46,47] introduced another direction of uncertainty by defining neutrosophic numbers (NN), which represent indeterminate and incomplete information in a new way. A NN consists of a determinate component and an indeterminate component. Thus, the NNs are more applicable to deal with indeterminate and incomplete information in real world problems. The NN is expressed as the function $N = p + qI$ in which $p$ is the determinate component and $qI$ is the indeterminate component. If $N = qI$, i.e., the indeterminate part reaches the maximum label, the worst situation occurs. If $N = p$, i.e., the indeterminate part does not appear, the best situation occurs. Thus, the application of NNs is more appropriate to deal with the indeterminate and incomplete information in real-world decision-making situations.

Information aggregation is an essential practice of accumulating relevant information from various sources. It is used to present aggregation between the min and max operators. The harmonic mean is usually used as a mathematical tool to accumulate the central tendency of information [48].

The harmonic mean (HM) is widely used in statistics to calculate the central tendency of a set of data. Park et al. [49] proposed multi-attribute group decision-making (MAGDM) strategy based on HM operators under uncertain linguistic environments. Wei [50] proposed a MAGDM strategy based on fuzzy-induced, ordered, weighted HM. In a fuzzy environment, Xu [48] studied a fuzzy-weighted HM operator, fuzzy ordered weighted HM operator, and a fuzzy hybrid HM operator, and employed them for MADM problems. Ye [51] proposed a multi-attribute decision-making (MADM) strategy based on harmonic averaging projection for a simplified neutrosophic sets (SNS) environment.

In a NN environment, Ye [52] proposed a MAGDM using de-neutrosophication strategy and a possibility degree ranking strategy for neutrosophic numbers. Liu and Liu [53] proposed a NN generalized weighted power averaging operator for MAGDM. Zheng et al. [54] proposed a MAGDM strategy based on a NN generalized hybrid weighted averaging operator. Pramanik et al. [55] studied a teacher selection strategy based on projection and bidirectional projection measures in a NN environment.

Only four [52–55] MCGDM strategies using NNs have been reported in the literature. Motivated from the works of Ye [52], Liu and Liu [53], Zheng et al. [54], and Pramanik et al. [55], we consider the proposed strategies to handle MCGDM problems in a NN environment.

The strategies [52–55] cannot deal with the situation when larger values other than arithmetic mean, geometric mean, and harmonic mean are necessary for experimental purposes. To fill the research gap, we propose two MCGDM strategies.

In this paper, we develop two new MCGDM strategies based on a NN harmonic mean operator (NNHMO) and a NN weighted harmonic mean operator (NNWHMO) to solve MCGDM problems. We define a cosine function to determine unknown weights of the criteria. To develop the proposed strategies, we define score and accuracy functions for ranking NNs for the first time in the literature.

The rest of the paper is structured as follows: Section 2 presents some preliminaries of NNs and score and accuracy functions of NNs. Section 3 devotes NN harmonic mean operator (NNHMO) and NN weighted harmonic mean operator (NNWHMO). Section 4 defines the cosine function to determine unknown criteria weights. Section 5 presents two novel decision-making strategies based on NNHMO and NNWHMO. In Section 6, a numerical example is presented to illustrate the proposed MCGDM strategies and the results show the feasibility of the proposed MCGDM strategies. Section 7 compares the obtained results derived from the proposed strategies and the existing strategies in NN environment. Finally, Section 8 concludes the paper with some remarks and future scope of research.

## 2. Preliminaries

In this section, definition of harmonic and weighted harmonic mean of positive real numbers, concepts of NNs, operations on NNs, score and accuracy functions of NNs are outlined.

### 2.1. Harmonic Mean and Weighted Harmonic Mean

Harmonic mean is a traditional average, which is generally used to determine central tendency of data. The harmonic mean is commonly considered as a fusion method of numerical data.

**Definition 1.** *[48]: The harmonic mean H of the positive real numbers $x_1, x_2, \ldots, x_n$ is defined as:* $\mathrm{H} = \frac{n}{\frac{1}{x_1}+\frac{1}{x_2}+\cdots+\frac{1}{x_n}} = \frac{n}{\sum_{i=1}^{n}\frac{1}{x_i}}$; $i = 1, 2, \ldots, n$.

**Definition 2.** *[49]: The weighted harmonic mean H of the positive real numbers $x_1, x_2, \ldots, x_n$ is defined as* $\mathrm{WH} = \frac{1}{\frac{w_1}{x_1}+\frac{w_2}{x_2}+\cdots+\frac{w_n}{x_n}} = \frac{1}{\sum_{i=1}^{n}\frac{w_i}{x_i}}$; $i = 1, 2, \ldots, n$.

Here, $\sum_{i=1}^{n} w_i = 1$.

2.2. NNs

A NN [46,47] consists of a determinate component $x$ and an indeterminate component $yI$, and is mathematically expressed as $z = x + yI$ for $x, y \in R$, where $I$ is indeterminacy interval and $R$ is the set of real numbers. A NN $z$ can be specified as a possible interval number, denoted by $z = [x + yI^L, x + yI^U]$ for $z \in Z$ (Z is set of all NNs) and $I \in [I^L, I^U]$. The interval $I \in [I^L, I^U]$ is considered as an indeterminate interval.

- If $yI = 0$, then z is degenerated to the determinate component $z = x$
- If $x = 0$, then z is degenerated to the indeterminate component $z = yI$
- If $I^L = I^U$, then z is degenerated to a real number.

Let two NNs be $z_1 = x_1 + y_1I$ and $z_2 = x_2 + y_2I$ for $z_1, z_2 \in Z$, and $I \in [I^L, I^U]$. Some basic operational rules for $z_1$ and $z_2$ are presented as follows:

(1) $I^2 = I$
(2) $I.0 = 0$
(3) $I/I$ = Undefined
(4) $z_1 + z_2 = x_1 + x_2 + (y_1 + y_2)I = [x_1 + x_2 + (y_1 + y_2)I^L, x_1 + x_2 + (y_1 + y_2)I^U]$
(5) $z_1 - z_2 = x_1 - x_2 + (y_1 - y_2)I = [x_1 - x_2 + (y_1 - y_2)I^L, x_1 - x_2 + (y_1 - y_2)I^U]$
(6) $z_1 \times z_2 = x_1x_2 + (x_1y_2 + x_2y_1)I + y_1y_2I^2 = x_1x_2 + (x_1y_2 + x_2y_1 + y_1y_2)I$
(7) $\frac{z_1}{z_2} = \frac{x_1+y_1I}{x_2+y_2I} = \frac{x_1}{x_2} + \frac{x_2y_1-x_1y_2}{x_2(x_2+y_2)}I;\ x_2 \neq 0,\ x_2 \neq -y_2$
(8) $\frac{1}{z_1} = \frac{1+0.I}{x_1+y_1I} = \frac{1}{x_1} + \frac{-y_1}{x_1(x_1+y_1)}I;\ x_1 \neq 0,\ x_1 \neq -y_1$
(9) $z_1^2 = x_1^2 + (2x_1y_1 + y_1^2)I$
(10) $\lambda z_1 = \lambda x_1 + \lambda y_1 I$

**Theorem 1.** *If z is a neutrosophic number then,* $\frac{1}{(z)^{-1}} = z,\ z \neq 0$.

**Proof.** Let $z = x + yI$. Then,

$$\frac{1}{z} = (z)^{-1} = \frac{1}{x} + \frac{-y}{x(x+y)}I;\ x \neq 0,\ x \neq -y$$

$$\begin{aligned}\frac{1}{(z)^{-1}} &= \frac{1}{1/x} + \frac{\frac{y}{x(x+y)}}{1/x\left(1/x - \frac{y}{x(x+y)}\right)}I;\ x \neq 0, x \neq -y \\ &= x + yI = z.\end{aligned}$$

□

**Definition 3.** *For any NN* $z = x + yI = [x + yI^L, x + yI^U]$, *(x and y not both zeroes), its score and accuracy functions are defined, respectively, as follows:*

$$Sc(z) = \left|\frac{x + y(I^U - I^L)}{2\sqrt{x^2 + y^2}}\right| \tag{1}$$

$$Ac(z) = 1 - \exp\left\langle -\left|x + y(I^U - I^L\right|\right\rangle \tag{2}$$

**Theorem 2.** *Both score function Sc(z) and accuracy function Ac(z) are bounded.*

**Proof.**
$x,\ y \in R$ and $I \in [0,\ 1]$
$\Rightarrow 0 \le \frac{x}{\sqrt{x^2+y^2}} \le 1,\ 0 \le \frac{y(I^U-I^L)}{\sqrt{x^2+y^2}} \le 1$
$\Rightarrow 0 \le \left|\frac{x+y(I^U-I^L)}{\sqrt{x^2+y^2}}\right| \le 2 \Rightarrow 0 \le \left|\frac{x+y(I^U-I^L)}{2\sqrt{x^2+y^2}}\right| \le 1 \Rightarrow 0 \le S(z) \le 1.$
Since $0 \le Sc(z) \le 1$, score function is bounded.
Again:
$0 \le \exp\langle -|x+y(I^U-I^L|\rangle \le 1$
$\Rightarrow -1 \le -\exp\langle -|x+y(I^U-I^L|\rangle \le 0$
$\Rightarrow 0 \le 1-\exp\langle -|x+y(I^U-I^L|\rangle \le 1$
Since $0 \le Ac(z) \le 1$, accuracy function is bounded. □

**Definition 4.** *Let two NNs be $z_1 = x_1 + y_1 I = [x_1 + y_1 I^L, x_1 + y_1 I^U]$, and $z_2 = x_2 + y_2 I = [x_2 + y_2 I^L, x_2 + y_2 I^U]$, then the following comparative relations hold:*

- *If $S(z_1) > S(z_2)$, then $z_1 > z_2$*
- *If $S(z_1) = S(z_2)$ and $A(z_1) < A(z_2)$, then $z_1 < z_2$*
- *If $S(z_1) = S(z_2)$ and $A(z_1) = A(z_2)$, then $z_1 = z_2$.*

**Example 1.** *Let three NNs be $z_1 = 10 + 2I$, $z_2 = 12$ and $z_3 = 12 + 5I$ and $I \in [0, 0.2]$. Then,*

*$S(z_1) = 0.5099$, $S(z_2) = 0.5$, $S(z_3) = 0.5577$, $A(z_1) = 0.999969$, $A(z_2) = 0.999994$, $A(z_3) = 0.999997$.*

*We see that, $S(z_1) \succ S(z_2) = S(z_3)$, and $A(z_3) \succ S(z_2)$.*
*Using Definition 2, we conclude that, $z_1 \succ z_3 \succ z_2$.*

## 3. Harmonic Mean Operators for NNs

In this section, we define harmonic mean operator and weighted harmonic mean operator for neutrosophic numbers.

### 3.1. NN-Harmonic Mean Operator (NNHMO)

**Definition 5.** *Let $z_i = x_i + y_i I$ $(i = 1, 2, \ldots, n)$ be a collection of NNs. Then the NNHMO is defined as follows:*

$$\mathrm{NNHMO}(z_1,\ z_2, \cdots, z_n) = n\left(\sum_{i=1}^{n} (z_i)^{-1}\right)^{-1} \tag{3}$$

**Theorem 3.** *Let $z_i = x_i + y_i I$ $(i = 1, 2, \ldots, n)$ be a collection of NNs. The aggregated value of the $\mathrm{NNHMO}(z_1,\ z_2, \cdots, z_n)$ operator is also a NN.*

**Proof.**
$\mathrm{NNHMO}(z_1,\ z_2, \cdots,\ z_n) = n\left(\sum_{i=1}^{n} (z_i)^{-1}\right)^{-1}$
$= n\left(\sum_{i=1}^{n} \frac{1}{x_i} + \sum_{i=1}^{n} \frac{-y_i}{x_i(x_i+y_i)} I\right)^{-1};\ x_i \ne 0,\ x_i \ne -y_i$
$= \frac{n}{\sum_{i=1}^{n}\frac{1}{x_i}} + \frac{-n\sum_{i=1}^{n}\frac{-y_i}{x_i(x_i+y_i)}}{\left(\sum_{i=1}^{n}\frac{1}{x_i}\right)\left(\sum_{i=1}^{n}\frac{1}{x_i}+\sum_{i=1}^{n}\frac{-y_i}{x_i(x_i+y_i)}\right)} I;\ \sum_{i=1}^{n}\frac{1}{x_i} \ne 0,\ \sum_{i=1}^{n}\frac{1}{x_i} \ne -\sum_{i=1}^{n}\frac{-y_i}{x_i(x_i+y_i)}$
This shows that NNHMO is also a NN. □

### 3.2. NN-Weighted Harmonic Mean Operator (NNWHMO)

**Definition 6.** *Let $z_i = x_i + y_iI$ ($i = 1, 2, \ldots, n$) be a collection of NNs and $w_i$ ($i = 1, 2, \ldots, n$) is the weight of $z_i$ ($i = 1, 2, \ldots, n$) and $\sum\limits_{i=1}^{n} w_i = 1$. Then the NN-weighted harmonic mean (NNWHMO) is defined as follows:*

$$\mathrm{NNWHMO}(z_1, z_2, \cdots, z_n) = \left(\sum_{i=1}^{n} \frac{w_i}{z_i}\right)^{-1}, \; z_i \neq 0 \tag{4}$$

**Theorem 4.** *Let $z_i = x_i + y_iI$ ($i = 1, 2, \ldots, n$) be a collection of NNs. The aggregated value of the $\mathrm{NNWHMO}(z_1, z_2, \cdots, z_n)$ operator is also a NN.*

**Proof.**

$$\mathrm{NNWHMO}(z_1, z_2, \cdots, z_n) = \left(\sum_{i=1}^{n} \frac{w_i}{z_i}\right)^{-1}, \; z_i \neq 0$$

$$= \left(\sum_{i=1}^{n} w_i\left(\tfrac{1}{x_i} + \tfrac{-y_i}{x_i(x_i+y_i)}I\right)\right)^{-1}; \; x_i \neq 0, \; x_i \neq -y_i$$

$$= \left(w_i \cdot \sum_{i=1}^{n} \tfrac{1}{x_i} + w_i \cdot \sum_{i=1}^{n} \tfrac{-y_i}{x_i(x_i+y_i)} I\right)^{-1}; \; x_i \neq 0, \; x_i \neq -y_i$$

$$= \frac{1}{w_i \cdot \sum\limits_{i=1}^{n} \frac{1}{x_i}} + \frac{-w_i \cdot \sum\limits_{i=1}^{n} \frac{-y_i}{x_i(x_i+y_i)}}{\left(w_i \cdot \sum\limits_{i=1}^{n} \frac{1}{x_i}\right)\left(w_i \cdot \sum\limits_{i=1}^{n} \frac{1}{x_i} + w_i \cdot \sum\limits_{i=1}^{n} \frac{-y_i}{x_i(x_i+y_i)}\right)} I; \; w_i \cdot \sum_{i=1}^{n} \tfrac{1}{x_i} \neq 0, \; w_i \cdot \sum_{i=1}^{n} \tfrac{1}{x_i} \neq -w_i \cdot \sum_{i=1}^{n} \tfrac{-y_i}{x_i(x_i+y_i)}; \; \sum_{i=1}^{n} w_i = 1$$

This shows that NNWHMO is also a NN. □

**Example 2.** *Let two NNs be $z_1 = 3 + 2I$ and $z_2 = 2 + I$ and $I \in [0, 0.2]$. Then:*

$$\mathrm{NNHMO}(z_1, z_2) = 2\left(\frac{1}{z_1} + \frac{1}{z_2}\right)^{-1} = 2\left(\frac{1}{3+2I} + \frac{1}{2+I}\right)^{-1} = 2.4 + 0.635I.$$

**Example 3.** *Let two NNs be $z_1 = 3 + 2I$ and $z_2 = 2 + I$, $I \in [0, 0.2]$ and $w_1 = 0.4$, $w_2 = 0.6$, then:*

$$\mathrm{NNWHMO}(z_1, z_2) = \left(w_1\frac{1}{z_1} + w_2\frac{1}{z_2}\right)^{-1} = \left(0.4\frac{1}{3+2I} + 0.6\frac{1}{2+I}\right)^{-1} = 2.308 + 1.370I.$$

The NNHMO operator and the NNWHMO operator satisfy the following properties.

*P1.* ***Idempotent law***: If $z_i = z$ for $i = 1, 2, \ldots, n$ then, $\mathrm{NNHMO}(z_1, z_2, \cdots, z_n) = z$ and $\mathrm{NNWHMO}(z_1, z_2, \cdots, z_n) = z$.

**Proof.** For, $z_i = z$, $\sum\limits_{i=1}^{n} w_i = 1$,

$$\mathrm{NNHMO}(z_1, z_2, \cdots, z_n) = n\left(\sum_{i=1}^{n} (z_i)^{-1}\right)^{-1} = n\left(\sum_{i=1}^{n} (z)^{-1}\right)^{-1} = \frac{n}{nz^{-1}} = z.$$

$$\mathrm{NNWHMO}(z_1, z_2, \cdots, z_n) = \left(\sum_{i=1}^{n} \frac{w_i}{z_i}\right)^{-1}, \; z_i \neq 0 = \left(\sum_{i=1}^{n} \frac{w_i}{z}\right)^{-1} = \left(\sum_{i=1}^{n} w_i\right)^{-1}\left(z^{-1}\right)^{-1} = z.$$

□

*P2.* ***Boundedness***: Both the operators are bounded.

**Proof.** Let $z_{\min} = \min(z_1, z_2, \cdots, z_n)$, $z_{\max} = \max(z_1, z_2, \cdots, z_n)$ for $i = 1, 2, \ldots, n$ then,
$z_{\min} \leq \text{NNHMO}(z_1, z_2, \cdots, z_n) \leq z_{\max}$ and $z_{\min} \leq \text{NNWHMO}(z_1, z_2, \cdots, z_n) \leq z_{\max}$.
Hence, both the operators are bounded. □

*P3.* ***Monotonicity***: If $z_i \leq z_i^*$ for $i = 1, 2, \ldots, n$ then, $\text{NNHMO}(z_1, z_2, \cdots, z_n) \leq \text{NNHMO}(z_1^*, z_2^*, \cdots, z_n^*)$ and $\text{NNWHMO}(z_1, z_2, \cdots, z_n) \leq \text{NNWHMO}(z_1^*, z_2^*, \cdots, z_n^*)$.

**Proof.** $\text{NNHMO}(z_1, z_2, \cdots, z_n) - \text{NNHMO}(z_1^*, z_2^*, \cdots, z_n^*) = \frac{n}{\frac{1}{z_1}+\frac{1}{z_2}+\cdots+\frac{1}{z_n}} - \frac{n}{\frac{1}{z_1^*}+\frac{1}{z_2^*}+\cdots+\frac{1}{z_n^*}} \leq 0$, since $z_i \leq z_i^*$ or $\frac{1}{z_i} \geq \frac{1}{z_i^*}$, for $i = 1, 2, \ldots, n$.
Again,
$\text{NNWHMO}(z_1, z_2, \cdots, z_n) - \text{NNWHMO}(z_1^*, z_2^*, \cdots, z_n^*) = \frac{1}{\frac{w_1}{z_1}+\frac{w_2}{z_2}+\cdots+\frac{w_n}{z_n}} - \frac{1}{\frac{w_1}{z_1^*}+\frac{w_2}{z_2^*}+\cdots+\frac{w_n}{z_n^*}} \leq 0$,
since $z_i \leq z_i^*$ or $\frac{1}{z_i} \geq \frac{1}{z_i^*}, \leq 0$, for $z_i \leq z_i^*$; $\sum\limits_{i=1}^{n} w_i = 1$; $(i = 1, 2, \ldots, n)$.
This proves the monotonicity of the functions $\text{NNHMO}(z_1, z_2, \cdots, z_n)$ and $\text{NNWHMO}(z_1, z_2, \cdots, z_n)$. □

*P4.* ***Commutativity***: If $(z_1^{\circ}, z_2^{\circ}, \cdots, z_n^{\circ})$ be any permutation of $(z_1, z_1, \cdots, z_n)$ then, $\text{NNHMO}(z_1, z_2, \cdots, z_n) = \text{NNHMO}(z_1^{\circ}, z_2^{\circ}, \cdots, z_n^{\circ})$ and $\text{NNWHMO}(z_1, z_2, \cdots, z_n) = \text{NNWHMO}(z_1^{\circ}, z_2^{\circ}, \cdots, z_n^{\circ})$.

**Proof.** $\text{NNHMO}(z_1, z_2, \cdots, z_n) - \text{NNHMO}(z_1^{\circ}, z_2^{\circ}, \cdots, z_n^{\circ}) = n\left(\sum\limits_{i=1}^{n} (z_i)^{-1}\right)^{-1} - n\left(\sum\limits_{i=1}^{n} (z_i^*)^{-1}\right)^{-1} = 0$,
because, $(z_1^{\circ}, z_2^{\circ}, \cdots, z_n^{\circ})$ is any permutation of $(z_1, z_2, \cdots, z_n)$.
Hence, we have $\text{NNHMO}(z_1, z_2, \cdots, z_n) = \text{NNHMO}(z_1^{\circ}, z_2^{\circ}, \cdots, z_n^{\circ})$.
Again:

$$\text{NNWHMO}(z_1, z_2, \cdots, z_n) - \text{NNWHMO}(z_1^{\circ}, z_2^{\circ}, \cdots, z_n^{\circ}) = \left(\sum_{i=1}^{n} w_i (z_i)^{-1}\right)^{-1} - \left(\sum_{i=1}^{n} w_i\, (z_i^*)^{-1}\right)^{-1} = 0,$$

because, $(z_1^{\circ}, z_2^{\circ}, \cdots, z_n^{\circ})$ is any permutation of $(z_1, z_2, \cdots, z_n)$.
Hence, we have $\text{NNWHMO}(z_1, z_2, \cdots, z_n) = \text{NNWHMO}(z_1^{\circ}, z_2^{\circ}, \cdots, z_n^{\circ})$. □

## 4. Cosine Function for Determining Unknown Criteria Weights

When criteria weights are completely unknown to decision-makers, the entropy measure [56] can be used to calculate criteria weights. Biswas et al. [57] employed entropy measure for MADM problems to determine completely unknown attribute weights of single valued neutrosophic sets (SVNSs). Literature review reflects that, strategy to determine unknown weights in the NN environment is yet to appear. In this paper, we propose a cosine function to determine unknown criteria weights.

**Definition 7.** *The cosine function of a NN* $P = x_{ij} + y_{ij}I = [x_{ij} + y_{ij}I^L, x_{ij} + y_{ij}I^U]$, $(i = 1, 2, \ldots, m;$ $j = 1, 2, \ldots, n)$ *is defined as follows:*

$$COS_j(P) = \frac{1}{n}\sum_{i=1}^{n} \cos\frac{\pi}{2}\left(\left|\frac{y_{ij}}{\sqrt{x_{ij}^2 + y_{ij}^2}}\right|\right),\ (x_{ij} \text{ and } y_{ij} \text{ are not both zeroes}) \tag{5}$$

*The weight structure is defined as follows:*

$$w_j = \frac{COS_j(P)}{\sum_{j=1}^{n} COS_j(P)};\ j = 1, 2, \cdots, n \;\&\; \sum_{j=1}^{n} w_j = 1 \tag{6}$$

*The cosine function* $COS_j(P)$ *satisfies the following properties:*

*P1.* $COS_j(P) = 1$, *if* $y_{ij} = 0$ and $x_{ij} \neq 0$.

*P2.* $COS_j(P) = 0$, *if* $x_{ij} = 0$ *and* $y_{ij} \neq 0$.

*P3.* $COS_j(P) \geq COS_j(Q)$, *if* $x_{ij}$ *of* $P > x_{ij}$ *of* $Q$ *or* $y_{ij}$ *of* $P < y_{ij}$ *of* $Q$ *or both.*

**Proof.**

*P1.* $y_{ij} = 0 \Rightarrow COS_j(P) = \frac{1}{n}\sum\limits_{i=1}^{n}[\cos 0] = 1$

*P2.* $x_{ij} = 0 \Rightarrow COS_j(P) = \frac{1}{n}\sum\limits_{i=1}^{n}\left[\cos \frac{\pi}{2}\right] = 0$

*P3.* For, $x_{ij}$ of $P > x_{ij}$ of $Q$

$\Rightarrow$ Determinate part of $P >$ Determinate part of $Q$

$\Rightarrow$ $COS_j(Q) < COS_j(P)$.

For, $y_{ij}$ of $P < y_{ij}$ of $Q$

$\Rightarrow$ Indeterminacy part of $P <$ Indeterminacy part of $Q$

$\Rightarrow$ $COS_j(Q) > COS_j(P)$.

For, $x_{ij}$ of $P > x_{ij}$ of $Q$ and $y_{ij}$ of $P < y_{ij}$ of $Q$

$\Rightarrow$ (Real part of $P >$ Real part of $Q$) & (Indeterminacy part of $P <$ Indeterminacy part of $Q$)

$\Rightarrow$ $COS_j(Q) > COS_j(P)$. □

**Example 4.** *Let two NNs be* $z_1 = 3 + 2I$, *and* $z_2 = 3 + 5I$, *then,* $COS(z_1) = 0.9066$, $COS(z_2) = 0.7817$.

**Example 5.** *Let two NNs be* $z_1 = 3 + I$, *and* $z_2 = 7 + I$, *then,* $COS(z_1) = 0.9693$, $COS(z_2) = 0.9938$.

**Example 6.** *Let two NNs be* $z_1 = 10 + 2I$, *and* $z_2 = 2 + 10I$, *then,* $COS(z_1) = 0.9882$, $COS(z_2) = 0.7178$.

## 5. Multi-Criteria Group Decision-Making Strategies Based on NNHMO and NNWHMO

Two MCGDM strategies using the NNHMO and NNWHMO respectively are developed in this section. Suppose that $A = \{A_1, A_2, \ldots, A_m\}$ is a set of alternatives, $C = \{C_1, C_2, \ldots, C_n\}$ is a set of criteria and $DM = \{DM_1, DM_2, \ldots, DM_k\}$ is a set of decision-makers. Decision-makers' assessment for each alternative $A_i$ will be based on each criterion $C_j$. All the assessment values are expressed by NNs. Steps of decision making strategies based on proposed NNHMO and NNWHMO to solve MCGDM problems are presented below.

### 5.1. MCGDM Strategy 1 (Based on NNHMO)

Strategy 1 is presented (see Figure 1) using the following six steps:

***Step 1.*** Determine the relation between alternatives and criteria.

Each decision-maker forms a NN decision matrix. The relation between the alternative $A_i$ ($i = 1, 2, \ldots, m$) and the criterion $C_j$ ($j = 1, 2, \ldots, n$) is presented in Equation (7).

$$DM_k[A|C] = \begin{matrix} \\ A_1 \\ A_2 \\ \vdots \\ A_m \end{matrix} \begin{pmatrix} C_1 & C_2 & \cdots & C_n \\ \langle x_{11}+y_{11}I\rangle_k & \langle x_{12}+y_{12}I\rangle_k & \cdots & \langle x_{1n}+y_{1n}I\rangle_k \\ \langle x_{21}+y_{21}I\rangle_k & \langle x_{22}+y_{22}I\rangle_k & \cdots & \langle x_{2n}+y_{2n}I\rangle_k \\ \vdots & \vdots & \ddots & \vdots \\ \langle x_{m1}+y_{m1}I\rangle_k & \langle x_{m2}+y_{m2}I\rangle_k & & \langle x_{mn}+y_{mn}I\rangle_k \end{pmatrix} \quad (7)$$

Note 1: Here, $\langle x_{ij} + y_{ij}I \rangle_k$ represents the NN rating value of the alternative $A_i$ with respect to the criterion $C_j$ for the decision-maker $DM_k$.

***Step 2.*** Using Equation (3), determine the aggregation values $(DM_k^{aggr}(A_i))$, $(i = 1, 2, \ldots, n)$ for all decision matrices.

***Step 3.*** To fuse all the aggregation values $(DM_k^{aggr}(A_i))$, corresponding to alternatives $A_i$, we define the averaging function as follows:

$$DM^{aggr}(A_i) = \sum_{t=1}^{k} w_t\,(DM_t^{aggr}(A_i));\ \sum_{t=1}^{k} w_t = 1.\ (i = 1,\ 2,\ \ldots, n;\ t = 1,\ 2,\ \ldots, k) \quad (8)$$

Here, $w_t$ $(t = 1, 2, \ldots, k)$ is the weight of the decision-maker $DM_t$.

***Step 4.*** Determine the preference ranking order.

Using Equation (1), determine the score values $Sc(z_i)$ (accuracy degrees $Ac(z_i)$, if necessary) $(i = 1, 2, \ldots, m)$ of all alternatives $A_i$. All the score values are arranged in descending order. The alternative corresponding to the highest score value (accuracy values) reflects the best choice.

***Step 5.*** Select the best alternative from the preference ranking order.

***Step 6.*** End.

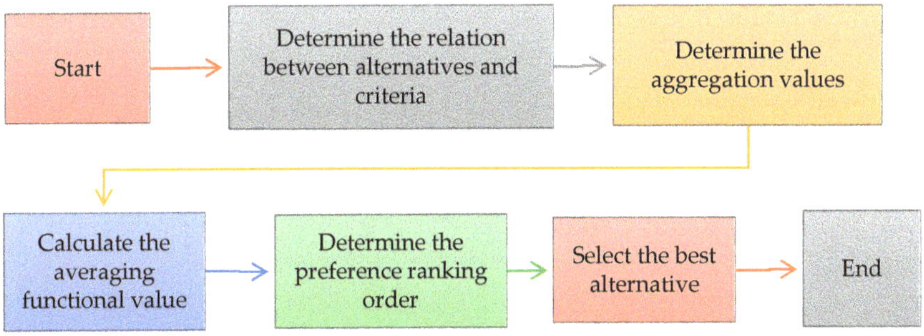


**Figure 1.** Steps of MCGDM Strategy 1 based on NNHMO.

### *5.2. MCGDM Strategy 2 (Based on NNWHMO)*

Strategy 2 is presented (see Figure 2) using the following seven steps:

***Step 1.*** This step is similar to the first step of Strategy 1.

***Step 2.*** Determine the criteria weights.

Using Equation (6), determine the criteria weights from decision matrices $(DM_t[A|C])$, $(t = 1, 2, \ldots, k)$.

***Step 3.*** Determine the weighted aggregation values $(DM_k^{waggr}(A_i))$.

Using Equation (4), determine the weighted aggregation values $(DM_k^{waggr}(A_i))$, $(i = 1, 2, \ldots, n)$ for all decision matrices.

***Step 4.*** Determine the averaging values.

To fuse all the weighted aggregation values ($DM_k^{waggr}(A_i)$), corresponding to alternatives $A_i$, we define the averaging function as follows:

$$DM^{waggr}(A_i) = \sum_{t=1}^{k} w_t(DM_t^{waggr}(A_i))(i = 1, 2, \ldots, n;\ t = 1, 2, \ldots, k) \tag{9}$$

Here, $w_t$ ($t$ = 1, 2, ... , $k$) is the weight of the decision maker $DM_t$.

***Step 5.*** Determine the ranking order.

Using Equation (1), determine the score values $S(z_i)$ (accuracy degrees $A(z_i)$, if necessary) ($i$ = 1, 2, ... , $m$) of all alternatives $A_i$. All the score values are arranged in descending order. The alternative corresponding to the highest score value (accuracy values) reflects the best choice.

***Step 6.*** Select the best alternative from the preference ranking order.

***Step 7.*** End.

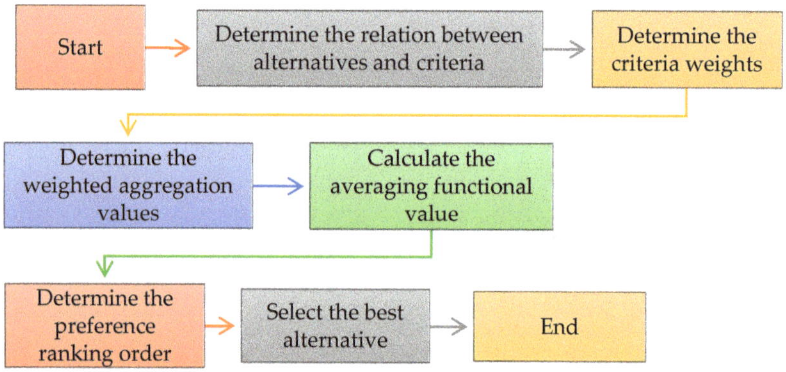


**Figure 2.** Steps of MCGDM strategy based on NNWHMO.

## 6. Simulation Results

We solve a numerical example studied by Zheng et al. [54]. An investment company desires to invest a sum of money in the best investment fund. There are four possible selection options to invest the money. Feasible selection options are namely, $A_1$: Car company (CARC); $A_2$: Food company (FOODC); $A_3$: Computer company (COMC); $A_4$: Arms company (ARMC). Decision-making must be based on the three criteria namely, risk analysis ($C_1$), growth analysis ($C_2$), environmental impact analysis ($C_3$). The four possible selection options/alternatives are to be selected under the criteria by the NN assessments provided by the three decision-makers $DM_1$, $DM_2$, and $DM_3$.

### 6.1. Solution Using MCGDM Strategy 1

***Step 1.*** Determine the relation between alternatives and criteria.

All assessment values are provided by the following three NN based decision matrices (shown in Equations (10)–(12).

$$DM_1[L|C] = \begin{matrix} \\ A_1 \\ A_2 \\ A_3 \\ A_4 \end{matrix} \begin{pmatrix} C_1 & C_2 & C_3 \\ 4+I & 5 & 3+I \\ 6 & 6 & 5 \\ 3 & 5+I & 6 \\ 7 & 6 & 4+I \end{pmatrix} \quad (10)$$

$$DM_2[L|C] = \begin{matrix} \\ A_1 \\ A_2 \\ A_3 \\ A_4 \end{matrix} \begin{pmatrix} C_1 & C_2 & C_3 \\ 5 & 4 & 4 \\ 5+I & 6 & 6 \\ 4 & 5 & 5+I \\ 6+I & 6 & 5 \end{pmatrix} \quad (11)$$

$$DM_3[L|C] = \begin{matrix} \\ A_1 \\ A_2 \\ A_3 \\ A_4 \end{matrix} \begin{pmatrix} C_1 & C_2 & C_3 \\ 4 & 5+I & 4 \\ 6 & 7 & 5+I \\ 4+I & 5 & 6 \\ 8 & 6 & 4+I \end{pmatrix} \quad (12)$$

Note 2: Here, $DM_1[L|C]$, $DM_2[L|C]$ and $DM_3[L|C]$ are the decision matrices for the decision makers $DM_1$, $DM_2$ and $DM_3$ respectively.

***Step 2.*** Determine the weighted aggregation values ($DM_k^{aggr}(A_i)$).

Using Equation (3), we calculate the aggregation values ($DM_k^{aggr}(A_i)$) as follows:

$DM_1^{aggr}(A_1) = 3.829 + 0.785I$; $DM_1^{aggr}(A_2) = 5.625$; $DM_1^{aggr}(A_3) = 4.285 + 0.214I$; $DM_1^{aggr}(A_4) = 5.362 + 0.514I$;

$DM_2^{aggr}(A_1) = 4.285$; $DM_2^{aggr}(A_2) = 5.206 + 0.415I$; $DM_2^{aggr}(A_3) = 4.196 + 0.532I$; $DM_2^{aggr}(A_4) = 5.234 + 0.618I$;

$DM_3^{aggr}(A_1) = 4.019 + 0.605I$; $DM_3^{aggr}(A_2) = 5.817 + 0.433I$; $DM_3^{aggr}(A_3) = 4.876 + 0.387I$; $DM_3^{aggr}(A_4) = 6.023 + 0.257I$.

***Step 3.*** Determine the averaging values.

Using Equation (8), we calculate the averaging values (Considering equal importance of all the decision makers) to fuse all the aggregation values corresponding to the alternative $A_i$.

$DM^{aggr}(A_1) = 4.044 + 0.463I$; $DM^{aggr}(A_2) = 5.549 + 0.282I$; $DM^{aggr}(A_3) = 4.452 + 0.378I$; $DM^{aggr}(A_4) = 5.539 + 0.463I$.

***Step 4.*** Using Equation (1), we calculate the score values $Sc(A_i)$ ($i$ = 1, 2, 3, 4). Sensitivity analysis and ranking order of alternatives are shown in Table 1 for different values of $I$.

**Table 1.** Sensitivity analysis and ranking order with variation of "$I$" on NNs for strategy 1.

| $I$ | $Sc(A_i)$ | Ranking Order |
|---|---|---|
| $I$ = [0, 0] | $S(A_1)$ = 0.4988, $S(A_2)$ = 0.4993, $S(A_3)$ = 0.4982, $S(A_4)$ = 0.4983 | $A_2 \succ A_1 \succ A_4 \succ A_3$ |
| $I \in$ [0, 0.2] | $S(A_1)$ = 0.5081, $S(A_2)$ = 0.5144, $S(A_3)$ = 0.5067, $S(A_4)$ = 0.5056 | $A_2 \succ A_1 \succ A_3 \succ A_4$ |
| $I \in$ [0, 0.4] | $S(A_1)$ = 0.5182, $S(A_2)$ = 0.5195, $S(A_3)$ = 0.5151, $S(A_4)$ = 0.5249 | $A_2 \succ A_1 \succ A_4 \succ A_3$ |
| $I \in$ [0, 0.6] | $S(A_1)$ = 0.5289, $S(A_2)$ = 0.5346, $S(A_3)$ = 0.5236, $S(A_4)$ = 0.5233 | $A_2 \succ A_1 \succ A_3 \succ A_4$ |
| $I \in$ [0, 0.8] | $S(A_1)$ = 0.5396, $S(A_2)$ = 0.5497, $S(A_3)$ = 0.5320, $S(A_4)$ = 0.5316 | $A_2 \succ A_1 \succ A_3 \succ A_4$ |
| $I \in$ [0, 1] | $S(A_1)$ = 0.5503, $S(A_2)$ = 0.5547, $S(A_3)$ = 0.5405, $S(A_4)$ = 0.5399 | $A_2 \succ A_1 \succ A_3 \succ A_4$ |

***Step 5.*** Food company (FOODC) is the best alternative for investment.

***Step 6.*** End.

Note 3: In Figure 3, we represent ranking order of alternatives with variation of "$I$" based on strategy 1. Figure 3 reflects that various values of $I$, ranking order of alternatives are different. However, the best choice is the same.

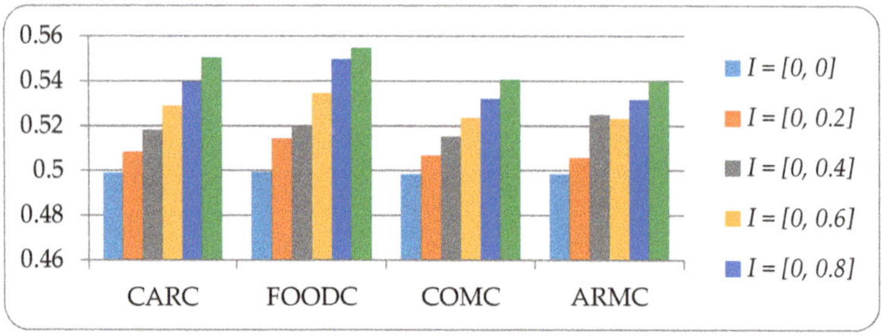


**Figure 3.** Ranking order with variation of "$I$" based on strategy 1.

### 6.2. Solution Using MCGDM Strategy 2

***Step 1.*** Determine the relation between alternatives and criteria.

This step is similar to the first step of strategy 1.

***Step 2.*** Determine the criteria weights.

Using Equations (5) and (6), criteria weights are calculated as follows:

$$[w_1 = 0.3265, w_2 = 0.3430, w_3 = 0.3305] \text{ for } DM_1,$$

$$[w_1 = 0.3332, w_2 = 0.3334, w_3 = 0.3334] \text{ for } DM_2,$$

$$[w_1 = 0.3333, w_2 = 0.3335, w_3 = 0.3332] \text{ for } DM_3.$$

***Step 3.*** Determine the weighted aggregation values ($DM_k^{waggr}(A_i)$).

Using Equation (4), we calculate the aggregation values ($DM_k^{aggr}(A_i)$) as follows:

$$DM_1^{aggr}(A_1) = 3.861 + 0.774I;\ DM_1^{aggr}(A_2) = 6.006;\ DM_1^{aggr}(A_3) = 4.307 + 0.234I;\ DM_1^{aggr}(A_4) = 5.399 + 0.541I;$$

$$DM_2^{aggr}(A_1) = 4.288;\ DM_2^{aggr}(A_2) = 5.219 + 0.429I;\ DM_2^{aggr}(A_3) = 4.206 + 0.541I;\ DM_2^{aggr}(A_4) = 5.251 + 0.629I;$$

$$DM_3^{aggr}(A_1) = 4.024 + 0.616I;\ DM_3^{aggr}(A_2) = 5.824 + 0.445I;\ DM_3^{aggr}(A_3) = 4.889 + 0.393I;\ DM_3^{aggr}(A_4) = 6.029 + 0.265I.$$

***Step 4.*** Determine the averaging values.

Using Equation (9), we calculate the averaging (Considering equal importance of all the decision makers to fuse all the aggregation values corresponding to the alternative $A_i$.

$$DM^{aggr}(A_1) = 4.057 + 0.463I;\ DM^{aggr}(A_2) = 5.568 + 0.291I;\ DM^{aggr}(A_3) = 4.467 + 0.389I;\ DM^{aggr}(A_4) = 5.559 + 0.478I.$$

***Step 5.*** Determine the ranking order.

Using Equation (1), we calculate the score values $Sc(A_i)$ ($i$ = 1, 2, 3, 4). Since scores values are different, accuracy values are not required. Sensitivity analysis and ranking order of alternatives are shown in Table 2 for different values of $I$.

**Table 2.** Sensitivity analysis and ranking order with variation of "$I$" on NNs for strategy 2.

| $I$ | $Sc(A_i)$ | Ranking Order |
|---|---|---|
| $I = 0$ | $S(A_1) = 0.4968$, $S(A_2) = 0.4993$, $S(A_3) = 0.4981$, $S(A_4) = 0.4982$ | $A_2 \succ A_4 \succ A_3 \succ A_1$ |
| $I \in [0, 0.2]$ | $S(A_1) = 0.5081$, $S(A_2) = 0.5095$, $S(A_3) = 0.5068$, $S(A_4) = 0.5067$ | $A_2 \succ A_1 \succ A_4 \succ A_3$ |
| $I \in [0, 0.4]$ | $S(A_1) = 0.5195$, $S(A_2) = 0.5198$, $S(A_3) = 0.5155$, $S(A_4) = 0.5153$ | $A_2 \succ A_1 \succ A_3 \succ A_4$ |
| $I \in [0, 0.6]$ | $S(A_1) = 0.5308$, $S(A_2) = 0.5350$, $S(A_3) = 0.5241$, $S(A_4) = 0.5239$ | $A_2 \succ A_1 \succ A_3 \succ A_4$ |
| $I \in [0, 0.8]$ | $S(A_1) = 0.5421$, $S(A_2) = 0.5502$, $S(A_3) = 0.5328$, $S(A_4) = 0.5324$ | $A_2 \succ A_1 \succ A_3 \succ A_4$ |
| $I \in [0, 1]$ | $S(A_1) = 0.5535$, $S(A_2) = 0.5654$, $S(A_3) = 0.5415$, $S(A_4) = 0.5410$ | $A_2 \succ A_1 \succ A_3 \succ A_4$ |

***Step 6.*** Food company (FOODC) is the best alternative for investment.

***Step 7.*** End.

Note 4: In Figure 4, we represent ranking order of alternatives with variation of "$I$" based on strategy 2. Figure 4 reflects that various values of $I$, ranking order of alternatives are different. However, the best choice is the same.

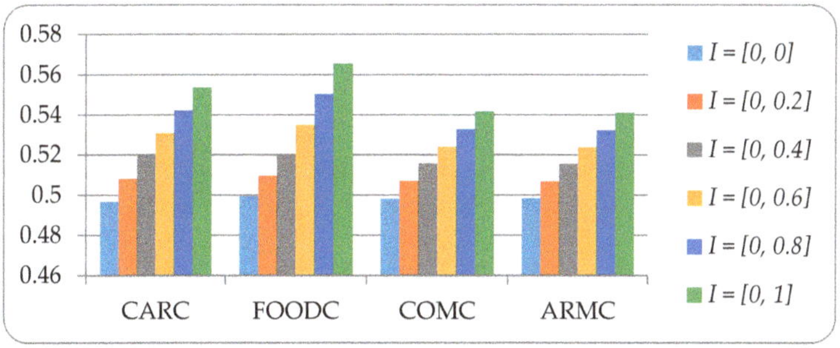


**Figure 4.** Ranking order with variation of "$I$" on NNs for Strategy 2.

## 7. Comparison Analysis and Contributions of the Proposed Approach

### *7.1. Comparison Analysis*

In this subsection, a comparison analysis is conducted between the proposed MCGDM strategies and the other existing strategies in the literature in NN environment. Table 1 reflects that $A_2$ is the best alternative for $I = 0$ and $I \neq 0$ i.e., for all cases considered. Table 2 reflects that $A_2$ is the best alternative for any values of $I$. Ranking order differs for different values of $I$.

The ranking results obtained from the existing strategies [52–54] are furnished in Table 3. The ranking orders of Ye [52] and Zheng et al. [54] are similar for all values of $I$ considered. When $I$ lies in [0, 0], [0, 0.2], [0, 0.4], $A_2$ is the best alternative for [52–54] and the proposed strategies. When $I$ lies in [0, 0.6], [0, 0.8], [0, 1], $A_4$ is the best alternative for [52,54], whereas $A_2$ is the best alternative for [53], and the proposed strategies.

**Table 3.** Comparison of ranking preference order with variation of "$I$" on NNs for different strategies.

| $I$ | Ye [52] | Zheng et al. [54] | Liu and Liu [53] | Proposed Strategy 1 | Proposed Strategy 2 |
|---|---|---|---|---|---|
| [0, 0] | $A_2 \succ A_4 \succ A_3 \succ A_1$ | $A_2 \succ A_4 \succ A_3 \succ A_1$ | $A_2 \succ A_4 \succ A_1 \succ A_3$ | $A_2 \succ A_1 \succ A_4 \succ A_3$ | $A_2 \succ A_4 \succ A_3 \succ A_1$ |
| [0, 0.2] | $A_2 \succ A_4 \succ A_3 \succ A_1$ | $A_2 \succ A_4 \succ A_3 \succ A_1$ | $A_2 \succ A_3 \succ A_1 \succ A_4$ | $A_2 \succ A_1 \succ A_3 \succ A_4$ | $A_2 \succ A_1 \succ A_4 \succ A_3$ |
| [0, 0.4] | $A_2 \succ A_4 \succ A_3 \succ A_1$ | $A_2 \succ A_4 \succ A_3 \succ A_1$ | $A_2 \succ A_3 \succ A_4 \succ A_1$ | $A_2 \succ A_1 \succ A_4 \succ A_3$ | $A_2 \succ A_1 \succ A_3 \succ A_4$ |
| [0, 0.6] | $A_4 \succ A_2 \succ A_3 \succ A_1$ | $A_4 \succ A_2 \succ A_3 \succ A_1$ | $A_2 \succ A_3 \succ A_4 \succ A_1$ | $A_2 \succ A_1 \succ A_3 \succ A_4$ | $A_2 \succ A_1 \succ A_3 \succ A_4$ |
| [0, 0.8] | $A_4 \succ A_2 \succ A_3 \succ A_1$ | $A_4 \succ A_2 \succ A_3 \succ A_1$ | $A_2 \succ A_3 \succ A_4 \succ A_1$ | $A_2 \succ A_1 \succ A_3 \succ A_4$ | $A_2 \succ A_1 \succ A_3 \succ A_4$ |
| [0, 1] | $A_4 \succ A_2 \succ A_3 \succ A_1$ | $A_4 \succ A_2 \succ A_3 \succ A_1$ | $A_2 \succ A_4 \succ A_3 \succ A_1$ | $A_2 \succ A_1 \succ A_3 \succ A_4$ | $A_2 \succ A_1 \succ A_3 \succ A_4$ |

In strategy [52], deneutrosophication process is analyzed. It does not recognize the importance of the aggregation information. MCGDM due to Liu and Liu [53] is based on NN generalized weighted power averaging operator. This strategy cannot deal the situation when larger value other than arithmetic mean, geometric mean, and harmonic mean is necessary for experimental purpose.

The strategy proposed by Zheng et al. [54] cannot be used when few observations contribute disproportionate amount to the arithmetic mean. The proposed two MCGDM strategies are free from these shortcomings.

### 7.2. Contributions of the Proposed Approach

- NNHMO and NNWHMO in NN environment are firstly defined in the literature. We have also proved their basic properties.
- We have proposed score and accuracy functions of NN numbers for ranking. If two score values are same, then accuracy function can be used for ranking purpose.
- The proposed two strategies can also be used when observations/experiments contribute is disproportionate amount to the arithmetic mean. The harmonic mean is used when sample values contain fractions and/or extreme values (either too small or too big).
- To calculate unknown weights structure of criteria in NN environment, we have proposed cosine function.
- Steps and calculations of the proposed strategies are easy to use.
- We have solved a numerical example to show the feasibility, applicability, and effectiveness of the proposed two strategies.

## 8. Conclusions

In the study, we have proposed NNHMO and NNWHMO. We have developed two strategies of ranking NNs based on proposed score and accuracy functions. We have proposed a cosine function to determine unknown weights of the criteria in a NN environment. We have developed two novel MCGDM strategies based on the proposed aggregation operators. We have solved a hypothetical case study and compared the obtained results with other existing strategies to demonstrate the effectiveness of the proposed MCGDM strategies. Sensitivity analysis for different values of $I$ is also conducted to show the influence of $I$ in preference ranking of the alternatives. The proposed MCGDM strategies can be applied in supply selection, pattern recognition, cluster analysis, medical diagnosis, etc.

**Acknowledgments:** The authors are very grateful to the anonymous reviewers for their insightful and constructive comments and suggestions that have led to an improved version of this paper.

**Author Contributions:** Kalyan Mondal and Surapati Pramanik conceived and designed the experiments; Kalyan Mondal, and Surapati Pramanik performed the experiments; Surapati Pramanik, Bibhas C. Giri, and Florentine Smarandache analyzed the data; Kalyan Mondal, Surapati Pramanik, Bibhas C. Giri, and Florentine Smarandache contributed to the analysis tools; and Kalyan Mondal and Surapati Pramanik wrote the paper.

**Conflicts of Interest:** The authors declare no conflict of interest.

## References

1. Hwang, L.; Lin, M.J. *Group Decision Making under Multiple Criteria: Methods and Applications*; Springer: Heidelberg, Germany, 1987.
2. Hwang, C.; Yoon, K. *Multiple Attribute Decision Making, Methods and Applications*; Springer: New York, NY, USA, 1981; Volume 186.
3. Chen, S.J.; Hwang, C.L. *Fuzzy Multiple Attribute Decision-Making, Methods and Applications*; Lecture Notes in Economics and Mathematical Systems; Springer: Berlin/Heidelberg, Germany, 1992; Volume 375.
4. Chang, T.H.; Wang, T.C. Using the fuzzy multi-criteria decision making approach for measuring the possibility of successful knowledge management. *Inf. Sci.* **2009**, *179*, 355–370. [CrossRef]
5. Krohling, R.A.; De Souza, T.T.M. Combining prospect theory and fuzzy numbers to multi-criteria decision making. *Exp. Syst. Appl.* **2012**, *39*, 11487–11493. [CrossRef]
6. Chen, C.T. Extension of the TOPSIS for group decision-making under fuzzy environment. *Fuzzy Sets Syst.* **2000**, *114*, 1–9. [CrossRef]
7. Zhang, G.; Lu, J. An integrated group decision-making method dealing with fuzzy preferences for alternatives and individual judgments for selection criteria. *Group Decis. Negot.* **2003**, *12*, 501–515. [CrossRef]
8. Krohling, R.A.; Campanharo, V.C. Fuzzy TOPSIS for group decision making: A case study for accidents with oil spill in the sea. *Exp. Syst. Appl.* **2011**, *38*, 4190–4197. [CrossRef]
9. Xia, M.; Xu, Z. A novel method for fuzzy multi-criteria decision making. *Int. J. Inf. Technol. Decis. Mak.* **2014**, *13*, 497–519. [CrossRef]
10. Mehlawat, M.K.; Guptal, P. A new fuzzy group multi-criteria decision making method with an application to the critical path selection. *Int. J. Adv. Manuf. Technol.* **2015**. [CrossRef]
11. Lin, L.; Yuan, X.H.; Xia, Z.Q. Multicriteria fuzzy decision-making based on intuitionistic fuzzy sets. *J. Comput. Syst. Sci.* **2007**, *73*, 84–88. [CrossRef]
12. Liu, H.W.; Wang, G.J. Multi-criteria decision-making methods based on intuitionistic fuzzy sets. *Eur. J. Oper. Res.* **2007**, *179*, 220–233. [CrossRef]
13. Pramanik, S.; Mondal, K. Weighted fuzzy similarity measure based on tangent function and its application to medical diagnosis. *Int. J. Innov. Res. Sci. Eng. Technol.* **2015**, *4*, 158–164.
14. Pramanik, S.; Mukhopadhyaya, D. Grey relational analysis based intuitionistic fuzzy multi-criteria group decision-making approach for teacher selection in higher education. *Int. J. Comput. Appl.* **2011**, *34*, 21–29. [CrossRef]
15. Mondal, K.; Pramanik, S. Intuitionistic fuzzy multi criteria group decision making approach to quality-brick selection problem. *J. Appl. Quant. Methods* **2014**, *9*, 35–50.
16. Dey, P.P.; Pramanik, S.; Giri, B.C. Multi-criteria group decision making in intuitionistic fuzzy environment based on grey relational analysis for weaver selection in Khadi institution. *J. Appl. Quant. Methods* **2015**, *10*, 1–14.
17. Ye, J. Multicriteria fuzzy decision-making method based on the intuitionistic fuzzy cross-entropy. In Proceedings of the International Conference on Intelligent Human-Machine Systems and Cybernetics, Hangzhou, China, 26–27 August 2009.
18. Chen, S.M.; Chang, C.H. A novel similarity measure between Atanassov's intuitionistic fuzzy sets based on transformation techniques with applications to pattern recognition. *Inf. Sci.* **2015**, *291*, 96–114. [CrossRef]
19. Chen, S.M.; Cheng, S.H.; Chiou, C.H. Fuzzy multi-attribute group decision making based on intuitionistic fuzzy sets and evidential reasoning methodology. *Inf. Fusion* **2016**, *27*, 215–227. [CrossRef]
20. Wang, J.Q.; Han, Z.Q.; Zhang, H.Y. Multi-criteria group decision making method based on intuitionistic interval fuzzy information. *Group Decis. Negot.* **2014**, *23*, 715–733. [CrossRef]
21. Yue, Z.L. TOPSIS-based group decision-making methodology in intuitionistic fuzzy setting. *Inf. Sci.* **2014**, *277*, 141–153. [CrossRef]
22. He, X.; Liu, W.F. An intuitionistic fuzzy multi-attribute decision-making method with preference on alternatives. *Oper. Res Manag. Sci.* **2013**, *22*, 36–40.
23. Mondal, K.; Pramanik, S. Intuitionistic fuzzy similarity measure based on tangent function and its application to multi-attribute decision making. *Glob. J. Adv. Res.* **2015**, *2*, 464–471.

24. Peng, H.G.; Wang, J.Q.; Cheng, P.F. A linguistic intuitionistic multi-criteria decision-making method based on the Frank Heronian mean operator and its application in evaluating coal mine safety. *Int. J. Mach. Learn. Cybern.* **2016**. [CrossRef]
25. Liang, R.X.; Wang, J.Q.; Zhang, H.Y. A multi-criteria decision-making method based on single-valued trapezoidal neutrosophic preference relations with complete weight information. *Neural Comput. Appl.* **2017**. [CrossRef]
26. Wang, J.Q.; Yang, Y.; Li, L. Multi-criteria decision-making method based on single-valued neutrosophic linguistic Maclaurin symmetric mean operators. *Neural Comput. Appl.* **2016**. [CrossRef]
27. Kharal, A. A neutrosophic multi-criteria decision making method. *New Math. Nat. Comput.* **2014**, *10*, 143–162. [CrossRef]
28. Ye, J. Multiple attribute group decision-making method with completely unknown weights based on similarity measures under single valued neutrosophic environment. *J. Intell. Fuzzy Syst.* **2014**, *27*, 2927–2935.
29. Mondal, K.; Pramanik, S. Multi-criteria group decision making approach for teacher recruitment in higher education under simplified neutrosophic environment. *Neutrosophic Sets Syst.* **2014**, *6*, 28–34.
30. Biswas, P.; Pramanik, S.; Giri, B.C. A new methodology for neutrosophic multi-attribute decision-making with unknown weight information. *Neutrosophic Sets Syst.* **2014**, *3*, 44–54.
31. Biswas, P.; Pramanik, S.; Giri, B.C. Cosine similarity measure based multi-attribute decision-making with trapezoidal fuzzy neutrosophic numbers. *Neutrosophic Sets Syst.* **2014**, *8*, 46–56.
32. Mondal, K.; Pramanik, S. Neutrosophic decision making model for clay-brick selection in construction field based on grey relational analysis. *Neutrosophic Sets Syst.* **2015**, *9*, 64–71.
33. Mondal, K.; Pramanik, S. Neutrosophic tangent similarity measure and its application to multiple attribute decision making. *Neutrosophic Sets Syst.* **2015**, *9*, 85–92.
34. Pramanik, S.; Biswas, P.; Giri, B.C. Hybrid vector similarity measures and their applications to multi-attribute decision making under neutrosophic environment. *Neural Comput. Appl.* **2015**. [CrossRef]
35. Sahin, R.; Küçük, A. Subsethood measure for single valued neutrosophic sets. *J. Intell. Fuzzy Syst.* **2015**, *29*, 525–530. [CrossRef]
36. Mondal, K.; Pramanik, S. Neutrosophic decision making model of school choice. *Neutrosophic Sets Syst.* **2015**, *7*, 62–68.
37. Ye, J. An extended TOPSIS method for multiple attribute group decision making based on single valued neutrosophic linguistic numbers. *J. Intell. Fuzzy Syst.* **2015**, *28*, 247–255.
38. Biswas, P.; Pramanik, S.; Giri, B.C. TOPSIS method for multi-attribute group decision-making under single-valued neutrosophic environment. *Neural Comput. Appl.* **2016**, *27*, 727–737. [CrossRef]
39. Biswas, P.; Pramanik, S.; Giri, B.C. Value and ambiguity index based ranking method of single-valued trapezoidal neutrosophic numbers and its application to multi-attribute decision making. *Neutrosophic Sets Syst.* **2016**, *12*, 127–138.
40. Biswas, P.; Pramanik, S.; Giri, B.C. Aggregation of triangular fuzzy neutrosophic set information and its application to multi-attribute decision making. *Neutrosophic Sets Syst.* **2016**, *12*, 20–40.
41. Smarandache, F.; Pramanik, S. (Eds.) *New Trends in Neutrosophic Theory and Applications*; Pons Editions: Brussels, Belgium, 2016; pp. 15–161; ISBN 978-1-59973-498-9.
42. Sahin, R.; Liu, P. Maximizing deviation method for neutrosophic multiple attribute decision making with incomplete weight information. *Neural Comput. Appl.* **2016**, *27*, 2017–2029. [CrossRef]
43. Pramanik, S.; Dalapati, S.; Alam, S.; Smarandache, S.; Roy, T.K. NS-cross entropy based MAGDM under single valued neutrosophic set environment. *Information* **2018**, *9*, 37. [CrossRef]
44. Sahin, R.; Liu, P. Possibility-induced simplified neutrosophic aggregation operators and their application to multi-criteria group decision-making. *J. Exp. Theor. Artif. Intell.* **2017**, *29*, 769–785. [CrossRef]
45. Biswas, P.; Pramanik, S.; Giri, B.C. Multi-attribute group decision making based on expected value of neutrosophic trapezoidal numbers. *New Math. Nat. Comput.* **2017**, in press.
46. Smarandache, F. *Introduction to Neutrosophic Measure, Neutrosophic Integral, and Neutrosophic, Probability*; Sitech & Education Publisher: Craiova, Romania, 2013.
47. Smarandache, F. *Introduction to Neutrosophic Statistics*; Sitech & Education Publishing: Columbus, OH, USA, 2014.
48. Xu, Z. Fuzzy harmonic mean operators. *Int. J. Intell. Syst.* **2009**, *24*, 152–172. [CrossRef]
49. Park, J.H.; Gwak, M.G.; Kwun, Y.C. Uncertain linguistic harmonic mean operators and their applications to multiple attribute group decision making. *Computing* **2011**, *93*, 47–64. [CrossRef]

50. Wei, G.W. FIOWHM operator and its application to multiple attribute group decision making. *Expert Syst. Appl.* **2011**, *38*, 2984–2989. [CrossRef]
51. Ye, J. Simplified neutrosophic harmonic averaging projection-based strategy for multiple attribute decision-making problems. *Int. J. Mach. Learn. Cybern.* **2017**, *8*, 981–987. [CrossRef]
52. Ye, J. Multiple attribute group decision making method under neutrosophic number environment. *J. Intell. Syst.* **2016**, *25*, 377–386. [CrossRef]
53. Liu, P.; Liu, X. The neutrosophic number generalized weighted power averaging operator and its application in multiple attribute group decision making. *Int. J. Mach. Learn. Cybern.* **2016**, 1–12. [CrossRef]
54. Zheng, E.; Teng, F.; Liu, P. Multiple attribute group decision-making strategy based on neutrosophic number generalized hybrid weighted averaging operator. *Neural Comput. Appl.* **2017**, *28*, 2063–2074. [CrossRef]
55. Pramanik, S.; Roy, R.; Roy, T.K. Teacher selection strategy based on bidirectional projection measure in neutrosophic number environment. In *Neutrosophic Operational Research*; Smarandache, F., Abdel-Basset, M., El-Henawy, I., Eds.; Pons Publishing House: Bruxelles, Belgium, 2017; Volume 2; ISBN 978-1-59973-537-5.
56. Majumdar, P.; Samanta, S.K. On similarity and entropy of neutrosophic sets. *J. Intell. Fuzzy Syst.* **2014**, *26*, 1245–1252.
57. Biswas, P.; Pramanik, S.; Giri, B.C. Entropy based grey relational analysis strategy for multi-attribute decision-making under single valued neutrosophic assessments. *Neutrosophic Sets. Syst.* **2014**, *2*, 102–110.

*axioms*

MDPI

*Article*

# Rough Neutrosophic Digraphs with Application

**Sidra Sayed [1], Nabeela Ishfaq [1], Muhammad Akram [1,*] and Florentin Smarandache [2]**

1. Department of Mathematics, University of the Punjab, New Campus, Lahore 54590, Pakistan; sidratulmuntha228@yahoo.com (S.S.); nabeelaishfaq123@gmail.com (N.I.)
2. Mathematics & Science Department, University of New Mexico, 705 Gurley Ave., Gallup, NM 87301, USA; fsmarandache@gmail.com

* Correspondence: m.akram@pucit.edu.pk; Tel.: +92-42-99231241

Received: 5 December 2017; Accepted: 15 January 2018; Published: 18 January 2018

**Abstract:** A rough neutrosophic set model is a hybrid model which deals with vagueness by using the lower and upper approximation spaces. In this research paper, we apply the concept of rough neutrosophic sets to graphs. We introduce rough neutrosophic digraphs and describe methods of their construction. Moreover, we present the concept of self complementary rough neutrosophic digraphs. Finally, we consider an application of rough neutrosophic digraphs in decision-making.

**Keywords:** rough neutrosophic sets; rough neutrosophic digraphs; decision-making

**MSC:** 03E72, 68R10, 68R05

---

## 1. Introduction

Smarandache [1] proposed the concept of neutrosophic sets as an extension of fuzzy sets [2]. A neutrosophic set has three components, namely, truth membership, indeterminacy membership and falsity membership, in which each membership value is a real standard or non-standard subset of the nonstandard unit interval $]0-, 1+[$ ([3]), where $0^- = 0 - \epsilon$, $1^+ = 1 + \epsilon$, $\epsilon$ is an infinitesimal number $> 0$. To apply neutrosophic set in real-life problems more conveniently, Smarandache [3] and Wang et al. [4] defined single-valued neutrosophic sets which takes the value from the subset of $[0, 1]$. Actually, the single valued neutrosophic set was introduced for the first time by Smarandache in 1998 in [3]. Ye [5] considered multicriteria decision-making method using the correlation coefficient under single-valued neutrosophic environment. Ye [6] also presented improved correlation coefficients of single valued neutrosophic sets and interval neutrosophic sets for multiple attribute decision making.

Rough set theory was proposed by Pawlak [7] in 1982. Rough set theory is useful to study the intelligence systems containing incomplete, uncertain or inexact information. The lower and upper approximation operators of rough sets are used for managing hidden information in a system. Therefore, many hybrid models have been built, such as soft rough sets, rough fuzzy sets, fuzzy rough sets, soft fuzzy rough sets, neutrosophic rough sets, andrough neutrosophic sets, for handling uncertainty and incomplete information effectively. Dubois and Prade [8] introduced the notions of rough fuzzy sets and fuzzy rough sets. Liu and Chen [9] have studied different decision-making methods. Broumi et al. [10] introduced the concept of rough neutrosophic sets. Yang et al. [11] proposed single valued neutrosophic rough sets by combining single valued neutrosophic sets and rough sets, and established an algorithm for decision-making problem based on single valued neutrosophic rough sets on two universes. Mordeson and Peng [12] presented operations on fuzzy graphs. Akram et al. [13–16] considered several new concepts of neutrosophic graphs with applications. Zafer and Akram [17] introduced a novel decision-making method based on rough fuzzy information. In this research study, we apply the concept of rough neutrosophic sets to graphs. We introduce rough neutrosophic digraphs and describe methods of their construction. Moreover, we present the concept of self complementary rough neutrosophic digraphs. We also present an application of rough neutrosophic digraphs in decision-making.

We have used standard definitions and terminologies in this paper. For other notations, terminologies and applications not mentioned in the paper, the readers are referred to [18–22].

## 2. Rough Neutrosophic Digraphs

**Definition 1.** *[4] Let Z be a nonempty universe. A neutrosophic set N on Z is defined as follows:*

$$N = \{< x : \mu_N(x), \sigma_N(x), \lambda_N(x) >, x \in Z\}$$

*where the functions* $\mu, \sigma, \lambda : Z \rightarrow [0,1]$ *represent the degree of membership, the degree of indeterminacy and the degree of falsity.*

**Definition 2.** *[7] Let Z be a nonempty universe and R an equivalence relation on Z.A pair* $(Z, R)$ *is called an approximation space. Let* $N^*$ *be a subset of Z and the lower and upper approximations of* $N^*$ *in the approximation space* $(Z, R)$ *denoted by* $\underline{R}N^*$ *and* $\overline{R}N^*$ *are defined as follows:*

$$\underline{R}N^* = \{x \in Z | [x]_R \subseteq N^*\},$$
$$\overline{R}N^* = \{x \in Z | [x]_R \subseteq N^*\},$$

*where* $[x]_R$ *denotes the equivalence class of R containing x. A pair* $(\underline{R}N^*, \overline{R}N^*)$ *is called a rough set.*

**Definition 3.** *[10] Let Z be a nonempty universe and R an equivalence relation on Z. Let N be a neutrosophic set(NS) on Z. The lower and upper approximations of N in the approximation space* $(Z, R)$ *denoted by* $\underline{R}N$ *and* $\overline{R}N$ *are defined as follows:*

$$\underline{R}N = \{< x, \mu_{\underline{R}(N)}(x), \sigma_{\underline{R}(N)}(x), \lambda_{\underline{R}(N)}(x) >: y \in [x]_R, x \in Z\},$$
$$\overline{R}N = \{< x, \mu_{\overline{R}(N)}(x), \sigma_{\overline{R}(N)}(x), \lambda_{\overline{R}(N)}(x) >: y \in [x]_R, x \in Z\},$$

*where,*

$$\mu_{\underline{R}(N)}(x) = \bigwedge_{y \in [x]_R} \mu_N(y), \quad \mu_{\overline{R}(N)}(x) = \bigvee_{y \in [x]_R} \mu_N(y),$$
$$\sigma_{\underline{R}(N)}(x) = \bigwedge_{y \in [x]_R} \sigma_N(y), \quad \sigma_{\overline{R}(N)}(x) = \bigvee_{y \in [x]_R} \sigma_N(y,$$
$$\lambda_{\underline{R}(N)}(x) = \bigvee_{y \in [x]_R} \lambda_N(y), \quad \lambda_{\overline{R}(N)}(x) = \bigwedge_{y \in [x]_R} \lambda_N(y).$$

*A pair* $(\underline{R}N, \overline{R}N)$ *is called a rough neutrosophic set.*

We now define the concept of rough neutrosophic digraph.

**Definition 4.** *Let* $V^*$ *be a nonempty set and R an equivalence relation on* $V^*$*. Let V be a NS on* $V^*$*, defined as*

$$V = \{< x, \mu_V(x), \sigma_V(x), \lambda_V(x) >: x \in V^*\}.$$

*Then, the lower and upper approximations of V represented by* $\underline{R}V$ *and* $\overline{R}V$*, respectively, are characterized as NSs in* $V^*$ *such that* $\forall\, x \in V^*$,

$$\underline{R}(V) = \{< x, \mu_{\underline{R}(V)}(x), \sigma_{\underline{R}(V)}(x), \lambda_{\underline{R}(V)}(x) >: y \in [x]_R\},$$
$$\overline{R}(V) = \{< x, \mu_{\overline{R}(V)}(x), \sigma_{\overline{R}(V)}(x), \lambda_{\overline{R}(V)}(x) >: y \in [x]_R\},$$

*where,*

$$\mu_{\underline{R}(V)}(x) = \bigwedge_{y \in [x]_R} \mu_V(y), \quad \mu_{\overline{R}(V)}(x) = \bigvee_{y \in [x]_R} \mu_V(y),$$
$$\sigma_{\underline{R}(V)}(x) = \bigwedge_{y \in [x]_R} \sigma_V(y), \quad \sigma_{\overline{R}(V)}(x) = \bigvee_{y \in [x]_R} \sigma_V(y),$$
$$\lambda_{\underline{R}(V)}(x) = \bigvee_{y \in [x]_R} \lambda_V(y), \quad \lambda_{\overline{R}(V)}(x) = \bigwedge_{y \in [x]_R} \lambda_V(y).$$

*Let* $E^* \subseteq V^* \times V^*$ *and* $S$ *an equivalence relation on* $E^*$ *such that*

$$((x_1, x_2), (y_1, y_2)) \in S \Leftrightarrow (x_1, y_1), (x_2, y_2) \in R.$$

*Let* $E$ *be a neutrosophic set on* $E^* \subseteq V^* \times V^*$ *defined as*

$$E = \{< xy, \mu_E(xy), \sigma_E(xy), \lambda_E(xy) >: xy \in V^* \times V^*\},$$

*such that*

$$\begin{aligned}
\mu_E(xy) &\leq \min\{\mu_{\underline{R}V}(x), \mu_{\underline{R}V}(y)\}, \\
\sigma_E(xy) &\leq \min\{\sigma_{\underline{R}V}(x), \sigma_{\underline{R}V}(y)\}, \\
\lambda_E(xy) &\leq \max\{\lambda_{\overline{R}V}(x), \lambda_{\overline{R}V}(y)\} \quad \forall x, y \in V^*.
\end{aligned}$$

*Then, the lower and upper approximations of* $E$ *represented by* $\underline{S}E$ *and* $\overline{S}E$*, respectively, are defined as follows*

$$\begin{aligned}
\underline{S}E &= \{< xy, \mu_{\underline{S}E}(xy), \sigma_{\underline{S}E}(xy), \lambda_{\underline{S}E}(xy) >: wz \in [xy]_S, xy \in V^* \times V^*\}, \\
\overline{S}E &= \{< xy, \mu_{\overline{S}E}(xy), \sigma_{\overline{S}E}(xy), \lambda_{\overline{S}E}(xy) >: wz \in [xy]_S, xy \in V^* \times V^*\},
\end{aligned}$$

*where,*

$$\begin{aligned}
\mu_{\underline{S}(E)}(xy) &= \bigwedge_{wz \in [xy]_S} \mu_E(wz), & \mu_{\overline{S}(E)}(xy) &= \bigvee_{wz \in [xy]_S} \mu_E(wz), \\
\sigma_{\underline{S}(E)}(xy) &= \bigwedge_{wz \in [xy]_S} \sigma_E(wz), & \sigma_{\overline{S}(E)}(xy) &= \bigvee_{wz \in [xy]_S} \sigma_E(wz), \\
\lambda_{\underline{S}(E)}(xy) &= \bigvee_{wz \in [xy]_S} \lambda_E(wz), & \lambda_{\overline{S}(E)}(xy) &= \bigwedge_{wz \in [xy]_S} \lambda_E(wz).
\end{aligned}$$

*A pair* $SE = (\underline{S}E, \overline{S}E)$ *is called a rough neutrosophic relation.*

**Definition 5.** *A rough neutrosophic digraph on a nonempty set* $V^*$ *is a four-ordered tuple* $G = (R, RV, S, SE)$ *such that*

(a) $R$ *is an equivalence relation on* $V^*$;
(b) $S$ *is an equivalence relation on* $E^* \subseteq V^* \times V^*$;
(c) $RV = (\underline{R}V, \overline{R}V)$ *is a rough neutrosophic set on* $V^*$;
(d) $SE = (\underline{S}E, \overline{S}E)$ *is a rough neutrosophic relation on* $V^*$ *and*
(e) $(RV, SE)$ *is a neutrosophic digraph where* $\underline{G} = (\underline{R}V, \underline{S}E)$ *and* $\overline{G} = (\overline{R}V, \overline{S}E)$ *are lower and upper approximate neutrosophic digraphs of* $G$ *such that*

$$\mu_{\underline{S}E}(xy) \leq \min\{\mu_{\underline{R}V}(x), \mu_{\underline{R}V}(y)\},$$

$$\sigma_{\underline{S}E}(xy) \leq \min\{\sigma_{\underline{R}V}(x), \sigma_{\underline{R}V}(y)\},$$

$$\lambda_{\underline{S}E}(xy) \leq \max\{\lambda_{\underline{R}V}(x), \lambda_{\underline{R}V}(y)\},$$

*and*

$$\begin{aligned}
\mu_{\overline{S}E}(xy) &\leq \min\{\mu_{\overline{R}V}(x), \mu_{\overline{R}V}(y)\}, \\
\sigma_{\overline{S}E}(xy) &\leq \min\{\sigma_{\overline{R}V}(x), \sigma_{\overline{R}V}(y)\}, \\
\lambda_{\overline{S}E}(xy) &\leq \max\{\lambda_{\overline{R}V}(x), \lambda_{\overline{R}V}(y)\} \ \forall\, x, y \in V^*.
\end{aligned}$$

**Example 1.** *Let* $V^* = \{a, b, c\}$ *be a set and* $R$ *an equivalence relation on* $V^*$

$$R = \begin{bmatrix} 1 & 0 & 1 \\ 0 & 1 & 0 \\ 1 & 0 & 1 \end{bmatrix}.$$

*Let* $V = \{(a, 0.2, 0.3, 0.6), (b, 0.8, 0.6, 0.5), (c, 0.9, 0.1, 0.4)\}$ *be a neutrosophic set on* $V^*$. *The lower and upper approximations of* $V$ *are given by,*

$$\underline{R}V = \{(a, 0.2, 0.1, 0.6), (b, 0.8, 0.6, 0.5), (c, 0.2, 0.1, 0.6)\},$$
$$\overline{R}V = \{(a, 0.9, 0.3, 0.4), (b, 0.8, 0.6, 0.5), (c, 0.9, 0.3, 0.4)\}.$$

*Let* $E^* = \{aa, ab, ac, bb, ca, cb\} \subseteq V^* \times V^*$ *and* $S$ *an equivalence relation on* $E^*$ *defined as:*

$$S = \begin{bmatrix} 1 & 0 & 1 & 0 & 1 & 0 \\ 0 & 1 & 0 & 0 & 0 & 1 \\ 1 & 0 & 1 & 0 & 1 & 0 \\ 0 & 0 & 0 & 1 & 0 & 0 \\ 1 & 0 & 1 & 0 & 1 & 0 \\ 0 & 1 & 0 & 0 & 0 & 1 \end{bmatrix}.$$

*Let* $E = \{(aa, 0.2, 0.1, 0.4), (ab, 0.2, 0.1, 0.5), (ac, 0.1, 0.1, 0.5), (bb, 0.7, 0.5, 0.5), (ca, 0.1, 0.1, 0.3), (cb, 0.2, 0.1, 0.5)\}$ *be a neutrosophic set on* $E^*$ *and* $SE = (\underline{S}E, \overline{S}E)$ *a rough neutrosophic relation where* $\underline{S}E$ *and* $\overline{S}E$ *are given as*

$$\underline{S}E = \{(aa, 0.1, 0.1, 0.5), (ab, 0.2, 0.1, 0.5), (ac, 0.1, 0.1, 0.5), (bb, 0.7, 0.5, 0.5), (ca, 0.1, 0.1, 0.5), (cb, 0.2, 0.1, 0.5)\},$$
$$\overline{S}E = \{(aa, 0.2, 0.1, 0.3), (ab, 0.2, 0.1, 0.5), (ac, 0.2, 0.1, 0.3), (bb, 0.7, 0.5, 0.5), (ca, 0.2, 0.1, 0.3), (cb, 0.2, 0.1, 0.5)\}.$$

*Thus,* $\underline{G} = (\underline{R}V, \underline{S}E)$ *and* $\overline{G} = (\overline{R}V, \overline{S}E)$ *are neutrosophic digraphs as shown in Figure 1.*

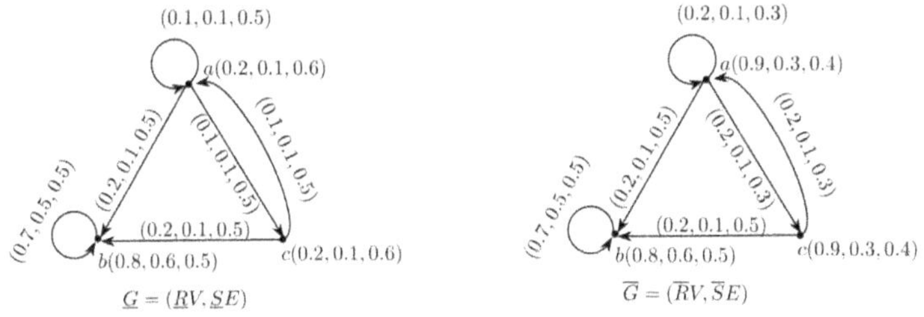


**Figure 1.** Rough neutrosophic digraph $G = (\underline{G}, \overline{G})$.

We now form new rough neutrosophic digraphs from old ones.

**Definition 6.** *Let* $G_1 = (\underline{G_1}, \overline{G_1})$ *and* $G_2 = (\underline{G_2}, \overline{G_2})$ *be two rough neutrosophic digraphs on a set* $V^*$. *Then, the intersection of* $G_1$ *and* $G_2$ *is a rough neutrosophic digraph* $G = G_1 \Cap G_2 = (\underline{G_1} \cap \underline{G_2}, \overline{G_1} \cap \overline{G_2})$, *where* $\underline{G_1} \cap \underline{G_2} = (\underline{R}V_1 \cap \underline{R}V_2, \underline{S}E_1 \cap \underline{S}E_2)$ *and* $\overline{G_1} \cap \overline{G_2} = (\overline{R}V_1 \cap \overline{R}V_2, \overline{S}E_1 \cap \overline{S}E_2)$ *are neutrosophic digraphs, respectively, such that*

$$
\begin{aligned}
(1)\ & \mu_{\underline{RV}_1\cap\underline{RV}_2}(x) = \min\{\mu_{\underline{RV}_1}(x), \mu_{\underline{RV}_2}(x)\},\\
& \sigma_{\underline{RV}_1\cap\underline{RV}_2}(x) = \min\{\sigma_{\underline{RV}_1}(x), \sigma_{\underline{RV}_2}(x)\},\\
& \lambda_{\underline{RV}_1\cap\underline{RV}_2}(x) = \max\{\lambda_{\underline{RV}_1}(x), \lambda_{\underline{RV}_2}(x)\} \quad \forall\, x \in \underline{RV}_1 \cap \underline{RV}_1,\\
& \mu_{\underline{SE}_1\cap\underline{SE}_2}(xy) = \min\{\mu_{\underline{SE}_1}(x), \mu_{\underline{SE}_2}(y)\},\\
& \sigma_{\underline{SE}_1\cap\underline{SE}_2}(xy) = \min\{\sigma_{\underline{SE}_1}(x), \sigma_{\underline{SE}_2}(y)\}\\
& \lambda_{\underline{SE}_1\cap\underline{SE}_2}(xy) = \max\{\lambda_{\underline{SE}_1}(x), \lambda_{\underline{SE}_2}(y)\} \quad \forall\, xy \in \underline{SE}_1 \cap \underline{SE}_2,
\end{aligned}
$$

$$
\begin{aligned}
(2)\ & \mu_{\overline{RV}_1\cap\overline{RV}_2}(x) = \min\{\mu_{\overline{RV}_1}(x), \mu_{\overline{RV}_2}(x)\},\\
& \sigma_{\overline{RV}_1\cap\overline{RV}_2}(x) = \min\{\sigma_{\overline{RV}_1}(x), \sigma_{\overline{RV}_2}(x)\},\\
& \lambda_{\overline{RV}_1\cap\overline{RV}_2}(x) = \max\{\lambda_{\overline{RV}_1}(x), \lambda_{\overline{RV}_2}(x)\} \quad \forall\, x \in \overline{RV}_1 \cap \overline{RV}_2,\\
& \mu_{\overline{SE}_1\cap\overline{SE}_2}(xy) = \min\{\mu_{\overline{SE}_1}(x), \mu_{\overline{SE}_2}(y)\}\\
& \sigma_{\overline{SE}_1\cap\overline{SE}_2}(xy) = \min\{\sigma_{\overline{SE}_1}(x), \sigma_{\overline{SE}_2}(y)\}\\
& \lambda_{\overline{SE}_1\cap\overline{SE}_2}(xy) = \max\{\lambda_{\overline{SE}_1}(x), \lambda_{\overline{SE}_2}(y)\} \quad \forall\, xy \in \overline{SE}_1 \cap \overline{SE}_2.
\end{aligned}
$$

**Example 2.** *Consider the two rough neutrosophic digraphs* $\underline{G}_1$ *and* $\overline{G}_2$ *as shown in Figures 1 and 2. The intersection of* $\underline{G}_1$ *and* $\overline{G}_2$ *is* $G = G_1 \Cap G_2 = (\underline{G}_1 \cap \underline{G}_2, \overline{G}_1 \cap \overline{G}_2)$ *where* $\underline{G}_1 \cap \underline{G}_2 = (\underline{RV}_1 \cap \underline{RV}_2, \underline{SE}_1 \cap \underline{SE}_2)$ *and* $\overline{G}_1 \cap \overline{G}_2 = (\overline{RV}_1 \cap \overline{RV}_2, \overline{SE}_1 \cap \overline{SE}_2)$ *are neutrosophic digraphs as shown in Figure 3.*

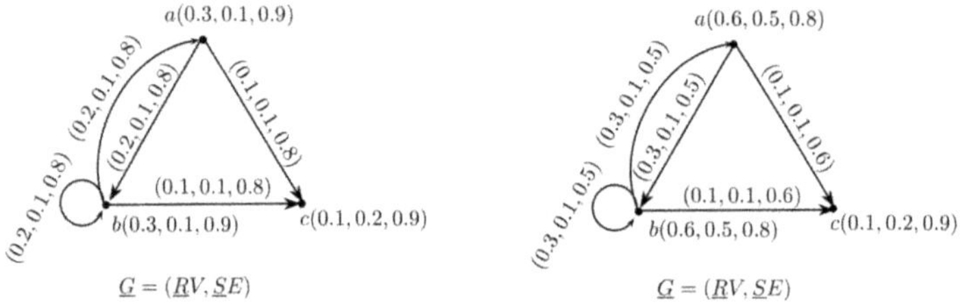


**Figure 2.** Rough neutrosophic digraph $G = (\underline{G}, \overline{G})$.

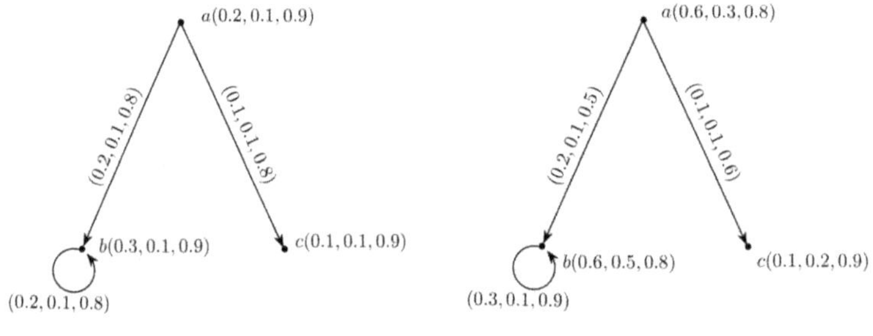


**Figure 3.** Rough neutrosophic digraph $G_1 \Cap G_2 = (\underline{G}_1 \cap \underline{G}_2, \overline{G}_1 \cap \overline{G}_2)$.

**Theorem 1.** *The intersection of two rough neutrosophic digraphs is a rough neutrosophic digraph.*

**Proof.** Let $G_1 = (\underline{G}_1, \overline{G}_1)$ and $G_2 = (\underline{G}_2, \overline{G}_2)$ be two rough neutrosophic digraphs. Let $G = G_1 \Cap G_2 = (\underline{G}_1 \cap \underline{G}_2, \overline{G}_1 \cap \overline{G}_2)$ be the intersection of $G_1$ and $G_2$, where $\underline{G}_1 \cap \underline{G}_2 = (\underline{RV}_1 \cap \underline{RV}_2, \underline{SE}_1 \cap, \underline{SE}_2)$ and $\overline{G}_1 \cap \overline{G}_2 = (\overline{RV}_1 \cap \overline{RV}_2, \overline{SE}_1 \cap \overline{SE}_2)$. To prove that $G = \underline{G}_1 \Cap \overline{G}_2$ is a rough neutrosophic digraph, it is enough to show that $\underline{S}E_1 \cap \underline{S}E_2$ a nd $\overline{S}E_1 \cap \overline{S}E_2$ are neutrosophic relation on $\underline{R}V_1 \cap \underline{R}V_2$ and $\overline{R}V_1 \cap \overline{R}V_2$, respectively. First, we show that $\underline{S}E_1 \cap \underline{S}E_2$ is a neutrosophic relation on $\underline{R}V_1 \cap \underline{R}V_2$.

$$\begin{aligned}
\mu_{\underline{S}E_1 \cap \underline{S}E_2}(xy) &= \mu_{\underline{S}E_1}(xy) \wedge \mu_{\underline{S}E_2}(xy)\\
&\leq (\mu_{\underline{R}V_1}(x) \wedge \mu_{\underline{R}V_2}(y)) \wedge (\mu_{\underline{R}V_1}(x) \wedge \mu_{\underline{R}V_2}(y))\\
&= (\mu_{\underline{R}V_1}(x) \wedge \mu_{\underline{R}V_2}(x)) \wedge (\mu_{\underline{R}V_1}(y) \wedge \mu_{\underline{R}V_2}(y)\\
&= \mu_{\underline{R}V_1 \cap \underline{R}V_2}(x) \wedge \mu_{\underline{R}V_1 \cap \underline{R}V_2}(y)\\
\mu_{\underline{S}E_1 \cap \underline{S}E_2}(xy) &\leq \min\{\mu_{\underline{R}V_1 \cap \underline{R}V_2}(x), \mu_{\underline{R}V_1 \cap \underline{R}V_2}(y)\}\\
\sigma_{\underline{S}E_1 \cap \underline{S}E_2}(xy) &= \sigma_{\underline{S}E_1}(xy) \wedge \sigma_{\underline{S}E_2}(xy)\\
&\leq (\sigma_{\underline{R}V_1}(x) \wedge \sigma_{\underline{R}V_2}(y)) \wedge (\sigma_{\underline{R}V_1}(x) \wedge \sigma_{\underline{R}V_2}(y))\\
&= (\sigma_{\underline{R}V_1}(x) \wedge \sigma_{\underline{R}V_2}(x)) \wedge (\sigma_{\underline{R}V_1}(y) \wedge \sigma_{\underline{R}V_2}(y)\\
&= \sigma_{\underline{R}V_1 \cap \underline{R}V_2}(x) \wedge \sigma_{\underline{R}V_1 \cap \underline{R}V_2}(y)\\
\sigma_{\underline{S}E_1 \cap \underline{S}E_2}(xy) &\leq \min\{\sigma_{\underline{R}V_1 \cap \underline{R}V_2}(x), \sigma_{\underline{R}V_1 \cap \underline{R}V_2}(y)\}\\
\lambda_{\underline{S}E_1 \cap \underline{S}E_2}(xy) &= \lambda_{\underline{S}E_1}(xy) \wedge \lambda_{\underline{S}E_2}(xy)\\
&\leq (\lambda_{\underline{R}V_1}(x) \vee \lambda_{\underline{R}V_2}(y)) \wedge (\lambda_{\underline{R}V_1}(x) \vee \lambda_{\underline{R}V_2}(y))\\
&= (\lambda_{\underline{R}V_1}(x) \wedge \lambda_{\underline{R}V_2}(x)) \vee (\lambda_{\underline{R}V_1}(y) \wedge \lambda_{\underline{R}V_2}(y)\\
&= \lambda_{\underline{R}V_1 \cap \underline{R}V_2}(x) \vee \lambda_{\underline{R}V_1 \cap \underline{R}V_2}(y)\\
\lambda_{\underline{S}E_1 \cap \underline{S}E_2}(xy) &\leq \max\{\lambda_{\underline{R}V_1 \cap \underline{R}V_2}(x), \lambda_{\underline{R}V_1 \cap \underline{R}V_2}(y)\}.
\end{aligned}$$

Thus, from above it is clear that $\underline{S}E_1 \cap \underline{S}E_2$ is a neutrosophic relation on $\underline{R}V_1 \cap \underline{R}V_2$.

Similarly, we can show that $\overline{S}E_1 \cap \overline{S}E_2$ is a neutrosophic relation on $\overline{R}V_1 \cap \overline{R}V_2$. Hence, $G$ is a rough neutrosophic digraph. □

**Definition 7.** *The Cartesian product of two neutrosophic digraphs $G_1$ and $G_2$ is a rough neutrosophic digraph $G = G_1 \ltimes G_2 = (\underline{G}_1 \ltimes \underline{G}_2, \overline{G}_1 \ltimes \overline{G}_2)$, where $\underline{G}_1 \ltimes \underline{G}_2 = (\underline{R}_1 \ltimes \underline{R}_2, \underline{S}E_1 \ltimes \underline{S}E_2$ and $\overline{G}_1 \ltimes \overline{G}_2 = (\overline{R}V_1 \ltimes \overline{R}V_2, \overline{S}E_1 \ltimes \overline{S}E_2)$ such that*

**(1)**
$$\begin{aligned}
\mu_{\underline{R}V_1 \ltimes \underline{R}V_2}(x_1, x_2) &= \min\{\mu_{\underline{R}V_1}(x_1), \mu_{\underline{R}V_2}(x_2)\},\\
\sigma_{\underline{R}V_1 \ltimes \underline{R}V_2}(x_1, x_2) &= \min\{\sigma_{\underline{R}V_1}(x_1), \mu_{\underline{R}V_2}(x_2)\},\\
\lambda_{\underline{R}V_1 \ltimes \underline{R}V_2}(x_1, x_2) &= \max\{\lambda_{\underline{R}V_1}(x_1), \mu_{\underline{R}V_2}(x_2)\}, \quad \forall (x_1, x_2) \in \underline{R}V_1 \ltimes \underline{R}V_2,\\
\mu_{\underline{S}E_1 \ltimes \underline{S}E_2}(x, x_2)(x, y_2) &= \min\{\mu_{\underline{R}V_1}(x), \mu_{\underline{S}E_2}(x_2, y_2)\},\\
\sigma_{\underline{S}E_1 \ltimes \underline{S}E_2}(x, x_2)(x, y_2) &= \min\{\sigma_{\underline{R}V_1}(x), \sigma_{\underline{S}E_2}(x_2, y_2)\},\\
\lambda_{\underline{S}E_1 \ltimes \underline{S}E_2}(x, x_2)(x, y_2) &= \max\{\lambda_{\underline{R}V_1}(x), \lambda_{\underline{S}E_2}(x_2, y_2)\} \quad \forall x \in \underline{R}V_1, x_2y_2 \in \underline{S}E_2,\\
\mu_{\underline{S}E_1 \ltimes \underline{S}E_2}(x_1, z)(y_1, z) &= \min\{\mu_{\underline{S}E_1}(x_1, y_1), \mu_{\underline{R}V_2}(z)\},\\
\sigma_{\underline{S}E_1 \ltimes \underline{S}E_2}(x_1, z)(y_1, z) &= \min\{\sigma_{\underline{S}E_1}(x_1, y_1), \sigma_{\underline{R}V_2}(z)\},\\
\lambda_{\underline{S}E_1 \ltimes \underline{S}E_2}(x_1, z)(y_1, z) &= \max\{\lambda_{\underline{S}E_1}(x_1, y_1), \lambda_{\underline{R}V_2}(z)\} \quad \forall x_1y_1 \in \underline{S}E_1, z \in \underline{R}V_2,
\end{aligned}$$

**(2)**
$$\begin{aligned}
\mu_{\overline{R}V_1 \ltimes \overline{R}V_2}(x_1, x_2) &= \min\{\mu_{\overline{R}V_1}(x_1), \mu_{\overline{R}V_2}(x_2)\},\\
\sigma_{\overline{R}V_1 \ltimes \overline{R}V_2}(x_1, x_2) &= \min\{\sigma_{\overline{R}V_1}(x_1), \mu_{\overline{R}V_2}(x_2)\},\\
\lambda_{\overline{R}V_1 \ltimes \overline{R}V_2}(x_1, x_2) &= \max\{\lambda_{\overline{R}V_1}(x_1), \mu_{\overline{R}V_2}(x_2)\} \quad \forall (x_1, x_2) \in \overline{R}V_1 \ltimes \overline{R}V_2,\\
\mu_{\overline{S}E_1 \ltimes \overline{S}E_2}(x, x_2)(x, y_2) &= \min\{\mu_{\overline{R}V_1}(x), \mu_{\overline{S}E_2}(x_2, y_2)\},\\
\sigma_{\overline{S}E_1 \ltimes \overline{S}E_2}(x, x_2)(x, y_2) &= \min\{\sigma_{\overline{R}V_1}(x), \sigma_{\overline{S}E_2}(x_2, y_2)\},\\
\lambda_{\overline{S}E_1 \ltimes \overline{S}E_2}(x, x_2)(x, y_2) &= \max\{\lambda_{\overline{R}V_1}(x), \lambda_{\overline{S}E_2}(x_2, y_2)\} \quad \forall x \in \overline{R}V_1, x_2y_2 \in \overline{S}E_2,
\end{aligned}$$

$$\mu_{\overline{SE}_1\times\overline{SE}_2}(x_1,z)(y_1,z)=\min\{\mu_{\overline{SE}_1}(x_1,y_1),\mu_{\overline{R}V_2}(z)\},$$
$$\sigma_{\overline{SE}_1\times\overline{SE}_2}(x_1,z)(y_1,z)=\min\{\sigma_{\overline{SE}_1}(x_1,y_1),\sigma_{\overline{R}V_2}(z)\},$$
$$\lambda_{\overline{SE}_1\times\overline{SE}_2}(x_1,z)(y_1,z)=\max\{\lambda_{\overline{SE}_1}(x_1,y_1),\lambda_{\overline{R}V_2}(z)\}\quad\forall\, x_1y_1\in\overline{S}E_1, z\in\overline{R}V_2,$$

**Example 3.** *Let $V^*=\{a,b,c,d\}$ be a set. Let $G_1=(\underline{G}_1,\overline{G}_1)$ and $G_2=(\underline{G}_2,\overline{G}_2)$ be two rough neutrosophic digraphs on $V^*$, as shown in Figures 4 and 5. The cartesian product of $G_1$ and $G_2$ is $G=(\underline{G}_1\times\underline{G}_2,\overline{G}_1\times\overline{G}_2)$, where $\underline{G}_1\times\underline{G}_2=(\underline{R}N_1\times\underline{R}N_2,\underline{S}E_1\times\underline{S}E_2)$ and $\overline{G}_1\times\overline{G}_2=(\overline{R}N_1\times\overline{R}N_2,\overline{S}E_1\times\overline{S}E_2)$ are neutrosophic digraphs, as shown in Figures 6 and 7, respectively.*

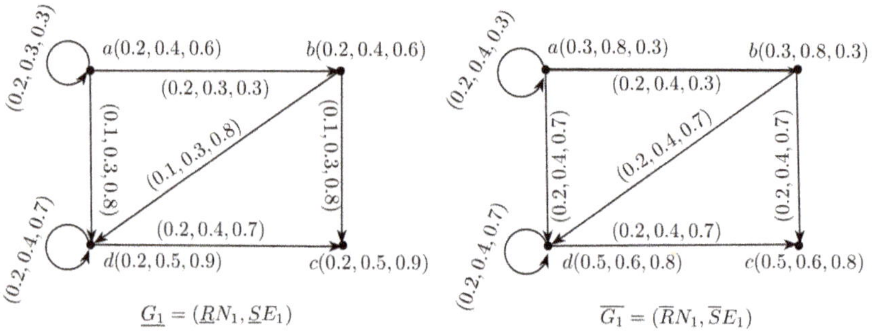


**Figure 4.** Rough neutrosophic digraph $G_1=(\underline{G}_1,\overline{G}_1)$.

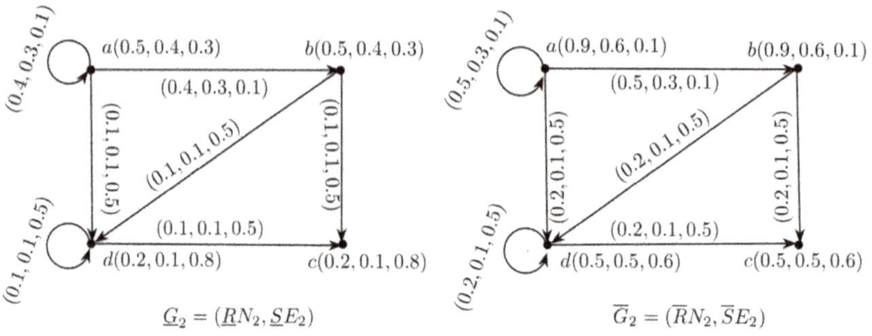


**Figure 5.** Rough neutrosophic digraph $G_2=(\underline{G}_2,\overline{G}_2)$.

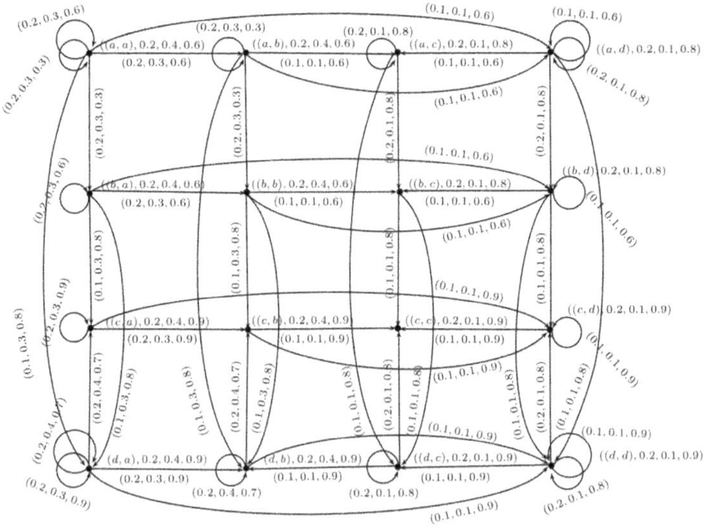


**Figure 6.** Neutrosophic digraph $\underline{G}_1 \times \underline{G}_2 = (\underline{R}N_1 \times \underline{R}N_2, \underline{S}E_1 \times \underline{S}E_2)$.

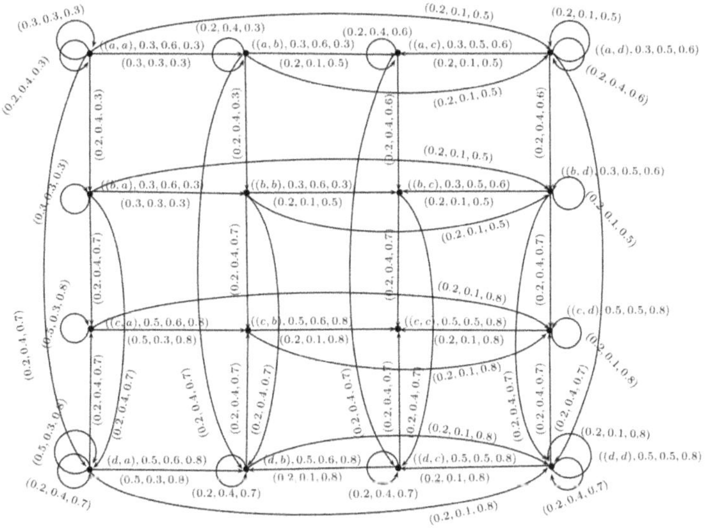


**Figure 7.** Neutrosophic digraph $\overline{G}_1 \times \overline{G}_2 = (\overline{R}N_1 \times \overline{R}N_2, \overline{S}E_1 \times \overline{S}E_2)$.

**Theorem 2.** *The Cartesian product of two rough neutrosophic digraphs is a rough neutrosophic digraph.*

**Proof.** Let $G_1 = (\underline{G}_1, \overline{G}_1)$ and $G_2 = (\underline{G}_2, \overline{G}_2)$ be two rough neutrosophic digraphs. Let $G = G_1 \ltimes G_2 = (\underline{G}_1 \ltimes \underline{G}_2, \overline{G}_1 \ltimes \overline{G}_2)$ be the Cartesian product of $G_1$ and $G_2$, where $\underline{G}_1 \ltimes \underline{G}_2 = (\underline{R}V_1 \ltimes \underline{R}V_2, \underline{S}E_1 \ltimes \underline{S}E_2)$ and $\overline{G}_1 \ltimes \overline{G}_2 = (\overline{R}V_1 \ltimes \overline{R}V_2, \overline{S}E_1 \ltimes \overline{S}E_2)$. To prove that $G = \underline{G}_1 \ltimes \overline{G}_2$ is a rough neutrosophic digraph, it is enough to show that $\underline{S}E_1 \ltimes \underline{S}E_2$ and $\overline{S}E_1 \ltimes \overline{S}E_2$ are neutrosophic relation on $\underline{R}V_1 \ltimes \underline{R}V_2$ and $\overline{R}V_1 \ltimes \overline{R}V_2$, respectively. First, we show that $\underline{S}E_1 \ltimes \underline{S}E_2$ is a neutrosophic relation on $\underline{R}V_1 \ltimes \underline{R}V_2$.

If $x \in \underline{R}V_1, x_2y_2 \in \underline{S}E_2$, then

$$\begin{aligned}
\mu_{\underline{S}E_1 \ltimes \underline{S}E_2}(x,x_2)(x,y_2) &= \mu_{\underline{R}V_1}(x) \wedge \mu_{\underline{S}E_2}(x_2,y_2) \\
&\leq \mu_{\underline{R}V_1}(x) \wedge (\mu_{\underline{R}V_2}(x_2) \wedge \mu_{\underline{R}V_2}(y_2)) \\
&= (\mu_{\underline{R}V_1}(x) \wedge \mu_{\underline{R}V_2}(x_2)) \wedge (\mu_{\underline{R}V_1}(x) \wedge \mu_{\underline{R}V_2}(y_2)) \\
&= \mu_{\underline{R}V_1 \ltimes \underline{R}V_2}(x,x_2) \wedge \mu_{\underline{R}V_1 \ltimes \underline{R}V_2}(x,y_2) \\
\mu_{\underline{S}E_1 \ltimes \underline{S}E_2}(x,x_2)(x,y_2) &\leq \min\{\mu_{\underline{R}V_1 \ltimes \underline{R}V_2}(x,x_2), \mu_{\underline{R}V_1 \ltimes \underline{R}V_2}(x,y_2)\}, \\
\sigma_{\underline{S}E_1 \ltimes \underline{S}E_2}(x,x_2)(x,y_2) &= \sigma_{\underline{R}V_1}(x) \wedge \sigma_{\underline{S}E_2}(x_2,y_2) \\
&\leq \sigma_{\underline{R}V_1}(x) \wedge (\sigma_{\underline{R}V_2}(x_2) \wedge \sigma_{\underline{R}V_2}(y_2)) \\
&= (\sigma_{\underline{R}V_1}(x) \wedge \sigma_{\underline{R}V_2}(x_2)) \wedge (\sigma_{\underline{R}V_1}(x) \wedge \sigma_{\underline{R}V_2}(y_2) \\
&= \sigma_{\underline{R}V_1 \ltimes \underline{R}V_2}(x,x_2) \wedge \sigma_{\underline{R}V_1 \ltimes \underline{R}V_2}(x,y_2) \\
\sigma_{\underline{S}E_1 \ltimes \underline{S}E_2}(x,x_2)(x,y_2) &\leq \min\{\sigma_{\underline{R}V_1 \ltimes \underline{R}V_2}(x,x_2), \sigma_{\underline{R}V_1 \ltimes \underline{R}V_2}(x,y_2)\}, \\
\lambda_{\underline{S}E_1 \ltimes \underline{S}E_2}(x,x_2)(x,y_2) &= \lambda_{\underline{R}V_1}(x) \vee \lambda_{\underline{S}E_2}(x_2,y_2) \\
&\leq \lambda_{\underline{R}V_1}(x) \vee (\lambda_{\underline{R}V_2}(x_2) \vee \lambda_{\underline{R}V_2}(y_2)) \\
&= (\lambda_{\underline{R}V_1}(x) \vee \lambda_{\underline{R}V_2}(x_2)) \vee (\lambda_{\underline{R}V_1}(x) \vee \lambda_{\underline{R}V_2}(y_2)) \\
&= \lambda_{\underline{R}V_1 \ltimes \underline{R}V_2}(x,x_2) \vee \lambda_{\underline{R}V_1 \ltimes \underline{R}V_2}(x,y_2) \\
\lambda_{\underline{S}E_1 \ltimes \underline{S}E_2}(x,x_2,x,y_2) &\leq \max\{\lambda_{\underline{R}V_1 \ltimes \underline{R}V_2}(x,x_2), \lambda_{\underline{R}V_1 \ltimes \underline{R}V_2}(x,y_2)\}.
\end{aligned}$$

If $x_1y_1 \in \underline{S}E_1, z \in \underline{R}V_2$, then

$$\begin{aligned}
\mu_{\underline{S}E_1 \ltimes \underline{S}E_2}(x_1,z)(y_1,z) &= \mu_{\underline{S}E_1}(x_1,y_1) \wedge \mu_{\underline{R}V_2}(z) \\
&\leq (\mu_{\underline{R}V_1}(x_1) \wedge \mu_{\underline{R}V_1}(y_1)) \wedge \mu_{\underline{R}V_2}(z) \\
&= (\mu_{\underline{R}V_1}(x_1) \wedge \mu_{\underline{R}V_2}(z)) \wedge (\mu_{\underline{R}V_1}(y_1) \wedge \mu_{\underline{R}V_2}(z)) \\
&= \mu_{\underline{R}V_1 \ltimes \underline{R}V_2}(x_1,z) \wedge \mu_{\underline{R}V_1 \ltimes \underline{R}V_2}(y_1,z) \\
\mu_{\underline{S}E_1 \ltimes \underline{S}E_2}(x_1,z)(y_1,z) &\leq \min\{\mu_{\underline{R}V_1 \ltimes \underline{R}V_2}(x_1,z), \mu_{\underline{R}V_1 \ltimes \underline{R}V_2}(y_1,z)\}, \\
\sigma_{\underline{S}E_1 \ltimes \underline{S}E_2}(x_1,z)(y_1,z) &= \sigma_{\underline{S}E_1}(x_1,y_1) \wedge \sigma_{\underline{R}V_2}(z) \\
&\leq (\sigma_{\underline{R}V_1}(x_1) \wedge \sigma_{\underline{R}V_1}(y_1)) \wedge \sigma_{\underline{R}V_2}(z) \\
&= (\sigma_{\underline{R}V_1}(x_1) \wedge \sigma_{\underline{R}V_2}(z)) \wedge (\sigma_{\underline{R}V_1}(y_1) \wedge \sigma_{\underline{R}V_2}(z)) \\
&= \sigma_{\underline{R}V_1 \ltimes \underline{R}V_2}(x_1,z) \wedge \sigma_{\underline{R}V_1 \ltimes \underline{R}V_2}(y_1,z) \\
\sigma_{\underline{S}E_1 \ltimes \underline{S}E_2}(x_1,z)(y_1,z) &\leq \min\{\sigma_{\underline{R}V_1 \ltimes \underline{R}V_2}(x_1,z), \sigma_{\underline{R}V_1 \ltimes \underline{R}V_2}(y_1,z)\}, \\
\lambda_{\underline{S}E_1 \ltimes \underline{S}E_2}(x_1,z)(y_1,z) &= \lambda_{\underline{S}E_1}(x_1,y_1) \vee \lambda_{\underline{R}V_2}(z) \\
&\leq (\lambda_{\underline{R}V_1}(x_1) \vee \lambda_{\underline{R}V_1}(y_1)) \vee \lambda_{\underline{R}V_2}(z) \\
&= (\lambda_{\underline{R}V_1}(x_1) \vee \lambda_{\underline{R}V_2}(z)) \vee (\lambda_{\underline{R}V_1}(y_1) \vee \lambda_{\underline{R}V_2}(z)) \\
&= \lambda_{\underline{R}V_1 \ltimes \underline{R}V_2}(x_1,z) \vee \lambda_{\underline{R}V_1 \ltimes \underline{R}V_2}(y_1,z) \\
\lambda_{\underline{S}E_1 \ltimes \underline{S}E_2}(x_1,z)(y_1,z) &\leq \max\{\lambda_{\underline{R}V_1 \ltimes \underline{R}V_2}(x_1,z), \lambda_{\underline{R}V_1 \ltimes \underline{R}V_2}(y_1,z)\}.
\end{aligned}$$

Thus, from above, it is clear that $\underline{S}E_1 \ltimes \underline{S}E_2$ is a neutrosophic relation on $\underline{R}V_1 \ltimes \underline{R}V_2$.
Similarly, we can show that $\overline{S}E_1 \ltimes \overline{S}E_2$ is a neutrosophic relation on $\overline{R}V_1 \ltimes \overline{R}V_2$. Hence, $G = (\underline{G}_1 \ltimes \underline{G}_2, \overline{G}_1 \ltimes \overline{G}_2)$ is a rough neutrosophic digraph. □

**Definition 8.** *The composition of two rough neutrosophic digraphs $G_1$ and $G_2$ is a rough neutrosophic digraph $G = G_1 \circ G_2 = (\underline{G}_1 \circ \underline{G}_2, \overline{G}_1 \circ \overline{G}_2)$, where $\underline{G}_1 \circ \underline{G}_2 = (\underline{R}V_1 \circ \underline{R}V_2, \underline{S}E_1 \circ \underline{S}E_2)$ and $\overline{G}_1 \circ \overline{G}_2 = (\overline{R}V_1 \circ \overline{R}V_2, \overline{S}E_1 \circ \overline{S}E_2)$ are neutrosophic digraphs, respectively, such that*

(1) $\mu_{\underline{R}V_1 \circ \underline{R}V_2}(x_1,x_2) = \min\{\mu_{\underline{R}V_1}(x_1), \mu_{\underline{R}V_2}(x_2)\},$

$$
\begin{aligned}
\sigma_{\underline{RV}_1\circ\underline{RV}_2}(x_1,x_2) &= \min\{\sigma_{\underline{RV}_1}(x_1),\mu_{\underline{RV}_2}(x_2)\},\\
\lambda_{\underline{RV}_1\circ\underline{RV}_2}(x_1,x_2) &= \max\{\lambda_{\underline{RV}_1}(x_1),\mu_{\underline{RV}_2}(x_2)\} \quad \forall\,(x_1,x_2)\in\underline{RV}_1\times\underline{RV}_2,\\
\mu_{\underline{SE}_1\circ\underline{SE}_2}(x,x_2)(x,y_2) &= \min\{\mu_{\underline{RV}_1}(x),\mu_{\underline{SE}_2}(x_2,y_2)\},\\
\sigma_{\underline{SE}_1\circ\underline{SE}_2}(x,x_2)(x,y_2) &= \min\{\sigma_{\underline{RV}_1}(x),\sigma_{\underline{SE}_2}(x_2,y_2)\},\\
\lambda_{\underline{SE}_1\circ\underline{SE}_2}(x,x_2)(x,y_2) &= \max\{\lambda_{\underline{RV}_1}(x),\lambda_{\underline{SE}_2}(x_2,y_2)\} \quad \forall\,x\in\underline{RV}_1,x_2y_2\in\underline{SE}_2,\\
\mu_{\underline{SE}_1\circ\underline{SE}_2}(x_1,z)(y_1,z) &= \min\{\mu_{\underline{SE}_1}(x_1,y_1),\mu_{\underline{RV}_2}(z)\},\\
\sigma_{\underline{SE}_1\circ\underline{SE}_2}(x_1,z)(y_1,z) &= \min\{\sigma_{\underline{SE}_1}(x_1,y_1),\sigma_{\underline{RV}_2}(z)\},\\
\lambda_{\underline{SE}_1\circ\underline{SE}_2}(x_1,z)(y_1,z) &= \max\{\lambda_{\underline{SE}_1}(x_1,y_1),\lambda_{\underline{RV}_2}(z)\} \quad \forall\,x_1y_1\in\underline{SE}_1,z\in\underline{RV}_2,\\
\mu_{\underline{SE}_1\circ\underline{SE}_2}(x_1,x_2)(y_1,y_2) &= \min\{\mu_{\underline{SE}_1}(x_1,y_1),\mu_{\underline{RV}_2}(x_2),\mu_{\underline{RV}_2}(y_2)\},\\
\sigma_{\underline{SE}_1\circ\underline{SE}_2}(x_1,x_2)(y_1,y_2) &= \min\{\sigma_{\underline{SE}_1}(x_1,y_1),\sigma_{\underline{RV}_2}(x_2),\sigma_{\underline{RV}_2}(y_2)\},\\
\lambda_{\underline{SE}_1\circ\underline{SE}_2}(x_1,x_2)(y_1,y_2) &= \max\{\lambda_{\underline{SE}_1}(x_1,y_1),\lambda_{\underline{RV}_2}(x_2),\lambda_{\underline{RV}_2}(y_2)\}\\
&\qquad \forall\,x_1y_1\in\underline{SE}_1,x_2,y_2\in\underline{RV}_2,x_2\neq y_2.
\end{aligned}
$$

$$
\begin{aligned}
(\mathbf{2})\qquad \mu_{\overline{RV}_1\circ\overline{RV}_2}(x_1,x_2) &= \min\{\mu_{\overline{RV}_1}(x_1),\mu_{\overline{RV}_2}(x_2)\},\\
\sigma_{\overline{RV}_1\circ\overline{RV}_2}(x_1,x_2) &= \min\{\sigma_{\overline{RV}_1}(x_1),\mu_{\overline{RV}_2}(x_2)\},\\
\lambda_{\overline{RV}_1\circ\overline{RV}_2}(x_1,x_2) &= \max\{\lambda_{\overline{RV}_1}(x_1),\mu_{\overline{RV}_2}(x_2)\} \quad \forall\,(x_1,x_2)\in\overline{RV}_1\times\overline{RV}_2,\\
\mu_{\overline{SE}_1\circ\overline{SE}_2}(x,x_2)(x,y_2) &= \min\{\mu_{\overline{RV}_1}(x),\mu_{\overline{SE}_2}(x_2,y_2)\},\\
\sigma_{\overline{SE}_1\circ\overline{SE}_2}(x,x_2)(x,y_2) &= \min\{\sigma_{\overline{RV}_1}(x),\sigma_{\overline{SE}_2}(x_2,y_2)\},\\
\lambda_{\overline{SE}_1\circ\overline{SE}_2}(x,x_2)(x,y_2) &= \max\{\lambda_{\overline{RV}_1}(x),\lambda_{\overline{SE}_2}(x_2,y_2)\} \quad \forall\,x\in\overline{RV}_1,x_2y_2\in\overline{SE}_2,\\
\mu_{\overline{SE}_1\circ\overline{SE}_2}(x_1,z)(y_1,z) &= \min\{\mu_{\overline{SE}_1}(x_1,y_1),\mu_{\overline{RV}_2}(z)\},\\
\sigma_{\overline{SE}_1\circ\overline{SE}_2}(x_1,z)(y_1,z) &= \min\{\sigma_{\overline{SE}_1}(x_1,y_1),\sigma_{\overline{RV}_2}(z)\},\\
\lambda_{\overline{SE}_1\circ\overline{SE}_2}(x_1,z)(y_1,z) &= \max\{\lambda_{\overline{SE}_1}(x_1,y_1),\lambda_{\overline{RV}_2}(z)\} \quad \forall\,x_1y_1\in\overline{SE}_1,z\in\overline{RV}_2,\\
\mu_{\overline{SE}_1\circ\overline{SE}_2}(x_1,x_2)(y_1,y_2) &= \min\{\mu_{\overline{SE}_1}(x_1,y_1),\mu_{\overline{RV}_2}(x_2),\mu_{\overline{RV}_2}(y_2)\},\\
\sigma_{\overline{SE}_1\circ\overline{SE}_2}(x_1,x_2)(y_1,y_2) &= \min\{\sigma_{\overline{SE}_1}(x_1,y_1),\sigma_{\overline{RV}_2}(x_2),\sigma_{\overline{RV}_2}(y_2)\},\\
\lambda_{\overline{SE}_1\circ\overline{SE}_2}(x_1,x_2)(y_1,y_2) &= \max\{\lambda_{\overline{SE}_1}(x_1,y_1),\lambda_{\overline{RV}_2}(x_2),\lambda_{\overline{RV}_2}(y_2)\}\\
&\qquad \forall\,x_1y_1\in\overline{SE}_1,x_2,y_2\in\overline{RV}_2,x_2\neq y_2
\end{aligned}
$$

**Example 4.** *Let $V^* = \{p,q,r\}$ be a set. Let $G_1 = (\underline{G}_1,\overline{G}_1)$ and $G_2 = (\underline{G}_2,\overline{G}_2)$ be two RND on $V^*$, where $\underline{G}_1 = (\underline{RV}_1,\underline{SE}_1)$ and $\overline{G}_1 = (\overline{RV}_1,\overline{SE}_1)$ are ND, as shown in Figure 8. $\underline{G}_2 = (\underline{RV}_2,\underline{SE}_2)$ and $\overline{G}_2 = (\overline{RV}_2,\overline{SE}_2)$ are also ND, as shown in Figure 9.*

*The composition of $G_1$ and $G_2$ is $G = G_1\circ G_2 = (\underline{G}_1\circ\underline{G}_2,\overline{G}_1\circ\overline{G}_2)$ where $\underline{G}_1\circ\underline{G}_2 = (\underline{RV}_1\circ\underline{RV}_2,\underline{SE}_1\circ\underline{SE}_2)$ and $\overline{G}_1\circ\overline{G}_2 = (\overline{RV}_1\circ\overline{RV}_2,\overline{SE}_1\circ\overline{SE}_2)$ are NDs, as shown in Figures 10 and 11.*

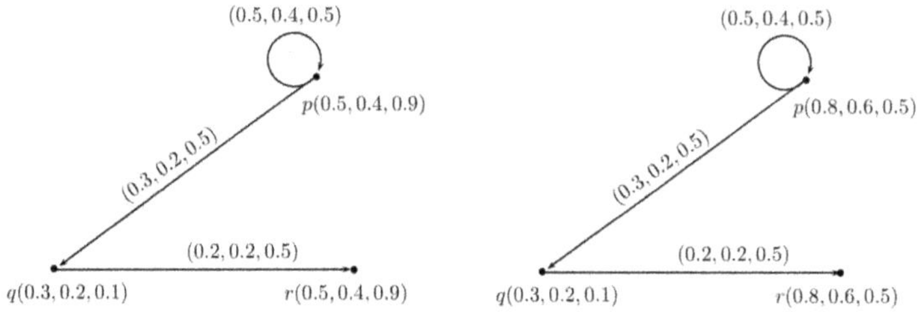


$\underline{G}_1 = (\underline{R}V_1, \underline{S}E_1)$ $\overline{G}_1 = (\overline{R}V_1, \overline{S}E_1)$

**Figure 8.** Rough neutrosophic digraph $G_1 = (\underline{G}_1, \overline{G}_1)$.

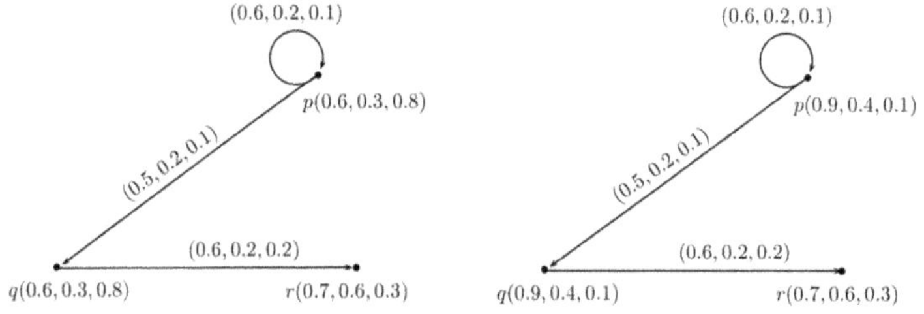


$\underline{G}_2 = (\underline{R}V_2, \underline{S}E_2)$ $\overline{G}_2 = (\overline{R}V_2, \overline{S}E_2$

**Figure 9.** Rough neutrosophic digraph $G_2 = (\underline{G}_2, \overline{G}_2)$.

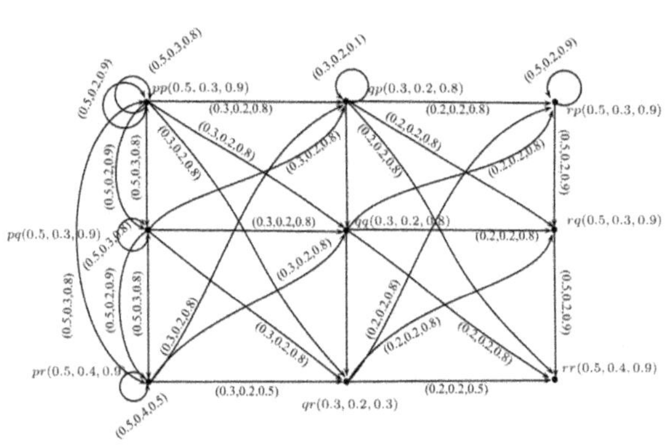


**Figure 10.** Neutrosophic digraph $\underline{G}_1 \circ \underline{G}_2 = (\underline{R}V_1 \circ \underline{R}V_2, \underline{S}E_1 \circ \underline{S}E_2)$.

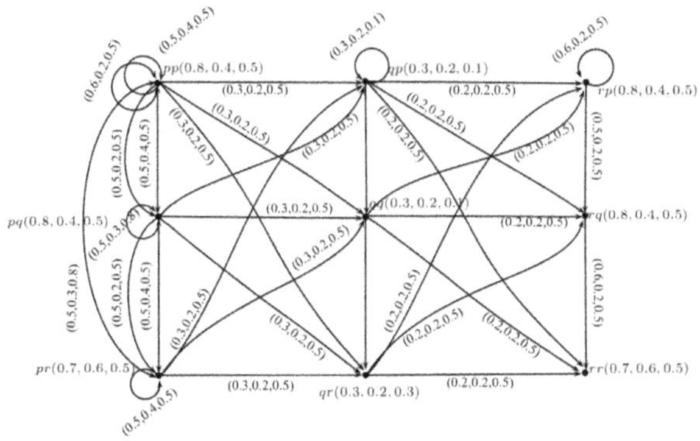


**Figure 11.** Neutrosophic digraph $\overline{G}_1 \circ \overline{G}_2 = (\underline{R}V_1 \circ \underline{R}V_2, \underline{S}E_1 \circ \underline{S}E_2)$.

**Theorem 3.** *The Composition of two rough neutrosophic digraphs is a rough neutrosophic digraph.*

**Proof.** Let $G_1 = (\underline{G}_1, \overline{G}_1)$ and $G_2 = (\underline{G}_2, \overline{G}_2)$ be two rough neutrosophic digraphs. Let $G = G_1 \circ G_2 = (\underline{G}_1 \circ \underline{G}_2, \overline{G}_1 \circ \overline{G}_2)$ be the *Composition* of $G_1$ and $G_2$, where $\underline{G}_1 \circ \underline{G}_2 = (\underline{R}V_1 \circ \underline{R}V_2, \underline{S}E_1 \circ \underline{S}E_2)$ and $\overline{G}_1 \circ \overline{G}_2 = (\overline{R}V_1 \circ \overline{R}V_2, \overline{S}E_1 \circ \overline{S}E_2)$. To prove that $G = \underline{G}_1 \circ \overline{G}_2$ is a rough neutrosophic digraph, it is enough to show that $\underline{S}E_1 \circ \underline{S}E_2$ and $\overline{S}E_1 \circ \overline{S}E_2$ are neutrosophic relations on $\underline{R}V_1 \circ \underline{R}V_2$ and $\overline{R}V_1 \circ \overline{R}V_2$, respectively. First, we show that $\underline{S}E_1 \circ \underline{S}E_2$ is a neutrosophic relation on $\underline{R}V_1 \circ \underline{R}V_2$.

If $x \in \underline{R}V_1, x_2y_2 \in \underline{S}E_2$, then

$$\begin{aligned}
\mu_{\underline{S}E_1 \circ \underline{S}E_2}(x, x_2)(x, y_2) &= \mu_{\underline{R}V_1}(x) \wedge \mu_{\underline{S}E_2}(x_2, y_2) \\
&\leq \mu_{\underline{R}V_1}(x) \wedge (\mu_{\underline{R}V_2}(x_2) \wedge \mu_{\underline{R}V_2}(y_2)) \\
&= (\mu_{\underline{R}V_1}(x) \wedge \mu_{\underline{R}V_2}(x_2)) \wedge (\mu_{\underline{R}V_1}(x) \wedge \mu_{\underline{R}V_2}(y_2)) \\
&= \mu_{\underline{R}V_1 \circ \underline{R}V_2}(x, x_2) \wedge \mu_{\underline{R}V_1 \circ \underline{R}V_2}(x, y_2) \\
\mu_{\underline{S}E_1 \circ \underline{S}E_2}(x, x_2)(x, y_2) &\leq \min\{\mu_{\underline{R}V_1 \circ \underline{R}V_2}(x, x_2), \mu_{\underline{R}V_1 \circ \underline{R}V_2}(x, y_2)\}, \\
\sigma_{\underline{S}E_1 \circ \underline{S}E_2}(x, x_2)(x, y_2) &= \sigma_{\underline{R}V_1}(x) \wedge \sigma_{\underline{S}E_2}(x_2, y_2) \\
&\leq \sigma_{\underline{R}V_1}(x) \wedge (\sigma_{\underline{R}V_2}(x_2) \wedge \sigma_{\underline{R}V_2}(y_2)) \\
&= (\sigma_{\underline{R}V_1}(x) \wedge \sigma_{\underline{R}V_2}(x_2)) \wedge (\sigma_{\underline{R}V_1}(x) \wedge \sigma_{\underline{R}V_2}(y_2) \\
&= \sigma_{\underline{R}V_1 \circ \underline{R}V_2}(x, x_2) \wedge \sigma_{\underline{R}V_1 \circ \underline{R}V_2}(x, y_2) \\
\sigma_{\underline{S}E_1 \circ \underline{S}E_2}(x, x_2)(x, y_2) &\leq \min\{\sigma_{\underline{R}V_1 \circ \underline{R}V_2}(x, x_2), \sigma_{\underline{R}V_1 \circ \underline{R}V_2}(x, y_2)\}, \\
\lambda_{\underline{S}E_1 \circ \underline{S}E_2}(x, x_2)(x, y_2) &= \lambda_{\underline{R}V_1}(x) \vee \lambda_{\underline{S}E_2}(x_2, y_2) \\
&\leq \lambda_{\underline{R}V_1}(x) \vee (\lambda_{\underline{R}V_2}(x_2) \vee \lambda_{\underline{R}V_2}(y_2)) \\
&= (\lambda_{\underline{R}V_1}(x) \vee \lambda_{\underline{R}V_2}(x_2)) \vee (\lambda_{\underline{R}V_1}(x) \vee \lambda_{\underline{R}V_2}(y_2)) \\
&= \lambda_{\underline{R}V_1 \circ \underline{R}V_2}(x, x_2) \vee \lambda_{\underline{R}V_1 \circ \underline{R}V_2}(x, y_2) \\
\lambda_{\underline{S}E_1 \circ \underline{S}E_2}(x, x_2, x, y_2) &\leq \max\{\lambda_{\underline{R}V_1 \circ \underline{R}V_2}(x, x_2), \lambda_{\underline{R}V_1 \circ \underline{R}V_2}(x, y_2)\}.
\end{aligned}$$

If $x_1y_1 \in \underline{S}E_1, z \in \underline{R}V_2$, then

$$\begin{aligned}
\mu_{\underline{S}E_1 \circ \underline{S}E_2}(x_1, z)(y_1, z) &= \mu_{\underline{S}E_1}(x_1, y_1) \wedge \mu_{\underline{R}V_2}(z) \\
&\leq (\mu_{\underline{R}V_1}(x_1) \wedge \mu_{\underline{R}V_1}(y_1)) \wedge \mu_{\underline{R}V_2}(z) \\
&= (\mu_{\underline{R}V_1}(x_1) \wedge \mu_{\underline{R}V_2}(z)) \wedge (\mu_{\underline{R}V_1}(y_1) \wedge \mu_{\underline{R}V_2}(z)) \\
&= \mu_{\underline{R}V_1 \circ \underline{R}V_2}(x_1, z) \wedge \mu_{\underline{R}V_1 \circ \underline{R}V_2}(y_1, z) \\
\mu_{\underline{S}E_1 \circ \underline{S}E_2}(x_1, z)(y_1, z) &\leq \min\{\mu_{\underline{R}V_1 \circ \underline{R}V_2}(x_1, z), \mu_{\underline{R}V_1 \circ \underline{R}V_2}(y_1, z)\},
\end{aligned}$$

$$\begin{aligned}
\sigma_{\underline{S}E_1\circ\underline{S}E_2}(x_1,z)(y_1,z) &= \sigma_{\underline{S}E_1}(x_1,y_1)\wedge\sigma_{\underline{R}V_2}(z)\\
&\leq (\sigma_{\underline{R}V_1}(x_1)\wedge\sigma_{\underline{R}V_1}(y_1))\wedge\sigma_{\underline{R}V_2}(z)\\
&= (\sigma_{\underline{R}V_1}(x_1)\wedge\sigma_{\underline{R}V_2}(z))\wedge(\sigma_{\underline{R}V_1}(y_1)\wedge\sigma_{\underline{R}V_2}(z))\\
&= \sigma_{\underline{R}V_1\circ\underline{R}V_2}(x_1,z)\wedge\sigma_{\underline{R}V_1\circ\underline{R}V_2}(y_1,z)\\
\sigma_{\underline{S}E_1\circ\underline{S}E_2}(x_1,z)(y_1,z) &\leq \min\{\sigma_{\underline{R}V_1\circ\underline{R}V_2}(x_1,z),\sigma_{\underline{R}V_1\circ\underline{R}V_2}(y_1,z)\},\\
\lambda_{\underline{S}E_1\circ\underline{S}E_2}(x_1,z)(y_1,z) &= \lambda_{\underline{S}E_1}(x_1,y_1)\vee\lambda_{\underline{R}V_2}(z)\\
&\leq (\lambda_{\underline{R}V_1}(x_1)\vee\lambda_{\underline{R}V_1}(y_1))\vee\lambda_{\underline{R}V_2}(z)\\
&= (\lambda_{\underline{R}V_1}(x_1)\vee\lambda_{\underline{R}V_2}(z))\vee(\lambda_{\underline{R}V_1}(y_1)\vee\lambda_{\underline{R}V_2}(z))\\
&= \lambda_{\underline{R}V_1\circ\underline{R}V_2}(x_1,z)\vee\lambda_{\underline{R}V_1\circ\underline{R}V_2}(y_1,z)\\
\lambda_{\underline{S}E_1\circ\underline{S}E_2}(x_1,z)(y_1,z) &\leq \max\{\lambda_{\underline{R}V_1\circ\underline{R}V_2}(x_1,z),\lambda_{\underline{R}V_1\circ\underline{R}V_2}(y_1,z)\}.
\end{aligned}$$

If $x_1y_1\in\underline{S}E_1, x_2,y_2\in\underline{R}V_2$ such that $x_2\neq y_2$,

$$\begin{aligned}
\mu_{\underline{S}E_1\circ\underline{S}E_2}(x_1,x_2)(y_1,y_2) &= \mu_{\underline{S}E_1}(x_1y_1)\wedge\mu_{\underline{R}V_2}(x_2)\wedge\mu_{\underline{R}V_2}(y_2)\\
&\leq (\mu_{\underline{R}V_1}(x_1)\wedge\mu_{\underline{R}V_1}(y_1))\wedge\mu_{\underline{R}V_2}(x_2)\wedge\mu_{\underline{R}V_2}(y_2)\\
&= (\mu_{\underline{R}V_1}(x_1)\wedge\mu_{\underline{R}V_2}(x_2))\wedge(\mu_{\underline{R}V_1}(y_1))\wedge\mu_{\underline{R}V_2}(y_2))\\
&= \mu_{\underline{R}V_1\circ\underline{R}V_2}(x_1,x_2)\wedge\mu_{\underline{R}V_1\circ\underline{R}V_2}(y_1,y_2)\\
\mu_{\underline{S}E_1\circ\underline{S}E_2}(x_1,x_2)(y_1,y_2) &\leq \min\{\mu_{\underline{R}V_1\circ\underline{R}V_2}(x_1,x_2),\mu_{\underline{R}V_1\circ\underline{R}V_2}(y_1,y_2)\}\\
\sigma_{\underline{S}E_1\circ\underline{S}E_2}(x_1,x_2)(y_1,y_2) &= \sigma_{\underline{S}E_1}(x_1y_1)\wedge\sigma_{\underline{R}V_2}(x_2)\wedge\sigma_{\underline{R}V_2}(y_2)\\
&\leq (\sigma_{\underline{R}V_1}(x_1)\wedge\sigma_{\underline{R}V_1}(y_1))\wedge\sigma_{\underline{R}V_2}(x_2)\wedge\sigma_{\underline{R}V_2}(y_2)\\
&= (\sigma_{\underline{R}V_1}(x_1)\wedge\sigma_{\underline{R}V_2}(x_2))\wedge(\sigma_{\underline{R}V_1}(y_1))\wedge\sigma_{\underline{R}V_2}(y_2))\\
&= \sigma_{\underline{R}V_1\circ\underline{R}V_2}(x_1,x_2)\wedge\sigma_{\underline{R}V_1\circ\underline{R}V_2}(y_1,y_2)\\
\sigma_{\underline{S}E_1\circ\underline{S}E_2}(x_1,x_2)(y_1,y_2) &\leq \min\{\sigma_{\underline{R}V_1\circ\underline{R}V_2}(x_1,x_2),\sigma_{\underline{R}V_1\circ\underline{R}V_2}(y_1,y_2)\}\\
\lambda_{\underline{S}E_1\circ\underline{S}E_2}(x_1,x_2)(y_1,y_2) &= \lambda_{\underline{S}E_1}(x_1y_1)\vee\lambda_{\underline{R}V_2}(x_2)\vee\lambda_{\underline{R}V_2}(y_2)\\
&\leq (\lambda_{\underline{R}V_1}(x_1)\vee\lambda_{\underline{R}V_1}(y_1))\vee\lambda_{\underline{R}V_2}(x_2)\vee\lambda_{\underline{R}V_2}(y_2)\\
&= (\lambda_{\underline{R}V_1}(x_1)\vee\lambda_{\underline{R}V_2}(x_2))\vee(\lambda_{\underline{R}V_1}(y_1))\vee\lambda_{\underline{R}V_2}(y_2))\\
&= \lambda_{\underline{R}V_1\circ\underline{R}V_2}(x_1,x_2)\vee\lambda_{\underline{R}V_1\circ\underline{R}V_2}(y_1,y_2)\\
\lambda_{\underline{S}E_1\circ\underline{S}E_2}(x_1,x_2)(y_1,y_2) &\leq \max\{\lambda_{\underline{R}V_1\circ\underline{R}V_2}(x_1,x_2),\lambda_{\underline{R}V_1\circ\underline{R}V_2}(y_1,y_2)\}.
\end{aligned}$$

Thus, from above, it is clear that $\underline{S}E_1\circ\underline{S}E_2$ is a neutrosophic relation on $\underline{R}V_1\circ\underline{R}V_2$.

Similarly, we can show that $\overline{S}E_1\circ\overline{S}E_2$ is a neutrosophic relation on $\overline{R}V_1\circ\overline{R}V_2$. Hence, $G=(\underline{G}_1\circ\underline{G}_2,\overline{G}_1\circ\overline{G}_2)$ is a rough neutrosophic digraph. □

**Definition 9.** *Let $G=(\underline{G},\overline{G})$ be a RND. The complement of G, denoted by $G'=(\underline{G}',\overline{G}')$ is a rough neutrosophic digraph, where $\underline{G}'=((\underline{R}V)',(\underline{S}E)')$ and $\overline{G}'=((\overline{R}V)',(\overline{S}E)')$ are neutrosophic digraph such that*

$$\begin{aligned}
(1)\quad \mu_{(\underline{R}V)'}(x) &= \mu_{\underline{R}V}(x),\\
\sigma_{(\underline{R}V)'}(x) &= \sigma_{\underline{R}V}(x),\\
\lambda_{(\underline{R}V)'}(x) &= \lambda_{\underline{R}V}(x)\ \ \forall\ x\in V^*\\
\mu_{(\underline{S}E)'}(x,y) &= \min\{\mu_{\underline{R}V}(x),\mu_{\underline{R}V}(y)\}-\mu_{\underline{S}E}(xy)\\
\sigma_{(\underline{S}E)'}(x,y) &= \min\{\sigma_{\underline{R}V}(x),\sigma_{\underline{R}V}(y)\}-\sigma_{\underline{S}E}(xy)\\
\lambda_{(\underline{S}E)'}(x,y) &= \max\{\lambda_{\underline{R}V}(x),\lambda_{\underline{R}V}(y)\}-\lambda_{\underline{S}E}(xy)\ \ \forall\ x,y\in V^*.
\end{aligned}$$

$$
\begin{aligned}
\textbf{(2)} \quad \mu_{\overline{R}V'}(x) &= \mu_{\overline{R}V}(x),\\
\sigma_{\overline{R}V'}(x) &= \sigma_{\overline{R}V}(x),\\
\lambda_{\overline{R}V'}(x) &= \lambda_{\overline{R}V}(x), \ \forall\, x \in V^*\\
\mu_{(\overline{S}E)'}(x,y) &= \min\{\mu_{\overline{R}V}(x), \mu_{\overline{R}V}(y)\} - \mu_{\overline{S}E}(xy)\\
\sigma_{(\overline{S}E)'}(x,y) &= \min\{\sigma_{\overline{R}V}(x), \sigma_{\overline{R}V}(y)\} - \sigma_{\overline{S}E}(xy)\\
\lambda_{(\overline{S}E)'}(x,y) &= \max\{\lambda_{\overline{R}V}(x), \lambda_{\overline{R}V}(y)\} - \lambda_{\overline{S}E}(xy) \ \forall\, x,y \in V^*.
\end{aligned}
$$

**Example 5.** *Consider a rough neutrosophic digraph as shown in Figure 4. The lower and upper approximations of graph G are* $\underline{G} = (\underline{R}V, \underline{S}E)$ *and* $\overline{G} = (\overline{R}V, \overline{S}E)$*, respectively, where*

$$
\begin{aligned}
\underline{R}V &= \{(a,0.2,0.4,0.6),(b,0.2,0.4,0.6),(c,0.2,0.5,0.9),(d,0.2,0.5,0.9)\},\\
\overline{R}V &= \{(a,0.3,0.8,0.3).(b,0.3,0.8,0.3),(c,0.5,0.6,0.8),(d,0.5,0.6,0.8)\},
\end{aligned}
$$

$$
\begin{aligned}
\underline{S}E = \{&(aa,0.2,0.3,0.3),(ab,0.2,0.3,0.3),(ad,0.1,0.3,0.8),(bc,0.1,0.3,0.8),\\
&(bd,0.1,0.3,0.8),(dc,0.2,0.4,0.7),(dd,0.2,0.4,0.7)\},\\
\overline{S}E = \{&(aa,0.2,0.4,0.3),(ab,0.2,0.4,0.3),(ad,0.2,0.4,0.7),(bc,0.2,0.4,0.7),\\
&(bd,0.2,0.4,0.7),(dc,0.2,0.4,0.7),(dd,0.2,0.4,0.7)\}.
\end{aligned}
$$

*The complement of G is* $G' = (\underline{G}', \overline{G}')$*. By calculations, we have*

$$
\begin{aligned}
(\underline{R}V)' &= \{(a,0.2,0.4,0.6),(b,0.2,0.4,0.6),(c,0.2,0.5,0.9),(d,0.2,0.5,0.9)\},\\
(\overline{R}V)' &= \{(a,0.3,0.8,0.3).(b,0.3,0.8,0.3),(c,0.5,0.6,0.8),(d,0.5,0.6,0.8)\},
\end{aligned}
$$

$$
\begin{aligned}
(\underline{S}E)' = \{&(aa,0,0.1,0.3),(ab,0,0.1,0.3),(ac,0.2,0.4,0.9),(ad,0.1,0.1,0.1),(ba,0.2,0.4,0.6),(bb,0.2,0.4,0.6),\\
&(bc,0.1,0.1,0.1),(bd,0.1,0.1,0.1),(ca,0.2,0.4,0.9),(cb,0.2,0.4,0.9),(cc,0.2,0.5,0.9),(cd,0.2,0.5,0.9),\\
&(da,0.2,0.4,0.9),(db,0.2,0.4,0.9),(dc,0,0.1,0.2),(dd,0,0.1,0.2)\},\\
(\overline{S}E)' = \{&(aa,0.1,0.4,0),(ab,0.1,0.4,0),(ac,0.3,0.6,0.8),(ad,0.1,0.2,0.1),(ba,0.3,0.8,0.3),(bb,0.3,0.8,0.3),\\
&(bc,0.1,0.2,0.1),(bd,0.1,0.2,0.1),(ca,0.3,0.6,0.8),(cb,0.3,0.6,0.8),(cc,0.5,0.6,0.8),(cd,0.5,0.6,0.8),\\
&(da,0.3,0.6,0.8),(db,0.3,0.6,0.8),(dc,0.3,0.2,0.1),(dd,0.3,0.2,0.1)\}.
\end{aligned}
$$

*Thus,* $\underline{G}' = ((\underline{R}V)', (\underline{S}E)')$ *and* $\overline{G}' = ((\overline{R}V)', (\overline{S}E)')$ *are neutrosophic digraph, as shown in Figure 12.*

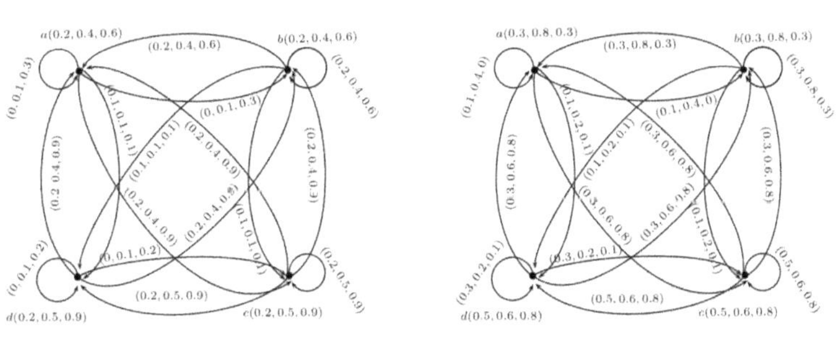


$\underline{G}' = ((\underline{R}N)', (\underline{S}E)')$ $\qquad$ $\overline{G}' = ((\overline{R}N)', (\overline{S}E)')$

**Figure 12.** Rough neutrosophic digraph $G' = (\underline{G}', \overline{G}')$.

**Definition 10.** *A rough neutrosophic digraph* $G = (\underline{G}, \overline{G})$ *is self complementary if G and G′ are isomorphic, that is,* $\underline{G} \cong \underline{G}'$ *and* $\overline{G} \cong \overline{G}'$*.*

**Example 6.** *Let $V^* = \{a, b, c\}$ be a set and $R$ an equivalence relation on $V^*$ defined as:*

$$R = \begin{bmatrix} 1 & 0 & 1 \\ 0 & 1 & 0 \\ 1 & 0 & 1 \end{bmatrix}.$$

*Let $V = \{(a, 0.2, 0.4, 0.8), (b, 0.2, 0.4, 0.8), (c, 0.4, 0.6, 0.4)\}$ be a neutrosophic set on $V^*$. The lower and upper approximations of $V$ are given as,*
$\underline{R}V = \{(a, 0.2, 0.4, 0.8), (b, 0.2, 0.4, 0.8), (c, 0.2, 0.4, 0.8)\}$,
$\overline{R}V = \{(a, 0.4, 0.6, 0.4), (b, 0.2, 0.4, 0.8), (c, 0.4, 0.6, 0.4)\}$.

*Let $E^* = \{aa, ab, ac, ba\} \subseteq V^* \times V^*$ and $S$ an equivalence relation on $E^*$ defined as*

$$S = \begin{bmatrix} 1 & 0 & 1 & 0 \\ 0 & 1 & 0 & 0 \\ 1 & 0 & 1 & 0 \\ 0 & 0 & 0 & 1 \end{bmatrix}.$$

*Let $E = \{(aa, 0.1, 0.3, 0.2), (ab, 0.1, 0.2, 0.4), (ac, 0.2, 0.2, 0.4), (ba, 0.1, 0.2, 0.4)\}$ be a neutrosophic set on $E^*$ and $SE = (\underline{S}E, \overline{S}E)$ a RNR where $\underline{S}E$ and $\overline{S}E$ are given as*
$\underline{S}E = \{(aa, 0.1, 0.2, 0.4), (ab, 0.1, 0.2, 0.4), (ac, 0.1, 0.2, 0.4), (ba, 0.1, 0.2, 0.4)\}$,
$\overline{S}E = \{(aa, 0.2, 0.3, 0.2), (ab, 0.1, 0.2, 0.4), (ac, 0.2, 0.3, 0.2), (ba, 0.1, 0.2, 0.4)\}$.

*Thus, $\underline{G} = (\underline{R}V, \underline{S}E)$ and $\overline{G} = (\overline{R}V, \overline{S}E)$ are neutrosophic digraphs, as shown in Figure 13. The complement of $G$ is $G' = (\underline{G}', \overline{G}')$, where $\underline{G}' = \underline{G}$ and $\overline{G}' = \overline{G}$ are neutrosophic digraphs, as shown in Figure 13, and it can be easily shown that $G$ and $G'$ are isomorphic. Hence, $G = (\underline{G}, \overline{G})$ is a self complementary RND.*

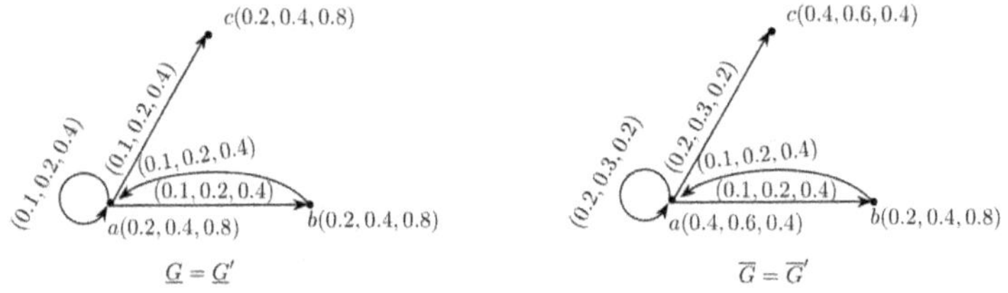


**Figure 13.** Self complementary RND $G = (\underline{G}, \overline{G})$.

**Theorem 4.** *Let $G = (\underline{G}, \overline{G})$ be a self complementary rough neutrosophic digraph. Then,*

$$\sum_{w,z \in V^*} \mu_{\underline{S}E}(wz) = \frac{1}{2} \sum_{w,z \in V^*} (\mu_{\underline{R}V}(w) \wedge \mu_{\underline{R}V}(z))$$

$$\sum_{w,z \in V^*} \sigma_{\underline{S}E}(wz) = \frac{1}{2} \sum_{w,z \in V^*} (\sigma_{\underline{R}V}(w) \wedge \sigma_{\underline{R}V}(z))$$

$$\sum_{w,z \in V^*} \lambda_{\underline{S}E}(wz) = \frac{1}{2} \sum_{w,z \in V^*} (\lambda_{\underline{R}V}(w) \vee \lambda_{\underline{R}V}(z))$$

$$\sum_{w,z \in V^*} \mu_{\overline{S}E}(wz) = \frac{1}{2} \sum_{w,z \in V^*} (\mu_{\overline{R}V}(w) \wedge \mu_{\overline{R}V}(z))$$

$$\sum_{w,z \in V^*} \sigma_{\overline{S}E}(wz) = \frac{1}{2} \sum_{w,z \in V^*} (\sigma_{\overline{R}V}(w) \wedge \sigma_{\overline{R}V}(z))$$

$$\sum_{w,z \in V^*} \lambda_{\overline{S}E}(wz) = \frac{1}{2} \sum_{w,z \in V^*} (\lambda_{\overline{R}V}(w) \vee \lambda_{\overline{R}V}(z)).$$

**Proof.** Let $G = (\underline{G}, \overline{G})$ be a self complementary rough neutrosophic digraph. Then, there exist two isomorphisms $\underline{g} : V^* \longrightarrow V^*$ and $\overline{g} : V^* \longrightarrow V^*$, respectively, such that

$$\begin{aligned}
\mu_{(\underline{RV})'}(\underline{g}(w)) &= \mu_{\underline{RV}}(w), \\
\sigma_{(\underline{RV})'}(\underline{g}(w)) &= \sigma_{\underline{RV}}(w), \\
\lambda_{(\underline{RV})'}(\underline{g}(w)) &= \lambda_{\underline{RV}}(w), \ \forall\, w \in V^* \\
\mu_{(\underline{SE})'}(\underline{g}(w)\underline{g}(z)) &= \mu_{(\underline{SE})}(wz), \\
\sigma_{(\underline{SE})'}(\underline{g}(w)\underline{g}(z)) &= \sigma_{(\underline{SE})}(wz), \\
\lambda_{(\underline{SE})'}(\underline{g}(w)\underline{g}(z)) &= \lambda_{(\underline{SE})}(wz) \ \forall\, w, z \in V^*.
\end{aligned}$$

and

$$\begin{aligned}
\mu_{(\overline{RV})'}(\overline{g}(w)) &= \mu_{\overline{RV}}(w), \\
\sigma_{(\overline{RV})'}(\overline{g}(w)) &= \sigma_{\overline{RV}}(w), \\
\lambda_{(\overline{RV})'}(\overline{g}(w)) &= \lambda_{\overline{RV}}(w), \ \forall\, w \in V^* \\
\mu_{(\overline{SE})'}(\overline{g}(w)\overline{g}(z)) &= \mu_{(\overline{SE})}(wz), \\
\sigma_{(\overline{SE})'}(\overline{g}(w)\overline{g}(z)) &= \sigma_{(\overline{SE})}(wz), \\
\lambda_{(\overline{SE})'}(\overline{g}(w)\overline{g}(z)) &= \lambda_{(\overline{SE})}(wz) \ \forall\, w, z \in V^*.
\end{aligned}$$

By Definition 7, we have

$$\begin{aligned}
\mu_{(\underline{SE})'}(\underline{g}(w)\underline{g}(z)) &= (\mu_{\underline{RV}}(w) \wedge \mu_{\underline{RV}}(z)) - \mu_{(\underline{SE})}(wz) \\
\mu_{(\underline{SE})}(wz) &= (\mu_{\underline{RV}}(w) \wedge \mu_{\underline{RV}}(z)) - \mu_{(\underline{SE})}(wz) \\
\sum_{w,z \in V^*} \mu_{(\underline{SE})}(wz) &= \sum_{w,z \in V^*} (\mu_{\underline{RV}}(w) \wedge \mu_{\underline{RV}}(z)) - \sum_{w,z \in V^*} \mu_{(\underline{SE})}(wz) \\
2 \sum_{w,z \in V^*} \mu_{(\underline{SE})}(wz) &= \sum_{w,z \in V^*} (\mu_{\underline{RV}}(w) \wedge \mu_{\underline{RV}}(z)) \\
\sum_{w,z \in V^*} \mu_{(\underline{SE})}(wz) &= \frac{1}{2} \sum_{w,z \in V^*} (\mu_{\underline{RV}}(w) \wedge \mu_{\underline{RV}}(z))
\end{aligned}$$

$$\begin{aligned}
\sigma_{(\underline{SE})'}(\underline{g}(w)\underline{g}(z)) &= (\sigma_{\underline{RV}}(w) \wedge \sigma_{\underline{RV}}(z)) - \sigma_{(\underline{SE})}(wz) \\
\sigma_{(\underline{SE})}(wz) &= (\sigma_{\underline{RV}}(w) \wedge \sigma_{\underline{RV}}(z)) - \sigma_{(\underline{SE})}(wz) \\
\sum_{w,z \in V^*} \sigma_{(\underline{SE})}(wz) &= \sum_{w,z \in V^*} (\sigma_{\underline{RV}}(w) \wedge \sigma_{\underline{RV}}(z)) - \sum_{w,z \in V^*} \sigma_{(\underline{SE})}(wz) \\
2 \sum_{w,z \in V^*} \sigma_{(\underline{SE})}(wz) &= \sum_{w,z \in V^*} (\sigma_{\underline{RV}}(w) \wedge \sigma_{\underline{RV}}(z)) \\
\sum_{w,z \in V^*} \sigma_{(\underline{SE})}(wz) &= \frac{1}{2} \sum_{w,z \in V^*} (\sigma_{\underline{RV}}(w) \wedge \sigma_{\underline{RV}}(z))
\end{aligned}$$

$$\begin{aligned}
\lambda_{(\underline{SE})'}(\underline{g}(w)\underline{g}(z)) &= (\lambda_{\underline{RV}}(w) \vee \lambda_{\underline{RV}}(z)) - \lambda_{(\underline{SE})}(wz) \\
\lambda_{(\underline{SE})}(wz) &= (\lambda_{\underline{RV}}(w) \vee \lambda_{\underline{RV}}(z)) - \lambda_{(\underline{SE})}(wz) \\
\sum_{w,z \in V^*} \lambda_{(\underline{SE})}(wz) &= \sum_{w,z \in V^*} (\lambda_{\underline{RV}}(w) \vee \lambda_{\underline{RV}}(z)) - \sum_{w,z \in V^*} \lambda_{(\underline{SE})}(wz) \\
2 \sum_{w,z \in V^*} \lambda_{(\underline{SE})}(wz) &= \sum_{w,z \in V^*} (\lambda_{\underline{RV}}(w) \vee \lambda_{\underline{RV}}(z)) \\
\sum_{w,z \in V^*} \lambda_{(\underline{SE})}(wz) &= \frac{1}{2} \sum_{w,z \in V^*} (\lambda_{\underline{RV}}(w) \vee \lambda_{\underline{RV}}(z))
\end{aligned}$$

Similarly, it can be shown that

$$\begin{aligned}
\sum_{w,z\in V^*} \mu_{\overline{S}E}(wz) &= \frac{1}{2}\sum_{w,z\in V^*} (\mu_{\overline{R}V}(w) \wedge \mu_{\overline{R}V}(z)) \\
\sum_{w,z\in V^*} \sigma_{\overline{S}E}(wz) &= \frac{1}{2}\sum_{w,z\in V^*} (\sigma_{\overline{R}V}(w) \wedge \sigma_{\overline{R}V}(z)) \\
\sum_{w,z\in V^*} \lambda_{\overline{S}E}(wz) &= \frac{1}{2}\sum_{w,z\in V^*} (\lambda_{\overline{R}V}(w) \vee \lambda_{\overline{R}V}(z)).
\end{aligned}$$

This completes the proof. □

## 3. Application

Investment is a very good way of getting profit and wisely invested money surely gives certain profit. The most important factors that influence individual investment decision are: company's reputation, corporate earnings and price per share. In this application, we combine these factors into one factor, i.e. company's status in industry, to describe overall performance of the company. Let us consider an individual Mr. Shahid who wants to invest his money. For this purpose, he considers some private companies, which are Telecommunication company (*TC*), Carpenter company (*CC*), Real Estate business (*RE*), Vehicle Leasing company (*VL*), Advertising company (*AD*), and Textile Testing company (*TT*). Let $V^*=\{TC, CC, RE, VL, AD, TT\}$ be a set. Let $T$ be an equivalence relation defined on $V^*$ as follows:

$$T = \begin{bmatrix} 1 & 0 & 1 & 0 & 1 & 0 \\ 0 & 1 & 0 & 0 & 0 & 0 \\ 1 & 0 & 1 & 0 & 1 & 0 \\ 0 & 0 & 0 & 1 & 0 & 1 \\ 1 & 0 & 1 & 0 & 1 & 0 \\ 0 & 0 & 0 & 1 & 0 & 1 \end{bmatrix}.$$

Let $V = \{(TC, 0.3, 0.4, 0.1), (CC, 0.8, 0.1, 0.5), (RE, 0.1, 0.2, 0.6), (VL, 0.9, 0.6, 0.1), (AD, 0.2, 0.5, 0.2), (TT, 0.8, 0.6, 0.5)\}$ be a neutrosophic set on $V^*$ with three components corresponding to each company, which represents its status in the industry and $TV = (\underline{T}V, \overline{T}V)$ a rough neutrosophic set, where $\underline{T}V$ and $\overline{T}V$ are lower and upper approximations of $V$, respectively, as follows:

$$\begin{aligned}
\underline{T}V &= \{(TC,0.1,0.2,0.6),(CC,0.8,0.1,0.5),(RE,0.1,0.2,0.6),(VL,0.8,0.6,0.5),(AD,\\
&\quad 0.1,0.2,0.6),(TT,0.8,0.6,0.5)\},\\
\overline{T}V &= \{(TC,0.3,0.5,0.1),(CC,0.8,0.1,0.5),(RE,0.3,0.5,0.1),(VL,0.9,0.6,0.1),(AD,\\
&\quad 0.3,0.5,0.1),(TT,0.9,0.6,0.1)\}.\\
\text{Let } E^* &= \{(TC,CC),(TC,AD),(TC,RE),(CC,VL),(CC,TT),(AD,RE),(TT,VL)\},
\end{aligned}$$

be the set of edges and $S$ an equivalence relation on $E^*$ defined as follows:

$$S = \begin{bmatrix} 1 & 0 & 0 & 0 & 0 & 0 & 0 \\ 0 & 1 & 1 & 0 & 0 & 1 & 0 \\ 0 & 1 & 1 & 0 & 0 & 1 & 0 \\ 0 & 0 & 0 & 1 & 1 & 0 & 0 \\ 0 & 1 & 0 & 1 & 1 & 0 & 0 \\ 0 & 0 & 1 & 0 & 0 & 1 & 0 \\ 0 & 0 & 0 & 0 & 0 & 0 & 1 \end{bmatrix}.$$

$$\text{Let } E = \{((TC,CC),0.1,0.1,01),((TC,AD),0.1,0.2,0.1),((TC,RE),0.1,0.2,0.1), ((CC,VL),0.8,0.1,0.5),((CC,TT),0.8,0.1,0.5),((AD,RE),0.1,0.2,0.1), ((TT,VL),0.8,0.6,0.1)\}$$

be a neutrosophic set on $E^*$ which represents relationship between companies and $SE = (\underline{S}E, \overline{S}E)$ a rough neutrosophic relation, where $\underline{S}E$ and $\overline{S}E$ are lower and upper upper approximations of $E$, respectively, as follows:

$$\underline{S}E = \{((TC,CC),0.1,0.1,0.1),((TC,AD),0.1,0.2,0.1),((TC,RE),0.1,0.2,0.1), ((CC,VL),0.8,0.1,0.5),((CC,TT),0.8,0.1,0.5),((AD,RE),0.1,0.2,0.1), ((TT,VL),0.8,0.6,0.1)\},$$

$$\overline{S}E = \{((TC,CC),0.1,0.1,0.1),((TC,AD),0.1,0.2,0.1),((TC,RE),0.1,0.2,0.1), ((CC,VL),0.8,0.1,0.5),((CC,TT),0.8,0.1,0.5),((AD,RE)0.1,0.2,0.1), ((TT,VL),0.8,0.6,0.1)\}.$$

Thus, $\underline{G} = (\underline{T}V, \underline{S}E)$ and $\overline{G} = (\overline{T}V, \overline{S}E)$ is a rough neutrosophic digraph as shown in Figure 14.

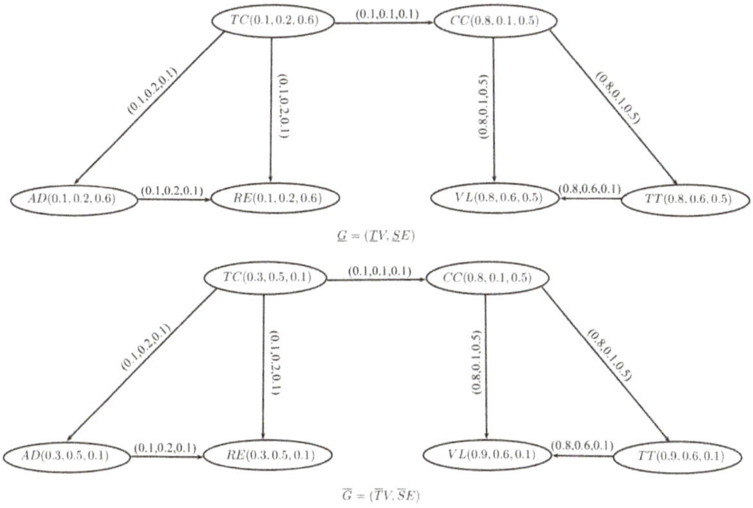


**Figure 14.** Rough neutrosophic digraph $G = (\underline{G}, \overline{G})$.

To find out the most suitable investment company, we define the score values

$$S(v_i) = \sum_{v_iv_j \in E^*} \frac{T(v_j) + I(v_j) - F(v_j)}{3 - (T(v_iv_j) + I(v_iv_j) - F(v_iv_j))},$$

where

$$T(v_j) = \frac{\underline{T}(v_j) + \overline{T}(v_j)}{2},$$
$$I(v_j) = \frac{\underline{I}(v_j) + \overline{I}(v_j)}{2},$$
$$F(v_j) = \frac{\underline{F}(v_j) + \overline{F}(v_j)}{2},$$

and

$$\begin{aligned} T(v_i v_j) &= \frac{\underline{T}(v_i v_j)+\overline{T}(v_i v_j)}{2}, \\ I(v_i v_j) &= \frac{\underline{I}(v_i v_j)+\overline{I}(v_i v_j)}{2}, \\ F(v_i v_j) &= \frac{\underline{F}(v_i v_j)+\overline{F}(v_i v_j)}{2}. \end{aligned}$$

of each selected company and industry decision is $v_k$ if $v_k = \max_i S(v_i)$. By calculation, we have $S(TC) = 0.4926$, $S(CC) = 1.4038$, $S(RE) = 0.0667$, $S(VL) = 0.3833$, $S(AD) = 0.1429$ and $S(TT) = 1.3529$. Clearly, $CC$ is the optimal decision. Therefore, the carpenter company is selected to get maximum possible profit. We present our proposed method as an algorithm. This Algorithm 1 returns the optimal solution for the investment problem.

**Algorithm 1** Calculation of Optimal decision

1: Input the vertex set $V^*$.
2: Construct an equivalence relation $T$ on the set $V^*$.
3: Calculate the approximation sets $\underline{T}V$ and $\overline{T}V$.
4: Input the edge set $E^* \subseteq V^* \times V^*$.
5: Construct an equivalence relation $S$ on $E^*$.
6: Calculate the approximation sets $\underline{S}E$ and $\overline{S}E$.
7: Calculate the score value, by using formula

$$S(v_i) = \sum_{v_i v_j \in E^*} \frac{T(v_j) + I(v_j) - F(v_j)}{3 - (T(v_i v_j) + I(v_i v_j) - F(v_i v_j))}.$$

8: The decision is $S(v_k) = \max_{v_i \in V^*} S(v_i)$.
9: If $v_k$ has more than one value, then any one of $S(v_k)$ may be chosen.

## 4. Conclusions and Future Directions

Neutrosophic sets and rough sets are very important models to handle uncertainty from two different perspectives. A rough neutrosophic model is a hybrid model which is made by combining two mathematical models, namely, rough sets and neutrosophic sets. This hybrid model deals with soft computing and vagueness by using the lower and upper approximation spaces. A rough neutrosophic set model gives more precise results for decision-making problems as compared to neutrosophic set model. In this paper, we have introduced the notion of rough neutrosophic digraphs. This research work can be extended to: (1) rough bipolar neutrosophic soft graphs; (2) bipolar neutrosophic soft rough graphs; (3) interval-valued bipolar neutrosophic rough graphs; and (4) neutrosophic soft rough graphs.

**Acknowledgments:** The authors are very thankful to the Editor and referees for their valuable comments and suggestions for improving the paper.

**Author Contributions:** Sidra Sayed and Nabeela Ishfaq conceived and designed the experiments; Muhammad Akram performed the experiments; Florentin Smarandache contributed reagents/materials/analysis tools.

**Conflicts of Interest:** The authors declare that they have no conflict of interest regarding the publication of the research article.

## References

1. Smarandache, F. *A Unifying Field in Logics. Neutrosophy: Neutrosophic Probability, Set and Logic*; American Research Press: Rehoboth, NM, USA, 1999.
2. Zadeh, L.A. Fuzzy sets. *Inf. Control* **1965**, *8*, 338–353.
3. Smarandache, F. *Neutrosophy Neutrosophic Probability, Set, and Logic*; American Research Press: Rehoboth, NM, USA, 1998.

4. Wang, H.; Smarandache, F.; Zhang, Y.; Sunderraman, R. Single-valued neutrosophic sets. *Multispace Multistructure* **2010**, *4*, 410–413.
5. Ye, J. Multicriteria decision-making method using the correlation coefficient under single-valued neutrosophic environment. *Int. J. Gen. Syst.* **2013**, *42*, 386–394.
6. Ye, J. Improved correlation coefficients of single valued neutrosophic sets and interval neutrosophic sets for multiple attribute decision making. *J. Intell. Fuzzy Syst.* **2014**, *27*, 2453–2462.
7. Pawlak, Z. Rough sets. *Int. J. Comput. Inf. Sci.* **1982**, *11*, 341–356.
8. Dubois, D.; Prade, H. Rough fuzzy sets and fuzzy rough sets. *Int. J. Gen. Syst.* **1990**, *17*, 191–209.
9. Liu, P.; Chen, S.M. Group decision making based on Heronian aggregation operators of intuitionistic fuzzy numbers. *IEEE Trans. Cybern.* **2017**, *47*, 2514–2530.
10. Broumi, S.; Smarandache, F.; Dhar, M. Rough neutrosophic sets. *Neutrosophic Sets Syst.* **2014**, *3*, 62–67.
11. Yang, H.L.; Zhang, C.L.; Guo, Z.L.; Liu, Y.L.; Liao, X. A hybrid model of single valued neutrosophic sets and rough sets: Single valued neutrosophic rough set model. *Soft Comput.* **2016**, *21*, 6253–6267, doi:10.1007/s00500-016-2356-y.
12. Mordeson, J.N.; Peng, C.S. Operations on fuzzy graphs. *Inf. Sci.* **1994**, *79*, 159–170.
13. Akram, M.; Shahzadi, S. Neutrosophic soft graphs with applicatin. *J. Intell. Fuzzy Syst.* **2017**, *32*, 841–858.
14. Akram, M.; Sarwar, M. Novel multiple criteria decision making methods based on bipolar neutrosophic sets and bipolar neutrosophic graphs. *Ital. J. Pure Appl. Math.* **2017**, *38*, 368–389.
15. Akram, M.; Siddique, S. Neutrosophic competition graphs with applications. *J. Intell. Fuzzy Syst.* **2017**, *33*, 921–935.
16. Akram, M.; Sitara, M. Interval-valued neutrosophic graph structures. *Punjab Univ. J. Math.* **2018**, *50*, 113–137.
17. Zafer, F.; Akram, M. A novel decision-making method based on rough fuzzy information. *Int. J. Fuzzy Syst.* **2017**, doi:10.1007/s40815-017-0368-0.
18. Banerjee, M.; Pal, S.K. Roughness of a fuzzy set. *Inf. Sci.* **1996**, *93*, 235–246.
19. Liu, P.; Chen, S.M.; Junlin, L. Some intuitionistic fuzzy interaction partitioned Bonferroni mean operators and their application to multi-attribute group decision making. *Inf. Sci.* **2017**, *411*, 98–121.
20. Liu, P. Multiple attribute group decision making method based on interval-valued intuitionistic fuzzy power Heronian aggregation operators. *Comput. Ind. Eng.* **2017**, *108*, 199–212.
21. Zhang, X.; Dai, J.; Yu, Y. On the union and intersection operations of rough sets based on various approximation spaces. *Inf. Sci.* **2015**, *292*, 214–229.
22. Bao, Y.L.; Yang, H.L. On single valued neutrosophic refined rough set model and its application. *J. Intell. Fuzzy Syst.* **2017**, *33*, 1235–1248.

axioms

MDPI

Article

# Neutrosophic Positive Implicative $\mathcal{N}$-Ideals in $BCK$-Algebras

**Young Bae Jun [1], Florentin Smarandache [2], Seok-Zun Song [3,*] and Madad Khan [4]**

[1] Department of Mathematics Education, Gyeongsang National University, Jinju 52828, Korea; skywine@gmail.com
[2] Mathematics & Science Department, University of New Mexico, 705 Gurley Ave., Gallup, NM 87301, USA; fsmarandache@gmail.com
[3] Department of Mathematics, Jeju National University, Jeju 63243, Korea
[4] Department of Mathematics, COMSATS Institute of Information Technology, Abbottabad 45550, Pakistan; madadmath@yahoo.com
* Correspondence: szsong@jejunu.ac.kr

Received: 30 October 2017; Accepted: 13 January 2018; Published: 15 January 2018

**Abstract:** The notion of a neutrosophic positive implicative $\mathcal{N}$-ideal in $BCK$-algebras is introduced, and several properties are investigated. Relations between a neutrosophic $\mathcal{N}$-ideal and a neutrosophic positive implicative $\mathcal{N}$-ideal are discussed. Characterizations of a neutrosophic positive implicative $\mathcal{N}$-ideal are considered. Conditions for a neutrosophic $\mathcal{N}$-ideal to be a neutrosophic positive implicative $\mathcal{N}$-ideal are provided. An extension property of a neutrosophic positive implicative $\mathcal{N}$-ideal based on the negative indeterminacy membership function is discussed.

**Keywords:** neutrosophic $\mathcal{N}$-structure; neutrosophic $\mathcal{N}$-ideal; neutrosophic positive implicative $\mathcal{N}$-ideal

**MSC:** 06F35; 03G25; 03B52

## 1. Introduction

There are many real-life problems which are beyond a single expert. It is because of the need to involve a wide domain of knowledge. As a generalization of the intuitionistic fuzzy set, paraconsistent set and intuitionistic set, the neutrosophic logic and set is introduced by F. Smarandache [1] and it is a useful tool to deal with uncertainty in several social and natural aspects. Neutrosophy provides a foundation for a whole family of new mathematical theories with the generalization of both classical and fuzzy counterparts. In a neutrosophic set, an element has three associated defining functions such as truth membership function ($T$), indeterminate membership function ($I$) and false membership function ($F$) defined on a universe of discourse $X$. These three functions are independent completely. The neutrosophic set has vast applications in various fields (see [2–6]).

In order to provide mathematical tool for dealing with negative information, Y. B. Jun, K. J. Lee and S. Z. Song [7] introduced the notion of negative-valued function, and constructed $\mathcal{N}$-structures. M. Khan, S. Anis, F. Smarandache and Y. B. Jun [8] introduced the notion of neutrosophic $\mathcal{N}$-structures, and it is applied to semigroups (see [8]) and $BCK/BCI$-algebras (see [9]). S. Z. Song, F. Smarandache and Y. B. Jun [10] studied a neutrosophic commutative $\mathcal{N}$-ideal in $BCK$-algebras. As well-known, $BCK$-algebras originated from two different ways: one of them is based on set theory, and another is from classical and non-classical propositional calculi (see [11]). The bounded commutative BCK-algebras are precisely MV-algebras. For MV-algebras, see [12]. The background of this study is displayed in the second section. In the third section, we introduce the notion of a neutrosophic positive implicative $\mathcal{N}$-ideal in $BCK$-algebras, and investigate several properties. We discuss relations between a neutrosophic $\mathcal{N}$-ideal and a neutrosophic positive implicative $\mathcal{N}$-ideal, and provide conditions for a neutrosophic $\mathcal{N}$-ideal to be a neutrosophic positive implicative $\mathcal{N}$-ideal. We consider characterizations of a neutrosophic positive implicative $\mathcal{N}$-ideal. We establish an extension property of a neutrosophic positive implicative $\mathcal{N}$-ideal based on the negative indeterminacy membership function. Conclusions are provided in the final section.

## 2. Preliminaries

By a *BCI-algebra* we mean a set $X$ with a binary operation "$*$" and a special element "0" in which the following conditions are satisfied:

(I) $((x * y) * (x * z)) * (z * y) = 0$,
(II) $(x * (x * y)) * y = 0$,
(III) $x * x = 0$,
(IV) $x * y = y * x = 0 \Rightarrow x = y$

for all $x, y, z \in X$. By a *BCK*-algebra, we mean a *BCI*-algebra $X$ satisfying the condition

$$(\forall x \in X)(0 * x = 0).$$

A partial ordering $\preceq$ on $X$ is defined by

$$(\forall x, y \in X)\,(x \preceq y \Rightarrow x * y = 0).$$

Every $BCK/BCI$-algebra $X$ verifies the following properties.

$$(\forall x \in X)\ (x * 0 = x), \tag{1}$$
$$(\forall x, y, z \in X)\ ((x * y) * z = (x * z) * y). \tag{2}$$

Let $I$ be a subset of a $BCK/BCI$-algebra. Then $I$ is called an *ideal* of $X$ if it satisfies the following conditions.

$$0 \in I, \tag{3}$$
$$(\forall x, y \in X)\,(x * y \in I,\ y \in I \Rightarrow x \in I). \tag{4}$$

Let $I$ be a subset of a $BCK$-algebra. Then $I$ is called a *positive implicative ideal* of $X$ if the Condition (3) holds and the following assertion is valid.

$$(\forall x, y, z \in X)\,((x * y) * z \in I,\ y * z \in I \Rightarrow x * z \in I). \tag{5}$$

Any positive implicative ideal is an ideal, but the converse is not true (see [13]).

**Lemma 1** ([13]). *A subset $I$ of a BCK-algebra $X$ is a positive implicative ideal of $X$ if and only if $I$ is an ideal of $X$ which satisfies the following condition.*

$$(\forall x, y \in X)\,((x * y) * y \in I \Rightarrow x * y \in I). \tag{6}$$

We refer the reader to the books [13,14] for further information regarding BCK/BCI-algebras.
For any family $\{a_i \mid i \in \Lambda\}$ of real numbers, we define

$$\bigvee\{a_i \mid i \in \Lambda\} := \sup\{a_i \mid i \in \Lambda\}$$

and

$$\bigwedge\{a_i \mid i \in \Lambda\} := \inf\{a_i \mid i \in \Lambda\}.$$

We denote the collection of functions from a set X to $[-1,0]$ by $\mathcal{F}(X,[-1,0])$. An element of $\mathcal{F}(X,[-1,0])$ is called a *negative-valued function* from X to $[-1,0]$ (briefly, $\mathcal{N}$*-function* on X). An ordered pair $(X,f)$ of X and an $\mathcal{N}$-function $f$ on X is called an $\mathcal{N}$*-structure* (see [7]).

A *neutrosophic* $\mathcal{N}$*-structure* over a nonempty universe of discourse X (see [8]) is defined to be the structure

$$X_{\mathbf{N}} := \left\{ \frac{x}{(T_N(x), I_N(x), F_N(x))} \mid x \in X \right\} \tag{7}$$

where $T_N$, $I_N$ and $F_N$ are $\mathcal{N}$-functions on X which are called the *negative truth membership function*, the *negative indeterminacy membership function* and the *negative falsity membership function*, respectively, on X.

For the sake of simplicity, we will use the notation $X_{\mathbf{N}}$ or $X_{\mathbf{N}} := \frac{X}{(T_N, I_N, F_N)}$ instead of the neutrosophic $\mathcal{N}$-structure in (7).

Recall that every neutrosophic $\mathcal{N}$-structure $X_{\mathbf{N}}$ over X satisfies the following condition:

$$(\forall x \in X)\,(-3 \leq T_N(x) + I_N(x) + F_N(x) \leq 0).$$

## 3. Neutrosophic Positive Implicative $\mathcal{N}$-ideals

In what follows, let X denote a *BCK*-algebra unless otherwise specified.

**Definition 1** ([9])**.** *Let $X_{\mathbf{N}}$ be a neutrosophic $\mathcal{N}$-structure over X. Then $X_{\mathbf{N}}$ is called a neutrosophic $\mathcal{N}$-ideal of X if the following condition holds.*

$$(\forall x, y \in X) \begin{pmatrix} T_N(0) \leq T_N(x) \leq \bigvee\{T_N(x*y), T_N(y)\} \\ I_N(0) \geq I_N(x) \geq \bigwedge\{I_N(x*y), I_N(y)\} \\ F_N(0) \leq F_N(x) \leq \bigvee\{F_N(x*y), F_N(y)\} \end{pmatrix}. \tag{8}$$

**Definition 2.** *A neutrosophic $\mathcal{N}$-structure $X_{\mathbf{N}}$ over X is called a neutrosophic positive implicative $\mathcal{N}$-ideal of X if the following assertions are valid.*

$$(\forall x \in X)\,(T_N(0) \leq T_N(x),\ I_N(0) \geq I_N(x),\ F_N(0) \leq F_N(x)), \tag{9}$$

$$(\forall x, y, z \in X) \begin{pmatrix} T_N(x*z) \leq \bigvee\{T_N((x*y)*z), T_N(y*z)\} \\ I_N(x*z) \geq \bigwedge\{I_N((x*y)*z), I_N(y*z)\} \\ F_N(x*z) \leq \bigvee\{F_N((x*y)*z), F_N(y*z)\} \end{pmatrix}. \tag{10}$$

**Example 1.** *Let $X = \{0,1,2,3,4\}$ be a BCK-algebra with the Cayley table in Table 1.*

**Table 1.** Cayley table for the binary operation "$*$".

| * | 0 | 1 | 2 | 3 | 4 |
|---|---|---|---|---|---|
| 0 | 0 | 0 | 0 | 0 | 0 |
| 1 | 1 | 0 | 0 | 1 | 0 |
| 2 | 2 | 2 | 0 | 2 | 0 |
| 3 | 3 | 3 | 3 | 0 | 3 |
| 4 | 4 | 4 | 4 | 4 | 0 |

*Let*

$$X_{\mathbf{N}} = \left\{ \frac{0}{(-0.9,-0.2,-0.7)}, \frac{1}{(-0.7,-0.6,-0.7)}, \frac{2}{(-0.5,-0.7,-0.6)}, \frac{3}{(-0.1,-0.4,-0.4)}, \frac{4}{(-0.3,-0.8,-0.2)} \right\}$$

*be a neutrosophic $\mathcal{N}$-structure over X. Then $X_{\mathbf{N}}$ is a neutrosophic positive implicative $\mathcal{N}$-ideal of X.*

If we take $z = 0$ in (10) and use (1), then we have the following theorem.

**Theorem 1.** *Every neutrosophic positive implicative $\mathcal{N}$-ideal is a neutrosophic $\mathcal{N}$-ideal.*

The following example shows that the converse of Theorem 1 does not holds.

**Example 2.** *Let $X = \{0, a, b, c\}$ be a BCK-algebra with the Cayley table in Table 2.*

**Table 2.** Cayley table for the binary operation "$*$".

| $*$ | **0** | $a$ | $b$ | $c$ |
|---|---|---|---|---|
| 0 | 0 | 0 | 0 | 0 |
| $a$ | $a$ | 0 | 0 | $a$ |
| $b$ | $b$ | $a$ | 0 | $b$ |
| $c$ | $c$ | $c$ | $c$ | 0 |

*Let*

$$X_{\mathbf{N}} = \left\{ \frac{0}{(t_0, i_2, f_0)}, \frac{a}{(t_1, i_1, f_2)}, \frac{b}{(t_1, i_1, f_2)}, \frac{c}{(t_2, i_0, f_1)} \right\}$$

*be a a neutrosophic $\mathcal{N}$-structure over $X$ where $t_0 < t_1 < t_2$, $i_0 < i_1 < i_2$ and $f_0 < f_1 < f_2$ in $[-1, 0]$. Then $X_{\mathbf{N}}$ is a neutrosophic $\mathcal{N}$-ideal of $X$. But it is not a neutrosophic positive implicative $\mathcal{N}$-ideal of $X$ since*

$$T_N(b * a) = T_N(a) = t_1 \not\leq t_0 = \bigvee\{T_N((b * a) * a), T_N(a * a)\},$$
$$I_N(b * a) = I_N(a) = i_1 \not\geq i_2 = \bigwedge\{I_N((b * a) * a), I_N(a * a)\},$$

*or*

$$F_N(b * a) = F_N(a) = f_2 \not\leq f_0 = \bigvee\{F_N((b * a) * a), F_N(a * a)\}.$$

Given a neutrosophic $\mathcal{N}$-structure $X_{\mathbf{N}}$ over $X$ and $\alpha, \beta, \gamma \in [-1, 0]$ with $-3 \leq \alpha + \beta + \gamma \leq 0$, we define the following sets.

$$T_N^{\alpha} := \{x \in X \mid T_N(x) \leq \alpha\},$$
$$I_N^{\beta} := \{x \in X \mid I_N(x) \geq \beta\},$$
$$F_N^{\gamma} := \{x \in X \mid F_N(x) \leq \gamma\}.$$

Then we say that the set

$$X_{\mathbf{N}}(\alpha, \beta, \gamma) := \{x \in X \mid T_N(x) \leq \alpha, I_N(x) \geq \beta, F_N(x) \leq \gamma\}$$

is the *$(\alpha, \beta, \gamma)$-level set* of $X_{\mathbf{N}}$ (see [9]). Obviously, we have

$$X_{\mathbf{N}}(\alpha, \beta, \gamma) = T_N^{\alpha} \cap I_N^{\beta} \cap F_N^{\gamma}.$$

**Theorem 2.** *If $X_{\mathbf{N}}$ is a neutrosophic positive implicative $\mathcal{N}$-ideal of $X$, then $T_N^{\alpha}$, $I_N^{\beta}$ and $F_N^{\gamma}$ are positive implicative ideals of $X$ for all $\alpha, \beta, \gamma \in [-1, 0]$ with $-3 \leq \alpha + \beta + \gamma \leq 0$ whenever they are nonempty.*

**Proof.** Assume that $T_N^{\alpha}$, $I_N^{\beta}$ and $F_N^{\gamma}$ are nonempty for all $\alpha, \beta, \gamma \in [-1, 0]$ with $-3 \leq \alpha + \beta + \gamma \leq 0$. Then $x \in T_N^{\alpha}$, $y \in I_N^{\beta}$ and $z \in F_N^{\gamma}$ for some $x, y, z \in X$. Thus $T_N(0) \leq T_N(x) \leq \alpha$, $I_N(0) \geq$

$I_N(y) \geq \beta$, and $F_N(0) \leq F_N(z) \leq \gamma$, that is, $0 \in T_N^\alpha \cap I_N^\beta \cap F_N^\gamma$. Let $(x * y) * z \in T_N^\alpha$ and $y * z \in T_N^\alpha$. Then $T_N((x * y) * z) \leq \alpha$ and $T_N(y * z) \leq \alpha$, which imply that

$$T_N(x * z) \leq \bigvee\{T_N((x * y) * z), T_N(y * z)\} \leq \alpha,$$

that is, $x * z \in T_N^\alpha$. If $(a * b) * c \in I_N^\beta$ and $b * c \in I_N^\beta$, then $I_N((a * b) * c) \geq \beta$ and $I_N(b * c) \geq \beta$. Thus

$$I_N(a * c) \geq \bigwedge\{I_N((a * b) * c), I_N(b * c)\} \geq \beta,$$

and so $a * c \in I_N^\beta$. Finally, suppose that $(u * v) * w \in F_N^\gamma$ and $v * w \in F_N^\gamma$. Then $F_N((u * v) * w) \leq \gamma$ and $F_N(v * w) \leq \gamma$. Thus

$$F_N(u * w) \leq \bigvee\{F_N((u * v) * w), F_N(v * w)\} \leq \gamma,$$

that is, $u * w \in F_N^\gamma$. Therefore $T_N^\alpha$, $I_N^\beta$ and $F_N^\gamma$ are positive implicative ideals of $X$. □

**Corollary 1.** *Let $X_{\mathbf{N}}$ be a neutrosophic $\mathcal{N}$-structure over $X$ and let $\alpha, \beta, \gamma \in [-1, 0]$ be such that $-3 \leq \alpha + \beta + \gamma \leq 0$. If $X_{\mathbf{N}}$ is a neutrosophic positive implicative $\mathcal{N}$-ideal of $X$, then the nonempty $(\alpha, \beta, \gamma)$-level set of $X_{\mathbf{N}}$ is a positive implicative ideal of $X$.*

**Proof.** Straightforward. □

The following example illustrates Theorem 2.

**Example 3.** *Let $X = \{0, 1, 2, 3, 4\}$ be a BCK-algebra with the Cayley table in Table 3.*

**Table 3.** Cayley table for the binary operation "$*$".

| * | 0 | 1 | 2 | 3 | 4 |
|---|---|---|---|---|---|
| 0 | 0 | 0 | 0 | 0 | 0 |
| 1 | 1 | 0 | 1 | 1 | 0 |
| 2 | 2 | 2 | 0 | 2 | 0 |
| 3 | 3 | 3 | 3 | 0 | 0 |
| 4 | 4 | 4 | 4 | 4 | 0 |

*Let*

$$X_{\mathbf{N}} = \left\{ \frac{0}{(-0.8,-0.3,-0.7)}, \frac{1}{(-0.7,-0.6,-0.4)}, \frac{2}{(-0.4,-0.4,-0.5)}, \frac{3}{(-0.3,-0.5,-0.6)}, \frac{4}{(-0.2,-0.9,-0.1)} \right\}$$

*be a neutrosophic $\mathcal{N}$-structure over $X$. Routine calculations show that $X_{\mathbf{N}}$ is a neutrosophic positive implicative $\mathcal{N}$-ideal of $X$. Then*

$$T_N^\alpha = \begin{cases} \emptyset & \text{if } \alpha \in [-1, -0.8), \\ \{0\} & \text{if } \alpha \in [-0.8, -0.7), \\ \{0, 1\} & \text{if } \alpha \in [-0.7, -0.4), \\ \{0, 1, 2\} & \text{if } \alpha \in [-0.4, -0.3), \\ \{0, 1, 2, 3\} & \text{if } \alpha \in [-0.3, -0.2), \\ X & \text{if } \alpha \in [-0.2, 0], \end{cases}$$

$$I_N^\beta = \begin{cases} \emptyset & \text{if } \beta \in (-0.3, 0], \\ \{0\} & \text{if } \beta \in (-0.4, -0.3], \\ \{0,2\} & \text{if } \beta \in (-0.5, -0.4], \\ \{0,2,3\} & \text{if } \beta \in (-0.6, -0.5], \\ \{0,1,2,3\} & \text{if } \beta \in (-0.9, -0.6], \\ X & \text{if } \beta \in [-1, -0.9], \end{cases}$$

*and*

$$F_N^\gamma = \begin{cases} \emptyset & \text{if } \gamma \in [-1, -0.7), \\ \{0\} & \text{if } \gamma \in [-0.7, -0.6), \\ \{0,3\} & \text{if } \gamma \in [-0.6, -0.5), \\ \{0,2,3\} & \text{if } \gamma \in [-0.5, -0.4), \\ \{0,1,2,3\} & \text{if } \gamma \in [-0.4, -0.1), \\ X & \text{if } \gamma \in [-0.1, 0], \end{cases}$$

*which are positive implicative ideals of X.*

**Lemma 2** ([9])**.** *Every neutrosophic $\mathcal{N}$-ideal $X_{\mathbf{N}}$ of $X$ satisfies the following assertions:*

$$(x, y \in X)\,(x \preceq y \ \Rightarrow\ T_N(x) \le T_N(y), I_N(x) \ge I_N(y), F_N(x) \le F_N(y)). \tag{11}$$

We discuss conditions for a neutrosophic $\mathcal{N}$-ideal to be a neutrosophic positive implicative $\mathcal{N}$-ideal.

**Theorem 3.** *Let $X_{\mathbf{N}}$ be a neutrosophic $\mathcal{N}$-ideal of X. Then $X_{\mathbf{N}}$ is a neutrosophic positive implicative $\mathcal{N}$-ideal of X if and only if the following assertion is valid.*

$$(\forall x, y \in X) \begin{pmatrix} T_N(x*y) \le T_N((x*y)*y), \\ I_N(x*y) \ge I_N((x*y)*y), \\ F_N(x*y) \le F_N((x*y)*y) \end{pmatrix}. \tag{12}$$

**Proof.** Assume that $X_{\mathbf{N}}$ is a neutrosophic positive implicative $\mathcal{N}$-ideal of $X$. If $z$ is replaced by $y$ in (10), then

$$\begin{aligned} T_N(x*y) &\le \bigvee\{T_N((x*y)*y), T_N(y*y)\} \\ &= \bigvee\{T_N((x*y)*y), T_N(0)\} = T_N((x*y)*y), \end{aligned}$$

$$\begin{aligned} I_N(x*y) &\ge \bigwedge\{I_N((x*y)*y), I_N(y*y)\} \\ &= \bigwedge\{I_N((x*y)*y), I_N(0)\} = I_N((x*y)*y), \end{aligned}$$

and

$$\begin{aligned} F_N(x*y) &\le \bigvee\{F_N((x*y)*y), F_N(y*y)\} \\ &= \bigvee\{F_N((x*y)*y), F_N(0)\} = F_N((x*y)*y) \end{aligned}$$

by (III) and (9).

Conversely, let $X_{\mathbf{N}}$ be a neutrosophic $\mathcal{N}$-ideal of $X$ satisfying (12). Since

$$((x*z)*z)*(y*z) \preceq (x*z)*y = (x*y)*z$$

for all $x, y, z \in X$, we have

$$(\forall x, y, z \in X) \left( \begin{array}{l} T_N(((x * z) * z) * (y * z)) \leq T_N((x * y) * z), \\ I_N(((x * z) * z) * (y * z)) \geq I_N((x * y) * z), \\ F_N(((x * z) * z) * (y * z)) \leq F_N((x * y) * z) \end{array} \right).$$

by Lemma 2. It follows from (8) and (12) that

$$\begin{aligned} T_N(x * z) &\leq T_N((x * z) * z) \\ &\leq \bigvee\{T_N(((x * z) * z) * (y * z)), T_N(y * z)\} \\ &\leq \bigvee\{T_N((x * y) * z), T_N(y * z)\}, \end{aligned}$$

$$\begin{aligned} I_N(x * z) &\geq I_N((x * z) * z) \\ &\geq \bigwedge\{I_N(((x * z) * z) * (y * z)), I_N(y * z)\} \\ &\geq \bigwedge\{I_N((x * y) * z), I_N(y * z)\}, \end{aligned}$$

and

$$\begin{aligned} F_N(x * z) &\leq F_N((x * z) * z) \\ &\leq \bigvee\{F_N(((x * z) * z) * (y * z)), F_N(y * z)\} \\ &\leq \bigvee\{F_N((x * y) * z), F_N(y * z)\}. \end{aligned}$$

Therefore $X_{\mathbf{N}}$ is a neutrosophic positive implicative $\mathcal{N}$-ideal of $X$. □

**Lemma 3** ([9]). *For any neutrosophic $\mathcal{N}$-ideal $X_{\mathbf{N}}$ of $X$, we have*

$$(\forall x, y, z \in X) \left( x * y \preceq z \Rightarrow \left\{ \begin{array}{l} T_N(x) \leq \bigvee\{T_N(y), T_N(z)\} \\ I_N(x) \geq \bigwedge\{I_N(y), I_N(z)\} \\ F_N(x) \leq \bigvee\{F_N(y), F_N(z)\} \end{array} \right. \right). \tag{13}$$

**Lemma 4.** *If a neutrosophic $\mathcal{N}$-structure $X_{\mathbf{N}}$ over $X$ satisfies the condition (13), then $X_{\mathbf{N}}$ is a neutrosophic $\mathcal{N}$-ideal of $X$.*

**Proof.** Since $0 * x \preceq x$ for all $x \in X$, we have $T_N(0) \leq T_N(x)$, $I_N(0) \geq I_N(x)$ and $F_N(0) \leq F_N(x)$ for all $x \in X$ by (13). Note that $x * (x * y) \preceq y$ for all $x, y \in X$. It follows from (13) that $T_N(x) \leq \bigvee\{T_N(x * y), T_N(y)\}$, $I_N(x) \geq \bigwedge\{I_N(x * y), I_N(y)\}$, and $F_N(x) \leq \bigvee\{F_N(x * y), F_N(y)\}$ for all $x, y \in X$. Therefore $X_{\mathbf{N}}$ is a neutrosophic $\mathcal{N}$-ideal of $X$. □

**Theorem 4.** *For any neutrosophic $\mathcal{N}$-structure $X_{\mathbf{N}}$ over $X$, the following assertions are equivalent.*

(1) *$X_{\mathbf{N}}$ is a neutrosophic positive implicative $\mathcal{N}$-ideal of $X$.*
(2) *$X_{\mathbf{N}}$ satisfies the following condition.*

$$((x * y) * y) * a \preceq b \Rightarrow \left\{ \begin{array}{l} T_N(x * y) \leq \bigvee\{T_N(a), T_N(b)\}, \\ I_N(x * y) \geq \bigwedge\{I_N(a), I_N(b)\}, \\ F_N(x * y) \leq \bigvee\{F_N(a), F_N(b)\}, \end{array} \right. \tag{14}$$

*for all $x, y, a, b \in X$.*

**Proof.** Suppose that $X_{\mathbf{N}}$ is a neutrosophic positive implicative $\mathcal{N}$-ideal of $X$. Then $X_{\mathbf{N}}$ is a neutrosophic $\mathcal{N}$-ideal of $X$ by Theorem 1. Let $x, y, a, b \in X$ be such that $((x * y) * y) * a \preceq b$. Then

$$\begin{aligned}
T_N(x*y) &\leq T_N(((x*y)*y)) \leq \bigvee\{T_N(a), T_N(b)\},\\
I_N(x*y) &\geq I_N(((x*y)*y)) \geq \bigwedge\{I_N(a), I_N(b)\},\\
F_N(x*y) &\leq F_N(((x*y)*y)) \leq \bigvee\{F_N(a), F_N(b)\}
\end{aligned}$$

by Theorem 3 and Lemma 3.

Conversely, let $X_{\mathbf{N}}$ be a neutrosophic $\mathcal{N}$-structure over $X$ that satisfies (14). Let $x, a, b \in X$ be such that $x * a \preceq b$. Then $((x * 0) * 0) * a \preceq b$, and so

$$\begin{aligned}
T_N(x) &= T_N(x*0) \leq \bigvee\{T_N(a), T_N(b)\},\\
I_N(x) &= I_N(x*0) \geq \bigwedge\{I_N(a), I_N(b)\},\\
F_N(x) &= F_N(x*y) \leq \bigvee\{F_N(a), F_N(b)\}.
\end{aligned}$$

Hence $X_{\mathbf{N}}$ is a neutrosophic $\mathcal{N}$-ideal of $X$ by Lemma 4. Since $((x * y) * y) * ((x * y) * y) \preceq 0$, it follows from (14) and (9) that

$$\begin{aligned}
T_N(x*y) &\leq \bigvee\{T_N((x*y)*y), T_N(0)\} = T_N((x*y)*y),\\
I_N(x*y) &\geq \bigwedge\{I_N((x*y)*y), I_N(0)\} = I_N((x*y)*y),\\
F_N(x*y) &\leq \bigvee\{F_N((x*y)*y), F_N(0)\} = F_N((x*y)*y),
\end{aligned}$$

for all $x, y \in X$. Therefore $X_{\mathbf{N}}$ is a neutrosophic positive implicative $\mathcal{N}$-ideal of $X$ by Theorem 3. □

**Lemma 5** ([9]). *Let $X_{\mathbf{N}}$ be a neutrosophic $\mathcal{N}$-structure over $X$ and assume that $T_N^\alpha$, $I_N^\beta$ and $F_N^\gamma$ are ideals of $X$ for all $\alpha, \beta, \gamma \in [-1, 0]$ with $-3 \leq \alpha + \beta + \gamma \leq 0$. Then $X_{\mathbf{N}}$ is a neutrosophic $\mathcal{N}$-ideal of $X$.*

**Theorem 5.** *Let $X_{\mathbf{N}}$ be a neutrosophic $\mathcal{N}$-structure over $X$ and assume that $T_N^\alpha$, $I_N^\beta$ and $F_N^\gamma$ are positive implicative ideals of $X$ for all $\alpha, \beta, \gamma \in [-1, 0]$ with $-3 \leq \alpha + \beta + \gamma \leq 0$. Then $X_{\mathbf{N}}$ is a neutrosophic positive implicative $\mathcal{N}$-ideal of $X$.*

**Proof.** If $T_N^\alpha$, $I_N^\beta$ and $F_N^\gamma$ are positive implicative ideals of $X$, then $T_N^\alpha$, $I_N^\beta$ and $F_N^\gamma$ are ideals of $X$. Thus $X_{\mathbf{N}}$ is a neutrosophic $\mathcal{N}$-ideal of $X$ by Lemma 5. Let $x, y \in X$ and $\alpha, \beta, \gamma \in [-1, 0]$ with $-3 \leq \alpha + \beta + \gamma \leq 0$ such that $T_N((x * y) * y) = \alpha$, $I_N((x * y) * y) = \beta$ and $F_N((x * y) * y) = \gamma$. Then $(x * y) * y \in T_N^\alpha \cap I_N^\beta \cap F_N^\gamma$. Since $T_N^\alpha \cap I_N^\beta \cap F_N^\gamma$ is a positive implicative ideal of $X$, it follows from Lemma 1 that $x * y \in T_N^\alpha \cap I_N^\beta \cap F_N^\gamma$. Hence

$$\begin{aligned}
T_N(x*y) &\leq \alpha = T_N((x*y)*y),\\
I_N(x*y) &\geq \beta = I_N((x*y)*y),\\
F_N(x*y) &\leq \gamma = F_N((x*y)*y).
\end{aligned}$$

Therefore $X_{\mathbf{N}}$ is a neutrosophic positive implicative $\mathcal{N}$-ideal of $X$ by Theorem 3. □

**Lemma 6** ([9]). *Let $X_{\mathbf{N}}$ be a neutrosophic $\mathcal{N}$-ideal of X. Then $X_{\mathbf{N}}$ satisfies the condition (12) if and only if it satisfies the following condition.*

$$(\forall x, y, z \in X)\left(\begin{array}{l} T_N((x*z)*(y*z)) \leq T_N((x*y)*z), \\ I_N((x*z)*(y*z)) \geq I_N((x*y)*z), \\ F_N((x*z)*(y*z)) \leq F_N((x*y)*z) \end{array}\right). \tag{15}$$

**Corollary 2.** *Let $X_{\mathbf{N}}$ be a neutrosophic $\mathcal{N}$-ideal of X. Then $X_{\mathbf{N}}$ is a neutrosophic positive implicative $\mathcal{N}$-ideal of X if and only if $X_{\mathbf{N}}$ satisfies (15).*

**Proof.** It follows from Theorem 3 and Lemma 6. □

**Theorem 6.** *For any neutrosophic $\mathcal{N}$-structure $X_{\mathbf{N}}$ over X, the following assertions are equivalent.*

(1) *$X_{\mathbf{N}}$ is a neutrosophic positive implicative $\mathcal{N}$-ideal of X.*
(2) *$X_{\mathbf{N}}$ satisfies the following condition.*

$$((x*y)*z)*a \preceq b \Rightarrow \left\{\begin{array}{l} T_N((x*z)*(y*z)) \leq \bigvee\{T_N(a), T_N(b)\}, \\ I_N((x*z)*(y*z)) \geq \bigwedge\{I_N(a), I_N(b)\}, \\ F_N((x*z)*(y*z)) \leq \bigvee\{F_N(a), F_N(b)\}, \end{array}\right. \tag{16}$$

*for all $x, y, z, a, b \in X$.*

**Proof.** Suppose that $X_{\mathbf{N}}$ is a neutrosophic positive implicative $\mathcal{N}$-ideal of X. Then $X_{\mathbf{N}}$ is a neutrosophic $\mathcal{N}$-ideal of X by Theorem 1. Let $x, y, z, a, b \in X$ be such that $((x*y)*z)*a \preceq b$. Using Corollary 2 and Lemma 3, we have

$$\begin{aligned} T_N((x*z)*(y*z)) &\leq T_N(((x*y)*z)) \leq \bigvee\{T_N(a), T_N(b)\}, \\ I_N((x*z)*(y*z)) &\geq I_N(((x*y)*z)) \geq \bigwedge\{I_N(a), I_N(b)\}, \\ F_N((x*z)*(y*z)) &\leq F_N(((x*y)*z)) \leq \bigvee\{F_N(a), F_N(b)\} \end{aligned}$$

for all $x, y, z, a, b \in X$.

Conversely, let $X_{\mathbf{N}}$ be a neutrosophic $\mathcal{N}$-structure over X that satisfies (16). Let $x, y, a, b \in X$ be such that $((x*y)*y)*a \preceq b$. Then

$$\begin{aligned} T_N(x*y) &= T_N((x*y)*(y*y)) \leq \bigvee\{T_N(a), T_N(b)\}, \\ I_N(x*y) &= I_N((x*y)*(y*y)) \geq \bigwedge\{I_N(a), I_N(b)\}, \\ F_N(x*y) &= F_N((x*y)*(y*y)) \leq \bigvee\{F_N(a), F_N(b)\} \end{aligned}$$

by (III), (1) and (16). It follows from Theorem 4 that $X_{\mathbf{N}}$ is a neutrosophic positive implicative $\mathcal{N}$-ideal of X. □

**Theorem 7.** *Let $X_{\mathbf{N}}$ be a neutrosophic $\mathcal{N}$-structure over X. Then $X_{\mathbf{N}}$ is a neutrosophic positive implicative $\mathcal{N}$-ideal of X if and only if $X_{\mathbf{N}}$ satisfies (9) and*

$$(\forall x, y, z \in X)\left(\begin{array}{l} T_N(x*y) \leq \bigvee\{T_N(((x*y)*y)*z), T_N(z)\}, \\ I_N(x*y) \geq \bigwedge\{I_N(((x*y)*y)*z), I_N(z)\}, \\ F_N(x*y) \leq \bigvee\{F_N(((x*y)*y)*z), F_N(z)\} \end{array}\right). \tag{17}$$

**Proof.** Assume that $X_{\mathbf{N}}$ is a neutrosophic positive implicative $\mathcal{N}$-ideal of $X$. Then $X_{\mathbf{N}}$ is a neutrosophic $\mathcal{N}$-ideal of $X$ by Theorem 1, and so the condition (9) is valid. Using (8), (III), (1), (2) and (15), we have

$$\begin{aligned} T_N(x*y) &\leq \bigvee\{T_N((x*y)*z), T_N(z)\} \\ &= \bigvee\{T_N(((x*z)*y)*(y*y)), T_N(z)\} \\ &\leq \bigvee\{T_N(((x*z)*y)*y), T_N(z)\} \\ &= \bigvee\{T_N(((x*y)*y)*z), T_N(z)\}, \end{aligned}$$

$$\begin{aligned} I_N(x*y) &\geq \bigwedge\{I_N((x*y)*z), I_N(z)\} \\ &= \bigwedge\{I_N(((x*z)*y)*(y*y)), I_N(z)\} \\ &\geq \bigwedge\{I_N(((x*z)*y)*y), I_N(z)\} \\ &= \bigwedge\{I_N(((x*y)*y)*z), I_N(z)\}, \end{aligned}$$

and

$$\begin{aligned} F_N(x*y) &\leq \bigvee\{F_N((x*y)*z), F_N(z)\} \\ &= \bigvee\{F_N(((x*z)*y)*(y*y)), F_N(z)\} \\ &\leq \bigvee\{F_N(((x*z)*y)*y), F_N(z)\} \\ &= \bigvee\{F_N(((x*y)*y)*z), F_N(z)\} \end{aligned}$$

for all $x, y, z \in X$. Therefore (17) is valid.

Conversely, if $X_{\mathbf{N}}$ is a neutrosophic $\mathcal{N}$-structure over $X$ satisfying two Conditions (9) and (17), then

$$\begin{aligned} T_N(x) &= T_N(x*0) \leq \bigvee\{T_N(((x*0)*0)*z), T_N(z)\} = \bigvee\{T_N(x*z), T_N(z)\}, \\ I_N(x) &= I_N(x*0) \geq \bigwedge\{I_N(((x*0)*0)*z), I_N(z)\} = \bigwedge\{I_N(x*z), I_N(z)\}, \\ F_N(x) &= F_N(x*0) \leq \bigvee\{F_N(((x*0)*0)*z), F_N(z)\} = \bigvee\{F_N(x*z), F_N(z)\} \end{aligned}$$

for all $x, z \in X$. Hence $X_{\mathbf{N}}$ is a neutrosophic $\mathcal{N}$-ideal of $X$. Now, if we take $z = 0$ in (17) and use (1), then

$$\begin{aligned} T_N(x*y) &\leq \bigvee\{T_N(((x*y)*y)*0), T_N(0)\} \\ &= \bigvee\{T_N((x*y)*y), T_N(0)\} = T_N((x*y)*y), \end{aligned}$$

$$\begin{aligned} I_N(x*y) &\geq \bigwedge\{I_N(((x*y)*y)*0), I_N(0)\} \\ &= \bigwedge\{I_N((x*y)*y), I_N(0)\} = I_N((x*y)*y), \end{aligned}$$

and

$$\begin{aligned} F_N(x*y) &\leq \bigvee\{F_N(((x*y)*y)*0), F_N(0)\} \\ &= \bigvee\{F_N((x*y)*y), F_N(0)\} = F_N((x*y)*y) \end{aligned}$$

for all $x, y \in X$. It follows from Theorem 3 that $X_{\mathbf{N}}$ is a neutrosophic positive implicative $\mathcal{N}$-ideal of $X$. □

Summarizing the above results, we have a characterization of a neutrosophic positive implicative $\mathcal{N}$-ideal.

**Theorem 8.** *For a neutrosophic $\mathcal{N}$-structure $X_{\mathbf{N}}$ over $X$, the following assertions are equivalent.*

(1) $X_{\mathbf{N}}$ *is a neutrosophic positive implicative $\mathcal{N}$-ideal of $X$.*
(2) $X_{\mathbf{N}}$ *is a neutrosophic $\mathcal{N}$-ideal of $X$ satisfying the condition (12).*
(3) $X_{\mathbf{N}}$ *is a neutrosophic $\mathcal{N}$-ideal of $X$ satisfying the condition (15).*
(4) $X_{\mathbf{N}}$ *satisfies two conditions (9) and (17).*
(5) $X_{\mathbf{N}}$ *satisfies the condition (14).*
(6) $X_{\mathbf{N}}$ *satisfies the condition (3).*

For any fixed numbers $\xi_T, \xi_F \in [-1,0), \xi_I \in (-1,0]$ and a nonempty subset $G$ of $X$, a neutrosophic $\mathcal{N}$-structure $X_{\mathbf{N}}^G$ over $X$ is defined to be the structure

$$X_{\mathbf{N}}^G := \frac{X}{(T_N^G, I_N^G, F_N^G)} = \left\{ \frac{x}{(T_N^G(x), I_N^G(x), F_N^G(x))} \mid x \in X \right\} \tag{18}$$

where $T_N^G$, $I_N^G$ and $F_N^G$ are $\mathcal{N}$-functions on $X$ which are given as follows:

$$T_N^G : X \to [-1,0],\ x \mapsto \begin{cases} \xi_T & \text{if } x \in G, \\ 0 & \text{otherwise,} \end{cases}$$

$$I_N^G : X \to [-1,0],\ x \mapsto \begin{cases} \xi_I & \text{if } x \in G, \\ -1 & \text{otherwise,} \end{cases}$$

and

$$F_N^G : X \to [-1,0],\ x \mapsto \begin{cases} \xi_F & \text{if } x \in G, \\ 0 & \text{otherwise.} \end{cases}$$

**Theorem 9.** *Given a nonempty subset $G$ of $X$, a neutrosophic $\mathcal{N}$-structure $X_{\mathbf{N}}^G$ over $X$ is a neutrosophic positive implicative $\mathcal{N}$-ideal of $X$ if and only if $G$ is a positive implicative ideal of $X$.*

**Proof.** Assume that $G$ is a positive implicative ideal of $X$. Since $0 \in G$, it follows that $T_N^G(0) = \xi_T \leq T_N^G(x)$, $I_N^G(0) = \xi_I \geq I_N^G(x)$, and $F_N^G(0) = \xi_F \leq F_N^G(x)$ for all $x \in X$. For any $x, y, z \in X$, we consider four cases:

Case 1. $(x * y) * z \in G$ and $y * z \in G$,
Case 2. $(x * y) * z \in G$ and $y * z \notin G$,
Case 3. $(x * y) * z \notin G$ and $y * z \in G$,
Case 4. $(x * y) * z \notin G$ and $y * z \notin G$.

Case 1 implies that $x * z \in G$, and thus

$$\begin{aligned} T_N^G(x * z) &= T_N^G((x * y) * z) = T_N^G(y * z) = \xi_T, \\ I_N^G(x * z) &= I_N^G((x * y) * z) = I_N^G(y * z) = \xi_I, \\ F_N^G(x * z) &= F_N^G((x * y) * z) = F_N^G(y * z) = \xi_F. \end{aligned}$$

Hence

$$\begin{aligned} T_N^G(x * z) &\leq \bigvee\{T_N^G((x * y) * z), T_N^G(y * z)\}, \\ I_N^G(x * z) &\geq \bigwedge\{I_N^G((x * y) * z), I_N^G(y * z)\}, \\ F_N^G(x * z) &\leq \bigvee\{F_N^G((x * y) * z), F_N^G(y * z)\}. \end{aligned}$$

If Case 2 is valid, then $T_N^G(y*z) = 0$, $I_N^G(y*z) = -1$ and $F_N^G(y*z) = 0$. Thus

$$\begin{aligned} T_N^G(x*z) &\leq 0 = \bigvee\{T_N^G((x*y)*z), T_N^G(y*z)\}, \\ I_N^G(x*z) &\geq -1 = \bigwedge\{I_N^G((x*y)*z), I_N^G(y*z)\}, \\ F_N^G(x*z) &\leq 0 = \bigvee\{F_N^G((x*y)*z), F_N^G(y*z)\}. \end{aligned}$$

For the Case 3, it is similar to the Case 2.
For the Case 4, it is clear that

$$\begin{aligned} T_N^G(x*z) &\leq \bigvee\{T_N^G((x*y)*z), T_N^G(y*z)\}, \\ I_N^G(x*z) &\geq \bigwedge\{I_N^G((x*y)*z), I_N^G(y*z)\}, \\ F_N^G(x*z) &\leq \bigvee\{F_N^G((x*y)*z), F_N^G(y*z)\}. \end{aligned}$$

Therefore $X_{\mathbf{N}}^G$ is a neutrosophic positive implicative $\mathcal{N}$-ideal of $X$.

Conversely, suppose that $X_{\mathbf{N}}^G$ is a neutrosophic positive implicative $\mathcal{N}$-ideal of $X$. Then $(T_N^G)^{\frac{\xi_T}{2}} = G$, $(I_N^G)^{\frac{\xi_I}{2}} = G$ and $(F_N^G)^{\frac{\xi_F}{2}} = G$ are positive implicative ideals of $X$ by Theorem 2. □

We consider an extension property of a neutrosophic positive implicative $\mathcal{N}$-ideal based on the negative indeterminacy membership function.

**Lemma 7** ([13]). *Let $A$ and $B$ be ideals of $X$ such that $A \subseteq B$. If $A$ is a positive implicative ideal of $X$, then so is $B$.*

**Theorem 10.** *Let*

$$X_{\mathbf{N}} := \frac{X}{(T_N, I_N, F_N)} = \left\{ \frac{x}{(T_N(x), I_N(x), F_N(x))} \mid x \in X \right\}$$

*and*

$$X_{\mathbf{M}} := \frac{X}{(T_M, I_M, F_M)} = \left\{ \frac{x}{(T_M(x), I_M(x), F_M(x))} \mid x \in X \right\}$$

*be neutrosophic $\mathcal{N}$-ideals of $X$ such that $X_{\mathbf{N}}(=,\leq,=)X_{\mathbf{M}}$, that is, $T_N(x) = T_M(x)$, $I_N(x) \leq I_M(x)$ and $F_N(x) = F_M(x)$ for all $x \in X$. If $X_{\mathbf{N}}$ is a neutrosophic positive implicative $\mathcal{N}$-ideal of $X$, then so is $X_{\mathbf{M}}$.*

**Proof.** Assume that $X_{\mathbf{N}}$ is a neutrosophic positive implicative $\mathcal{N}$-ideal of X. Then $T_N^\alpha$, $I_N^\beta$ and $F_N^\gamma$ are positive implicative ideals of $X$ for all $\alpha, \beta, \gamma \in [-1, 0]$ by Theorem 2. The condition $X_{\mathbf{N}}(=,\leq,=)X_{\mathbf{M}}$ implies that $T_N^{\xi_T} = T_M^{\xi_T}$, $I_N^{\xi_I} \subseteq I_M^{\xi_I}$ and $F_N^{\xi_F} = F_M^{\xi_F}$. It follows from Lemma 7 that $T_M^\alpha$, $I_M^\beta$ and $F_M^\gamma$ are positive implicative ideals of $X$ for all $\alpha, \beta, \gamma \in [-1, 0]$. Therefore $X_{\mathbf{M}}$ is a neutrosophic positive implicative $\mathcal{N}$-ideal of X by Theorem 5. □

## 4. Conclusions

The aim of this paper is to study neutrosophic $\mathcal{N}$-structure of positive implicative ideal in $BCK$-algebras, and to provide a mathematical tool for dealing with several informations containing uncertainty, for example, decision making problem, medical diagnosis, graph theory, pattern recognition, etc. As a more general platform which extends the concepts of the classic set and fuzzy set, intuitionistic fuzzy set and interval valued intuitionistic fuzzy set, F. Smarandache have developed neutrosophic set (NS) in [1,15]. In this manuscript, we have discussed the notion of a neutrosophic positive implicative $\mathcal{N}$-ideal in $BCK$-algebras, and investigated several properties. We have considered relations between a neutrosophic $\mathcal{N}$-ideal and a neutrosophic positive implicative $\mathcal{N}$-ideal. We have provided conditions for a neutrosophic $\mathcal{N}$-ideal to be a neutrosophic positive implicative $\mathcal{N}$-ideal, and considered characterizations of a neutrosophic positive implicative $\mathcal{N}$-ideal. We have established an extension property of a neutrosophic positive implicative $\mathcal{N}$-ideal based on the negative indeterminacy membership function.

Various sources of uncertainty can be a challenge to make a reliable decision. Based on the results in this paper, our future research will be focused to solve real-life problems under the opinions of experts in a neutrosophic set environment, for example, decision making problem, medical diagnosis etc. The future works also may use the study neutrosophic set theory on several related algebraic structures, $BL$-algebras, $MTL$-algebras, $R_0$-algebras, $MV$-algebras, $EQ$-algebras and lattice implication algebras etc.

**Acknowledgements:** The authors thank the academic editor for his valuable comments and suggestions and the anonymous reviewers for their valuable suggestions. The corresponding author, Seok-Zun Song, was supported by Basic Science Research Program through the National Research Foundation of Korea (NRF) funded by the Ministry of Education (No. 2016R1D1A1B02006812).

**Author Contributions:** This paper is a result of common work of the authors in all aspects.

**Conflicts of Interest:** The authors declare no conflict of interest

## References

1. Smarandache, F. *A Unifying Field in Logics: Neutrosophic Logic. Neutrosophy, Neutrosophic Set, Neutrosophic Probability*; American Reserch Press: Rehoboth, NM, USA, 1999.
2. Broumi, S.; Smarandache, F. Correlation coefficient of interval neutrosophic sets. *Appl. Mech. Mater.* **2013**, *436*, 511–517.
3. Cheng, H.D.; Guo, Y. A new neutrosophic approach to image thresholding. *New Math. Nat. Comput.* **2008**, *4*, 291–308.
4. Guo, Y.; Cheng, H.D. New nutrosophic approach to image segmentation. *Pat. Recognit.* **2009**, *42*, 587–595.
5. Kharal, A. A neutrosophic multicriteria decision making method. *New Math. Nat. Comput.* **2014**, *10*, 143–162.
6. Ye, J. Similarity measures between interval neutrosophic sets and their multicriteria decision-making method. *J. Intell. Fuzzy Syst.* **2014**, *26*, 165–172.
7. Jun, Y.B.; Lee, K.J.; Song, S.Z. $\mathcal{N}$-ideals of BCK/BCI-algebras. *J. Chungcheong Math. Soc.* **2009**, *22*, 417–437.
8. Khan, M.; Anis, S.; Smarandache, F.; Jun, Y.B. Neutrosophic $\mathcal{N}$-structures and their applications in semigroups. *Ann. Fuzzy Math. Inform.* **2017**, *14*, 583–598.
9. Jun, Y.B.; Smarandache, F.; Bordbar, H. Neutrosophic $\mathcal{N}$-structures applied to $BCK/BCI$-algebras. *Information* **2017**, *8*, 128.
10. Song, S.Z.; Smarandache, F.; Jun, Y.B. Neutrosophic commutative $\mathcal{N}$-ideals in $BCK$-algebras. *Information* **2017**, *8*, 130.
11. Hong, S.M.; Jun, Y.B.; Ozturk, M.A. Generalization of $BCK$-algebras. *Sci. Math. Jpn.* **2003**, *58*, 603–611.
12. Oner, T.; Senturk, I.; Oner, G. An independent set of axioms of MV-algebras and solutions of the Set-Theoretical Yang-Baxter equation. *Axioms* **2017**, *6*, 17
13. Meng, J.; Jun, Y.B. *BCK-Algebras*; Kyungmoon Sa Co.: Seoul, Korea, 1994.
14. Huang, Y.S. *BCI-Algebra*; Science Press: Beijing, China, 2006.
15. Smarandache, F. Neutrosophic set-a generalization of the intuitionistic fuzzy set. *Int. J. Pure Appl. Math.* **2005**, *24*, 287–297.

axioms

MDPI

*Article*

# Neutrosophic Hough Transform

**Ümit Budak [1], Yanhui Guo [2,*], Abdulkadir Şengür [3] and Florentin Smarandache [4]**

1 Department of Electrical-Electronics Engineering, Engineering Faculty, Bitlis Eren University, 13000 Bitlis, Turkey; umtbudak@gmail.com
2 Department of Computer Science, University of Illinois at Springfield, One University Plaza, Springfield, IL 62703, USA
3 Department of Electrical and Electronics Engineering, Technology Faculty, Firat University, Elazig 23119, Turkey; ksengur@gmail.com
4 Mathematics & Science Department, University of New Mexico, 705 Gurley Ave., Gallup, NM 87301, USA; fsmarandache@gmail.com
* Correspondence: yguo56@uis.edu; Tel.: +1-217-2068-170

Received: 22 November 2017; Accepted: 14 December 2017; Published: 18 December 2017

**Abstract:** Hough transform (HT) is a useful tool for both pattern recognition and image processing communities. In the view of pattern recognition, it can extract unique features for description of various shapes, such as lines, circles, ellipses, and etc. In the view of image processing, a dozen of applications can be handled with HT, such as lane detection for autonomous cars, blood cell detection in microscope images, and so on. As HT is a straight forward shape detector in a given image, its shape detection ability is low in noisy images. To alleviate its weakness on noisy images and improve its shape detection performance, in this paper, we proposed neutrosophic Hough transform (NHT). As it was proved earlier, neutrosophy theory based image processing applications were successful in noisy environments. To this end, the Hough space is initially transferred into the NS domain by calculating the NS membership triples (T, I, and F). An indeterminacy filtering is constructed where the neighborhood information is used in order to remove the indeterminacy in the spatial neighborhood of neutrosophic Hough space. The potential peaks are detected based on thresholding on the neutrosophic Hough space, and these peak locations are then used to detect the lines in the image domain. Extensive experiments on noisy and noise-free images are performed in order to show the efficiency of the proposed NHT algorithm. We also compared our proposed NHT with traditional HT and fuzzy HT methods on variety of images. The obtained results showed the efficiency of the proposed NHT on noisy images.

**Keywords:** Hough transform; fuzzy Hough transform; neutrosophy theory; line detection

## 1. Introduction

The Hough transform (HT), which is known as a popular pattern descriptor, was proposed in sixties by Paul Hough [1]. This popular and efficient shape based pattern descriptor was then introduced to the image processing and computer vision community in seventies by Duda and Hart [2]. The HT converts image domain features (e.g., edges) into another parameter space, namely Hough space. In Hough space, every points $(x_i, y_i)$ on a straight line in the image domain, corresponds a point $(\rho_i, \theta_i)$.

In the literature, there have been so many variations of Hough's proposal [3–14]. In [3], Chung et al. proposed a new HT based on affine transformation. Memory-usage efficiency was provided by proposed affine transformation, which was in the center of Chung's method. Authors targeted to detect the lines by using slope intercept parameters in the Hough space. In [4], Guo et al. proposed a methodology for an efficient HT by utilizing surround-suppression. Authors introduced a measure of isotropic surround-suppression, which suppresses the false peaks caused by the texture regions in Hough space. In traditional HT, image features are voted in the parameter space in order to construct the Hough space. Thus, if an image has more features (e.g., texture), construction of the Hough space becomes computationally expensive. In [5], Walsh et al. proposed the probabilistic HT. Probabilistic HT was proposed in order to reduce the computation-cost of the traditional HT by using a selection step. The selection step in probabilistic HT approach just considered the image features that contributed the Hough space. In [6], Xu et al. proposed another probabilistic HT algorithm, called randomized HT. Authors used a random selection mechanism to select a number of image domain features in order to construct the Hough space parameters. In other words, only the hypothesized parameter set was used to increment the accumulator. In [7], Han et al. used the fuzzy concept in HT (FHT) for detecting the shapes in noisy images. FHT enables detecting shapes in noisy environment by approximately fitting the image features avoiding the spurious detected using the traditional HT. The FTH can be obtained by one-dimensional (1-D) convolution of the rows of the Hough space. In [8], Montseny et al. proposed a new FHT method where edge orientations were considered. Gradient vectors were considered to enable edge orientations in FHT. Based on the stability properties of the gradient vectors, some relevant orientations were taken into consideration, which reduced the computation burden, dramatically. In [9], Mathavan et al. proposed an algorithm to detect cracks in pavement images based on FHT. Authors indicated that due to the intensity variations and texture content of the pavement images, FHT was quite convenient to detect the short lines (cracks). In [10], Chatzis et al. proposed a variation of HT. In the proposed method, authors first split the Hough space into fuzzy cells, which were defined as fuzzy numbers. Then, a fuzzy voting procedure was adopted to construct the Hough domain parameters. In [11], Suetake et al. proposed a generalized FHT (GFHT) for arbitrary shape detection in contour images. The GFHT was derived by fuzzifying the vote process in the HT. In [12], Chung et al. proposed another orientation based HT for real world applications. In the proposed method, authors filtered out those inappropriate edge pixels before performing their HT based methodology. In [13], Cheng et al. proposed an eliminating- particle-swarm-optimization (EPSO) algorithm to reduce the computation cost of HT. In [14], Zhu et al. proposed a new HT for reducing the computation complexity of the traditional HT. The authors called their new method as Probabilistic Convergent Hough Transform (PCHT). To enable the HT to detect an arbitrary object, the Generalized Hough Transform (GHT) is the modification of the HT using the idea of template matching [15]. The problem of finding the object is solved by finding the model's position, and the transformation's parameter in GHT.

The theory of neutrosophy (NS) was introduced by Smarandache as a new branch of philosophy [16,17]. Different from fuzzy logic, in NS, every event has not only a certain degree of truth, but also a falsity degree and an indeterminacy degree that have to be considered independently from each other [18]. Thus, an event or entity {A} is considered with its opposite {Anti-A} and the neutrality {Neut-A}. In this paper, we proposed a novel HT algorithm, namely NHT, to improve the line detection ability of the HT in noisy environments. Specifically, the neutrosophic theory was adopted to improve the noise consistency of the HT. To this end, the Hough space is initially transferred into the NS domain by calculating the NS memberships. A filtering mechanism is adopted that is based on the indeterminacy membership of the NS domain. This filter considers the neighborhood information in order to remove the indeterminacy in the spatial neighborhood of Hough space. A peak search algorithm on the Hough space is then employed to detect the potential lines in the image domain. Extensive experiments on noisy and noise-free images are performed in order to show the efficiency of the proposed NHT algorithm. We further compare our proposed NHT with traditional HT and FHT methods on a variety of images. The obtained results show the efficiency of the proposed NHT on both noisy and noise-free images.

The rest of the paper is structured as follows: Section 2 briefly reviews the Hough transform and fuzzy Hough transform. Section 3 describes the proposed method based on neutrosophic domain images and indeterminacy filtering. Section 4 describes the extensive experimental works and results, and conclusions are drawn in Section 5.

## 2. Previous Works

In the following sub-sections, we briefly re-visit the theories of HT and FHT. The interested readers may refer to the related references for more details about the methodologies.

### 2.1. Hough Transform

As it was mentioned in the introduction section, HT is a popular feature extractor for image based pattern recognition applications. In other definitions, HT constructs a specific parameter space, called Hough space, and then uses it to detect the arbitrary shapes in a given image. The construction of the Hough space is handled by a voting procedure. Let us consider a line given in an image domain as;

$$y = kx + b \tag{1}$$

where $k$ is the slope and $b$ is the intercept. Figure 1 depicts such a straight line with $\rho$ and $\theta$ in image domain. In Equation (1), if we replace $(k, b)$ with $(\rho, \theta)$, then we then obtain the following equation;

$$y = \left(-\frac{cos\theta}{sin\theta}\right)x + \frac{\rho}{sin\theta} \tag{2}$$

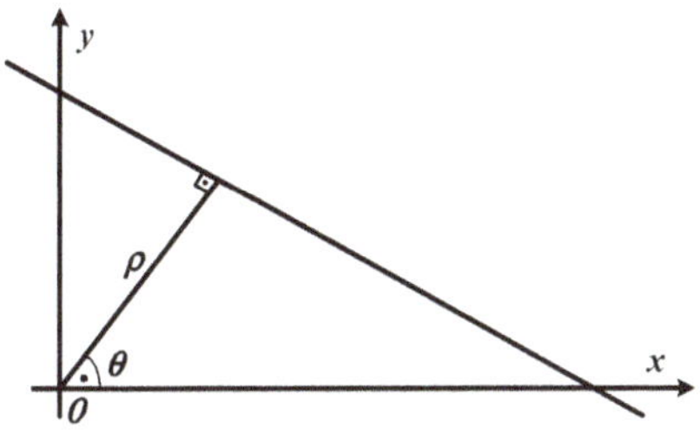


**Figure 1.** A straight line in an image domain.

If we re-write the Equation (2), as shown in Equation (3), the image domain data can be represented in the Hough space parameters.

$$\rho = xcos\theta + ysin\theta \tag{3}$$

Thus, a point in the Hough space can specify a straight line in the image domain. After Hough transform, an image $I$ in the image domain is transferred into Hough space, which is denoted as $I_{\mathrm{HT}}$.

### 2.2. Fuzzy Hough Transform

Introducing fuzzy concept into HT is arisen due to the lines that are not straight in the image domain [19]. In other words, traditional HT only considers the pixels that align on a straight line and non-straight lines detection with HT becomes challenging. Non-straight lines generally occur due to the noise or discrepancies in pre-processing operations, such as thresholding or edge detection. FHT aims to alleviate this problem by applying a strategy. In this strategy, closer the pixel to any line contributes more to the corresponding accumulator array bin. As it was mentioned in [7], the application of the FHT can be achieved by two successive steps. In the first step, the traditional HT is applied to obtain the Hough parameter space. In the second step, a 1-D Gaussian function is used to convolve the rows of the obtained Hough parameter space. The 1-D Gaussian function is given as following;

$$f(t) = \begin{cases} e^{-\frac{t^2}{\sigma^2}}, if\ t < R \\ 0, otherwise \end{cases} \tag{4}$$

where $R = \sigma$ defining the width of the 1-D Gaussian function.

## 3. Proposed Method

### 3.1. Neutrosophic Hough Space Image

An element in *NS* is defined as: let $A = \{A_1, A_2, \ldots, A_m\}$ as a set of alternatives in neutrosophic set. The alternative $A_i$ is $\{T(A_i), I(A_i), F(A_i)\}/A_i$, where $T(A_i)$, $I(A_i)$, and $F(A_i)$ are the membership values to the true, indeterminate, and false set.

A Hough space image $I_{HT}$ is mapped into neutrosophic set domain, denoted as $I_{NHT}$, which is interpreted using $T_{HT}$, $I_{HT}$, and $F_{HT}$. Given a pixel $P(\rho,\theta)$ in $I_{HT}$, it is interpreted as $P_{NHT}(\rho,\theta) = \{T_{HT}(\rho,\theta), I_{HT}(\rho,\theta), F_{HT}(\rho,\theta)\}$. $T_{HT}(\rho,\theta)$, $I_{HT}(\rho,\theta)$, and $F_{HT}(\rho,\theta)$ represent the memberships belonging to foreground, indeterminate set, and background, respectively [20–27].

Based on the Hough transformed value and local neighborhood information, the true membership and indeterminacy membership are used to describe the indeterminacy among local neighborhood as:

$$T_{HT}(\rho,\theta) = \frac{g(\rho,\theta) - g_{\min}}{g_{\max} - g_{\min}} \tag{5}$$

$$I_{HT}(\rho,\theta) = \frac{Gd(\rho,\theta) - Gd_{\min}}{Gd_{\max} - Gd_{\min}} \tag{6}$$

$$F_{HT}(\rho,\theta) = 1 - T_{HT}(\rho,\theta) \tag{7}$$

where $g(\rho,\theta)$ and $Gd(\rho,\theta)$ are the HT values and its gradient magnitude at the pixel of $P_{NHT}(\rho,\theta)$ on the image $I_{NHT}$.

### 3.2. Indeterminacy Filtering

A filter is defined based on the indeterminacy and is used to remove the effect of indeterminacy information for further segmentation, whose the kernel function is defined as follows:

$$G_I(\rho,\theta) = \frac{1}{2\pi\sigma_I^2} \exp\left(-\frac{\rho^2+\theta^2}{2\sigma_I^2}\right) \tag{8}$$

where $\sigma_I$ is the standard deviation value where is defined as a function $f(\cdot)$ that is associated to the indeterminacy degree. When the indeterminacy level is high, $\sigma_I$ is large and the filtering can make the current local neighborhood more smooth. When the indeterminacy level is low, $\sigma_I$ is small and the filtering takes a less smooth operation on the local neighborhood.

An indeterminate filtering is taken on $T_{HT}(\rho,\theta)$ to make it more homogeneous.

$$T'_{HT}(\rho,\theta) = T_{HT}(\rho,\theta) \oplus G_I(u,v) = \sum_{v=\theta-m/2}^{\theta+m/2} \sum_{u=\rho-m/2}^{\rho+m/2} T_{HT}(\rho-u,\theta-v) G_I(u,v) \tag{9}$$

$$G_I(u,v) = \frac{1}{2\pi\sigma_I^2} \exp\left(-\frac{u^2+v^2}{2\sigma_I^2}\right) \tag{10}$$

$$\sigma_I(\rho,\theta) = f(I_{HT}(\rho,\theta)) = aI_{HT}(\rho,\theta) + b \tag{11}$$

where $T'_{\mathrm{HT}}$ is the indeterminate filtering result, $a$ and $b$ are the parameters in the linear function to transform the indeterminacy level to standard deviation value.

### 3.3. Thresholding Based on Histogram in Neutrosophic Hough Image

After indeterminacy filtering, the new true membership set $T'_{\mathrm{HT}}$ in Hough space image become homogenous, and the clusters with high HT values become compact, which is suitable to be identified. A thresholding method is used on $T'_{\mathrm{HT}}$ to pick up the clustering with high HT values, which respond to the lines in original image domain. The thresholding value is automatic determined by the maximum peak on this histogram of the new Hough space image.

$$T_{BW}(\rho,\theta) = \begin{cases} 1 & T_{\mathrm{HT}}(\rho,\theta) > Th_{\mathrm{HT}} \\ 0 & otherwise \end{cases} \tag{12}$$

where $Th_{\mathrm{HT}}$ is the threshold value that is obtained from the histogram of $T'_{\mathrm{HT}}$.

In the binary image after thresholding, the object regions are detected and the coordinators of their center are identified as the parameters $(\rho,\theta)$ of the detected lines. Finally, the detected lines are found and recovered in the original images using the values of $\rho$ and $\theta$.

## 4. Experimental Results

In order to specify the efficiency of our proposed NHT, we conducted various experiments on a variety of noisy and noise-free images. Before illustration of the obtained results, we opted to show the effect of our proposed NHT on a synthetic image. All of the necessary parameters for NHT were fixed in the experiments where the sigma parameter and the window size of the indeterminacy filter were chosen as 0.5 and 7, respectively. The value of $a$ and $b$ in Equation (11) are set as 0.5 by the trial and error method.

In Figure 2, a synthetic image, its NS Hough space, detected peaks on NS Hough space and the detected lines illustration are illustrated below.

As seen in Figure 2a, we constructed a synthetic image, where four dotted lines were crossed in the center of image. The lines are designed not to be perfectly straight in order to test the proposed NHT. Figure 2b shows the NS Hough space of the input image. As seen in Figure 2b, each image point in the image domain corresponds to a sinus curve in NS Hough space, and the intersections of these sinus curves indicate the straight lines in the image domain. The intersected curves regions, as shown in Figure 2b, were detected based on the histogram thresholding method and the exact location of the peaks was determined by finding the center of the region. The obtained peaks locations are shown in Figure 2c. The obtained peak locations were then converted to lines and the detected lines were superimposed in the image domain, as shown in Figure 2d.

We performed the above experiment one more time when the input synthetic image was corrupted with noise. With this experiment, we can investigate the behavior of our proposed NHT on noisy images. To this end, the input synthetic image was degraded with a 10% salt & pepper noise. The degraded input image can be seen in Figure 3a. The corresponding NS Hough space is depicted in Figure 3b. As seen in Figure 3b, the NS Hough space becomes denser when compared with the Figure 2b, because the all of the noise points contributed the NS Hough space. The thresholded the NS Hough space is illustrated in Figure 3c, where four peaks are still visible. Finally, as seen in Figure 3d, the proposed method can detect the four lines correctly in a noisy image. We further experimented on some noisy and noise-free images, and the obtained results were shown in Figure 4. While the images in the Figure 4a column show the input noisy and noise-free images, in Figure 4b column, we give the obtained results. As seen in the results, the proposed NHT is quite effective for both noisy and noise-free images. All of the lines were detected successfully. In addition, as shown in the first row of Figure 4, the proposed NHT could detect the lines that were not straight perfectly.

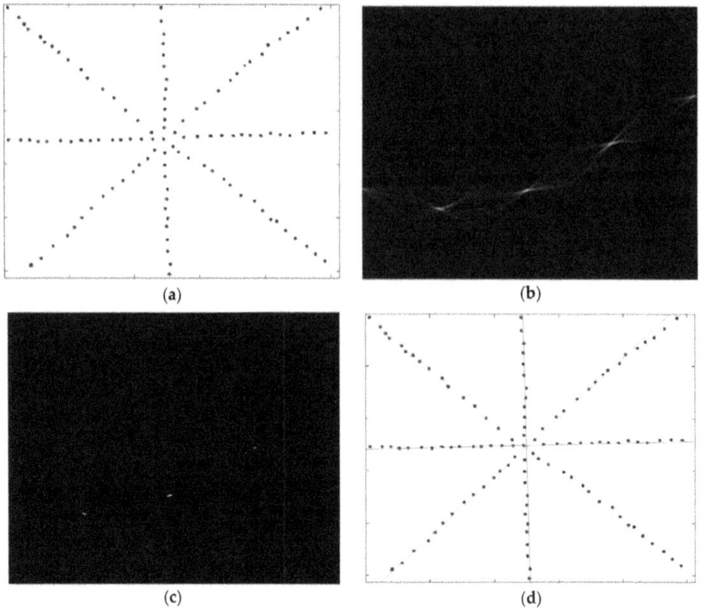


**Figure 2.** Application of the neutrosophic Hough transform (NHT) on a synthetic image, (**a**) input synthetic image; (**b**) neutrosophy (NS) Hough space; (**c**) Detected peaks in NS Hough space; and, (**d**) Detected lines.

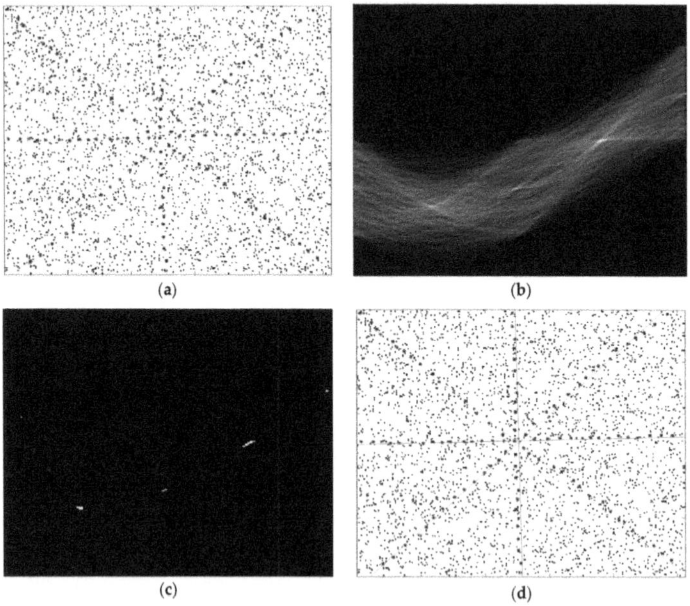


**Figure 3.** Application of the NHT on a noisy synthetic image, (**a**) input noisy synthetic image; (**b**) NS Hough space; (**c**) Detected peaks in NS Hough space; and, (**d**) Detected lines.

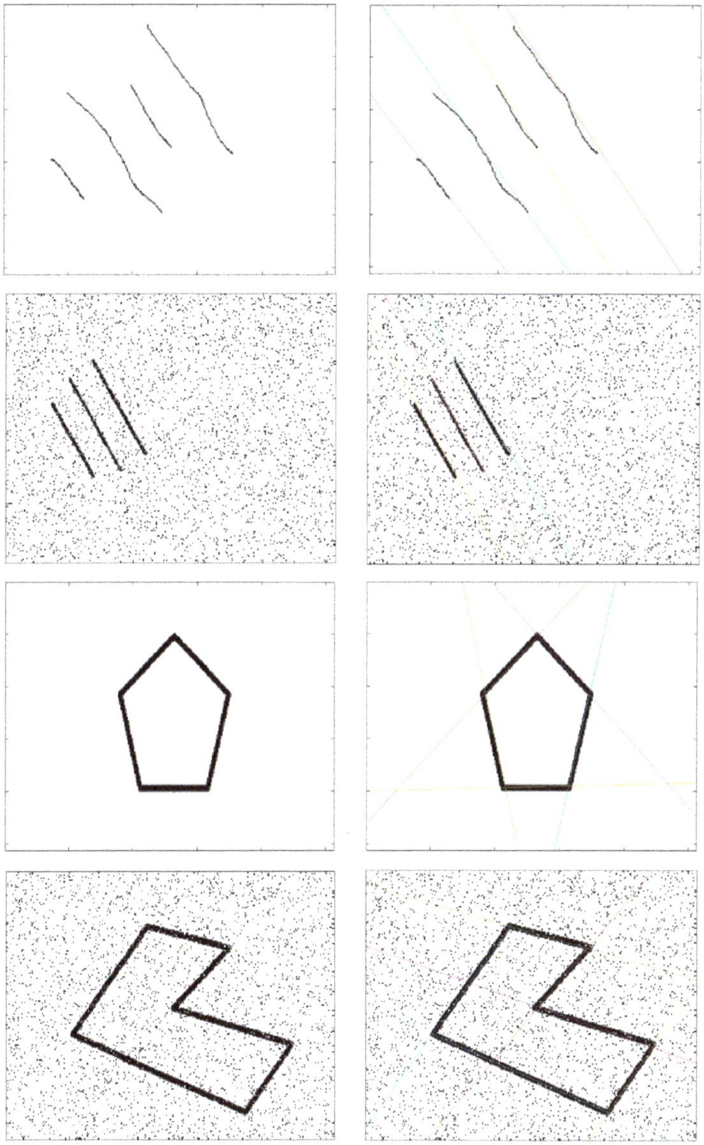

(a) (b)

**Figure 4.** Application of the NHT on a noisy synthetic image, (**a**) input noisy and noise-free synthetic image; (**b**) Detected lines with proposed NHT.

As the proposed NHT was quite good in detection of the lines in both noise and noise-free synthetic images, we performed some experiments on gray scale real-world images. The obtained results were indicated in Figure 5. Six images were used and obtained lines were superimposed on the input images. As seen from the results, the proposed NHT yielded successful results. For example, for the images of Figure 5a,c–f, the NHT performed reasonable lines. For the chessboard image (Figure 5b), only few lines were missed.

We further compared the proposed NHT with FHT and the traditional HT on a variety of noisy images and obtained results were given in Figure 6. The first column of Figure 6 shows the HT performance, the second column shows the FHT achievements, and final column shows the obtained results with proposed NHT.

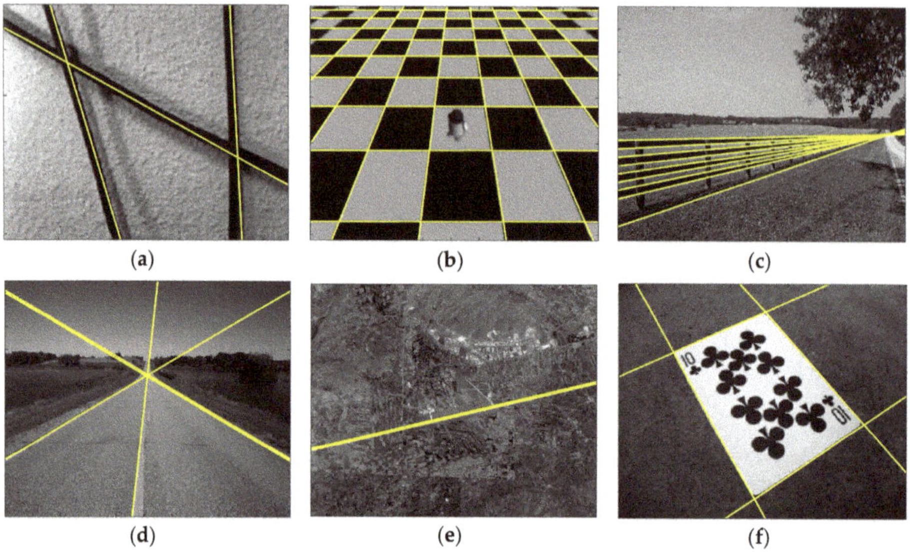


**Figure 5.** Various experimental results for gray scale real-world images (**a–f**).

With visual inspection, the proposed method obtained better results than HT and FHT. Generally, HT missed one or two lines in the noisy environment. Especially for the noisy images that were given in the third row of Figure 6, HT just detected one line. FHT generally produced better results than HT. FHT rarely missed lines but frequently detected the lines in the noisy images due to its fuzzy nature. In addition, the detected lines with FHT were not on the ground-truth lines, as seen in first and second rows of Figure 6. NHT detected all of the lines in all noisy images that were given Figure 6. The detected lines with NHT were almost superimposed on the ground-truth lines as we inspected it visually. In order to evaluate the comparison results quantitatively, we computed the *F-measure* values for each method, which is defined as:

$$F\text{-}measure = \frac{2 \times (precision \times recall)}{precision + recall} \tag{13}$$

where precision is the number of correct results divided by the number of all returned results. The recall is the number of correct results divided by the number of results that should have been returned.

To this end, the detected lines and the ground-truth lines were considered. A good line detector produces an *F-measure* percentage, which is close to 100%, while a poor one produces an *F-measure* value that is closer to 0%. As Figure 6 shows the comparison results, the corresponding *F-measure* values were tabulated in Table 1. As seen in Table 1, the highest average *F-measure* value was obtained by the proposed NHT method. The second highest average *F-measure* value was obtained by FHT. The worst *F-measure* values were produced by HT method.

**Table 1.** *F-measure* percentages for compared methods.

| Input Image | HT | FHT | NHT |
|---|---|---|---|
| First row (Figure 6) | 81.09 | 85.96 | 89.50 |
| Second row (Figure 6) | 71.87 | 86.75 | 98.28 |
| Third row (Figure 6) | 32.85 | 42.58 | 59.96 |
| Average | 61.94 | 71.76 | 82.58 |

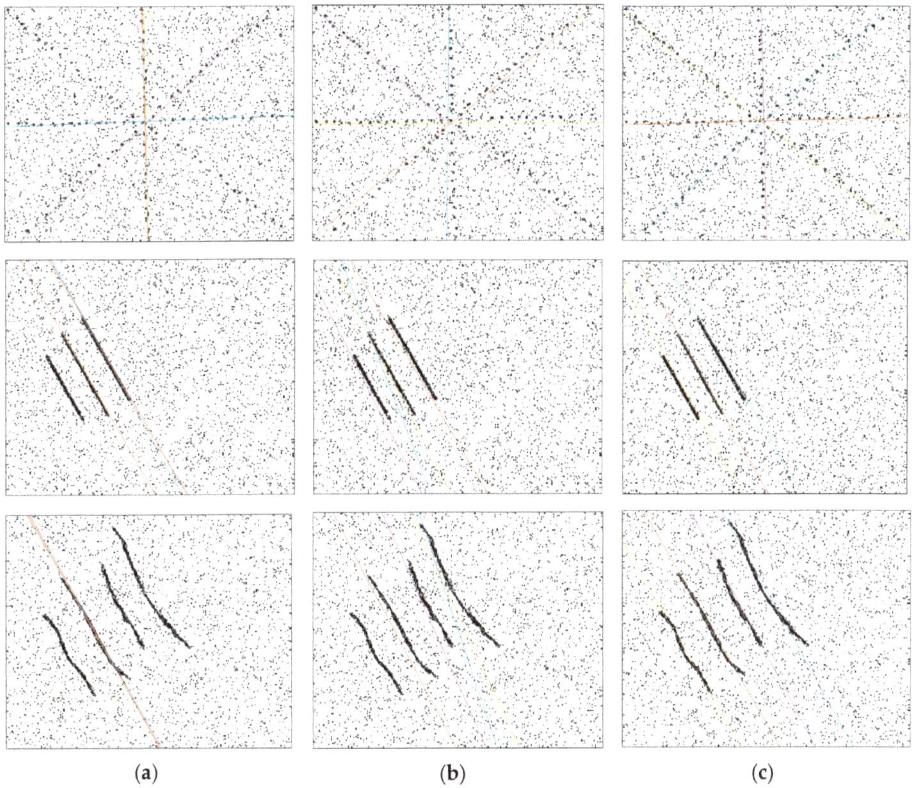

**Figure 6.** Comparison of NHT with Hough transform (HT) and FHT on noisy images (**a**) HT results; (**b**) FHT results and (**c**) NHT results.

We also experimented on various noisy real-world images and compared the obtained results with HT and FHT achievements. The real-world images were degraded with a 10% salt & pepper noise. The obtained results were given in Figure 7. The first, second, and third columns of Figure 7 show the HT, FHT, and NHT achievement, respectively. In addition, in the last column of Figure 7, the running times of each method for each image were given for comparison purposes.

As can be seen in Figure 7, the proposed NHT method is quite robust against noise and can able to find the most of the true lines in the given images. In addition, with a visual inspection, the proposed NHT achieved better results than the compared HT and FHT. For example, for the image, which is depicted in the first row of Figure 7, the NHT detected all of the lines. FHT almost detected the lines with error. However, HT only detected just one line and missed the other lines. Similar results were also obtained for the other images that were depicted in the second and third rows of the Figure 7.

In addition, for a comparison of the running times, it is seen that there is no significant differences between the compared methods running times.

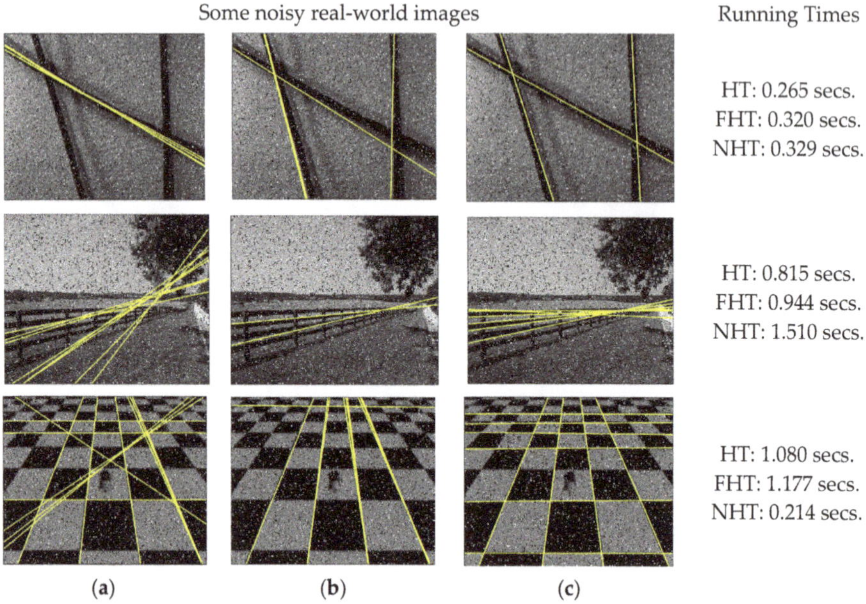


**Figure 7.** Comparison of NHT with HT and FHT on noisy images (**a**) HT results; (**b**) FHT results; and, (**c**) NHT results.

## 5. Conclusions

In this paper, a novel Hough transform, namely NHT, was proposed, which uses the NS theory in voting procedure of the HT algorithm. The proposed NHT is quite efficient in the detection of the lines in both noisy and noise-free images. We compared the performance achievement of proposed NHT with traditional HT and FHT methods on noisy images. NHT outperformed in line detection on noisy images. In future works, we are planning to extend the NHT on more complex images, such natural images, where there are so many textured regions. In addition, Neutrosophic circular HT will be investigated in our future works.

**Author Contributions:** Ümit Budak, Yanhui Guo, Abdulkadir Şengür and Florentin Smarandache conceived and worked together to achieve this work.

**Conflicts of Interest:** The authors declare no conflict of interest.

## References

1. Hough, P.V.C. Method and Means for Recognizing. US Patent 3,069,654, 25 March 1960.
2. Duda, R.O.; Hart, P.E. Use of the Hough transform to detect lines and curves in pictures. *Commun. ACM* **1972**, *15*, 11–15. [CrossRef]
3. Chung, K.L.; Chen, T.C.; Yan, W.M. New memory- and computation-efficient Hough transform for detecting lines. *Pattern Recognit.* **2004**, *37*, 953–963. [CrossRef]
4. Guo, S.; Pridmore, T.; Kong, Y.; Zhang, X. An improved Hough transform voting scheme utilizing surround suppression. *Pattern Recognit. Lett.* **2009**, *30*, 1241–1252. [CrossRef]
5. Walsh, D.; Raftery, A. Accurate and efficient curve detection in images: The importance sampling Hough transform. *Pattern Recognit.* **2002**, *35*, 1421–1431. [CrossRef]

6. Xu, L.; Oja, E. Randomized Hough transform (RHT): Basic mechanisms, algorithms, and computational complexities. *CVGIP Image Underst.* **1993**, *57*, 131–154. [CrossRef]
7. Han, J.H.; Koczy, L.T.; Poston, T. Fuzzy Hough transform. *Patterns Recognit. Lett.* **1994**, *15*, 649–658. [CrossRef]
8. Montseny, E.; Sobrevilla, P.; Marès Martí, P. Edge orientation-based fuzzy Hough transform (EOFHT). In Proceedings of the 3rd Conference of the European Society for Fuzzy Logic and Technology, Zittau, Germany, 10–12 September 2003.
9. Mathavan, S.; Vaheesan, K.; Kumar, A.; Chandrakumar, C.; Kamal, K.; Rahman, M.; Stonecliffe-Jones, M. Detection of pavement cracks using tiled fuzzy Hough transform. *J. Electron. Imaging* **2017**, *26*, 053008. [CrossRef]
10. Chatzis, V.; Ioannis, P. Fuzzy cell Hough transform for curve detection. *Pattern Recognit.* **1997**, *30*, 2031–2042. [CrossRef]
11. Suetake, N.; Uchino, E.; Hirata, K. Generalized fuzzy Hough transform for detecting arbitrary shapes in a vague and noisy image. *Soft Comput.* **2006**, *10*, 1161–1168. [CrossRef]
12. Chung, K.-L.; Lin, Z.-W.; Huang, S.-T.; Huang, Y.-H.; Liao, H.-Y.M. New orientation-based elimination approach for accurate line-detection. *Pattern Recognit. Lett.* **2010**, *31*, 11–19. [CrossRef]
13. Cheng, H.-D.; Guo, Y.; Zhang, Y. A novel Hough transform based on eliminating particle swarm optimization and its applications. *Pattern Recognit.* **2009**, *42*, 1959–1969. [CrossRef]
14. Zhu, L.; Chen, Z. Probabilistic Convergent Hough Transform. In Proceedings of the International Conference on Information and Automation, Changsha, China, 20–23 June 2008.
15. Ballard, D.H. Generalizing the Hough Transform to Detect Arbitrary Shapes. *Pattern Recognit.* **1981**, *13*, 111–122. [CrossRef]
16. Smarandache, F. *Neutrosophy: Neutrosophic Probability, Set, and Logic*; ProQuest Information & Learning; American Research Press: Rehoboth, DE, USA, 1998; 105p.
17. Smarandache, F. *Introduction to Neutrosophic Measure, Neutrosophic Integral and Neutrosophic Probability*; Sitech Education Publishing: Columbus, OH, USA, 2013; 55p.
18. Smarandache, F. A Unifying Field in Logics Neutrosophic Logic. In *Neutrosophy, Neutrosophic Set, Neutrosophic Probability*, 3rd ed.; American Research Press: Rehoboth, DE, USA, 2003.
19. Vaheesan, K.; Chandrakumar, C.; Mathavan, S.; Kamal, K.; Rahman, M.; Al-Habaibeh, A. Tiled fuzzy Hough transform for crack detection. In Proceedings of the Twelfth International Conference on Quality Control by Artificial Vision (SPIE 9534), Le Creusot, France, 3–5 June 2015.
20. Guo, Y.; Şengür, A. A novel image segmentation algorithm based on neutrosophic filtering and level set. *Neutrosophic Sets Syst.* **2013**, *1*, 46–49.
21. Guo, Y.; Xia, R.; Şengür, A.; Polat, K. A novel image segmentation approach based on neutrosophic c-means clustering and indeterminacy filtering. *Neural Comput. Appl.* **2017**, *28*, 3009–3019. [CrossRef]
22. Karabatak, E.; Guo, Y.; Sengur, A. Modified neutrosophic approach to color image segmentation. *J. Electron. Imaging* **2013**, *22*, 013005. [CrossRef]
23. Guo, Y.; Sengur, A. A novel color image segmentation approach based on neutrosophic set and modified fuzzy c-means. *Circuits Syst. Signal Process.* **2013**, *32*, 1699–1723. [CrossRef]
24. Sengur, A.; Guo, Y. Color texture image segmentation based on neutrosophic set and wavelet transformation. *Comput. Vis. Image Underst.* **2011**, *115*, 1134–1144. [CrossRef]
25. Guo, Y.; Şengür, A.; Tian, J.W. A novel breast ultrasound image segmentation algorithm based on neutrosophic similarity score and level set. *Comput. Methods Programs Biomed.* **2016**, *123*, 43–53. [CrossRef] [PubMed]
26. Guo, Y.; Sengur, A. NCM: Neutrosophic c-means clustering algorithm. *Pattern Recognit.* **2015**, *48*, 2710–2724. [CrossRef]
27. Guo, Y.; Şengür, A. A novel image segmentation algorithm based on neutrosophic similarity clustering. *Appl. Soft Comput.* **2014**, *25*, 391–398. [CrossRef]

MDPI
St. Alban-Anlage 66
4052 Basel
Switzerland
Tel. +41 61 683 77 34
Fax +41 61 302 89 18
www.mdpi.com

*Axioms* Editorial Office
E-mail: axioms@mdpi.com
www.mdpi.com/journal/axioms

www.ingramcontent.com/pod-product-compliance
Lightning Source LLC
Chambersburg PA
CBHW051849210326
41597CB00033B/5829
* 9 7 8 3 0 3 8 9 7 2 8 8 4 *